T0215343

Communications
in Computer and Information Science 1001

Commenced Publication in 2007
Founding and Former Series Editors:
Phoebe Chen, Alfredo Cuzzocrea, Xiaoyong Du, Orhun Kara, Ting Liu,
Krishna M. Sivalingam, Dominik Ślęzak, Takashi Washio, and Xiaokang Yang

More information about this series at http://www.springer.com/series/7899

Kevin Knight · Changshui Zhang ·
Geoff Holmes · Min-Ling Zhang (Eds.)

Artificial Intelligence

Second CCF International Conference, ICAI 2019
Xuzhou, China, August 22–23, 2019
Proceedings

 Springer

Editors
Kevin Knight
USC/Information Sciences Institute
University of Southern California
Marina Del Rey, CA, USA

Geoff Holmes
University of Waikato
Hamilton, New Zealand

Changshui Zhang
Tsinghua University
Beijing, China

Min-Ling Zhang
Southeast University
Nanjing, China

ISSN 1865-0929   ISSN 1865-0937 (electronic)
Communications in Computer and Information Science
ISBN 978-981-32-9297-0   ISBN 978-981-32-9298-7 (eBook)
https://doi.org/10.1007/978-981-32-9298-7

This Springer imprint is published by the registered company Springer Nature Singapore Pte Ltd.
The registered company address is: 152 Beach Road, #21-01/04 Gateway East, Singapore 189721, Singapore

Preface

On behalf of the Organizing Committee, it is our great pleasure to welcome you to the Second CCF International Conference on Artificial Intelligence (CCF-ICAI 2019) held in Xuzhou, China.

CCF-ICAI, organized by the China Computer Federation (CCF) and co-organized by the CCF Artificial Intelligence & Pattern Recognition Society, is an international conference in the field of artificial intelligence. It aims to provide a leading international forum for researchers, practitioners, and other potential users in artificial intelligence and related fields to share their new ideas, progresses, and achievements.

We received 97 submissions to CCF-ICAI 2019, where each of them was reviewed by at least three reviewers. Based on the reviewers' comments, 23 papers were finally accepted for presentation at the conference yielding an acceptance rate of 23.7%. The CCF-ICAI 2019 proceedings are published by Springer as a dedicated volume in the *Communications in Computer and Information Science* (CCIS) series.

We are very grateful to the many authors who submitted their work to CCF-ICAI 2019. We were delighted to feature three outstanding keynote speakers: Professor James Kwok from The Hong Kong University of Science and Technology, Dr. Bo An from Nanyang Technological University, and Professor Jianxin Wu from Nanjing University. The conference program was further enriched with oral and poster presentations of accepted papers.

We wish to sincerely thank all the Program Committee (PC) members for their invaluable efforts in ensuring a timely, fair, and highly effective paper review and selection procedure. We are also very thankful to the other Organizing Committee members: the publication co-chairs, Deyu Meng and Wenjian Wang, local arrangements co-chairs, Yong Zhou and Cunhua Li, publicity co-chairs, Shifei Ding and Xinzheng Xu, and finance co-chairs, Qiang Niu and Li Yuan. We appreciate Springer for publishing the conference proceedings, and particularly Celine Chang, Jane Li, and Sanja Evenson from Springer for their efforts and patience in collecting and editing the proceedings.

We appreciate the hosting organization, China University of Mining and Technology, and the co-host organization, HuaiHai Institute of Technology, for their institutional and financial support of CCF-ICAI 2019. Last but not least, our sincere thanks go to all the participants and volunteers of CCF-ICAI 2019—there would be no conference without you.

May 2019

Kevin Knight
Changshui Zhang
Geoff Holmes
Min-Ling Zhang

Organization

Organizing Committee

General Co-chairs

Kevin Knight   DiDi Labs; University of Southern California, USA
Changshui Zhang   Tsinghua University, China

Program Co-chairs

Geoff Holmes   University of Waikato, New Zealand
Min-Ling Zhang   Southeast University, China

Publication Co-chairs

Deyu Meng   Xi'an Jiaotong University, China
Wenjian Wang   Shanxi University, China

Local Arrangements Co-chairs

Yong Zhou   China University of Mining and Technology, China
Cunhua Li   HuaiHai Institute of Technology, China

Publicity Co-chairs

Shifei Ding   China University of Mining and Technology, China
Xinzheng Xu   China University of Mining and Technology, China

Finance Co-chairs

Qiang Niu   China University of Mining and Technology, China
Li Yuan   China University of Mining and Technology, China

Program Committee

Xiang Bai	Huazhong University of Science and Technology, China
Wei-Neng Chen	South China University of Technology, China
Chaoran Cui	Shandong University of Finance and Economics, China
Shifei Ding	China University of Mining and Technology, China
Hongbin Dong	Harbin Engineering University, China
Jun Dong	Institute of Intelligent Machines, CAS, China
Jingyang Gao	Beijing University of Chemical Technology, China
Xin Geng	Southeast University, China
Bing Han	Xidian University, China
Chenping Hou	National University of Defense Technology, China

Di Huang	Beihang University, China
Genlin Ji	Nanjing Normal University, China
Yichuan Jiang	Southeast University, China
Deyu Li	Shanxi University, China
Guo-Zheng Li	China Academy of Chinese Medical Science, China
Jinping Li	University of Jinan, China
Kewen Li	China University of Petroleum, China
Ming Li	Nanjing University, China
Qingyong Li	Beijing Jiaotong University, China
Wu-Jun Li	Nanjing University, China
Yu-Feng Li	Nanjing University, China
Meiqin Liu	Zhejiang University, China
Zhendong Liu	Shandong Jianzhu University, China
Jiwen Lu	Tsinghua University, China
Qianli Ma	South China University of Technology, China
Xinjun Mao	National University of Defense Technology, China
Xiushan Nie	Shandong University of Finance and Economics, China
Qiang Qi	Shandong University of Finance and Economics, China
Chao Qian	University of Science and Technology of China, China
Yuhua Qian	Shanxi University, China
Lin Shang	Nanjing University, China
Chuan Shi	Beijing University of Posts and Telecommunications, China
Zhenwei Shi	Beihang University, China
Shiliang Sun	East China Normal University, China
Xiaoyang Tan	Nanjing University of Aeronautics and Astronautics, China
Xiangrong Tong	Yantai University, China
Hongyuan Wang	Changzhou University, China
Jun Wang	Southwest University, China
Shangfei Wang	University of Science and Technology of China, China
Zhe Wang	East China University of Science and Technology, China
Yimin Wen	Guilin University of Electronic Technology, China
Jianxin Wu	Nanjing University, China
Fen Xiao	Xiangtan University, China
Jungang Xu	University of Chinese Academy of Sciences, China
Xin-Shun Xu	Shandong University, China
Xinzheng Xu	China University of Mining and Technology, China
Yong Xu	South China University of Technology, China
Hui Xue	Southeast University, China
Gongping Yang	Shandong University, China
Ming Yang	Nanjing Normal University, China
Yan Yang	Southwest Jiaotong University, China
Yu-Bin Yang	Nanjing University, China
Minghao Yin	Northeast Normal University, China

Guoxian Yu	Southwest University, China
Zhiwen Yu	South China University of Technology, China
Zhi-Hui Zhan	South China University of Technology, China
Li Zhang	Soochow University, China
Lijun Zhang	Nanjing University, China
Shihui Zhang	Yanshan University, China
Zhao Zhang	Hefei University of Technology, China
Jun Zhu	Tsinghua University, China
Fuzhen Zhuang	Institute of Computing Technology, CAS, China
Li Zou	Liaoning Normal University, China
Quan Zou	University of Electronic Science and Technology of China, China

Contents

NLP and Recommender System

Machine Learning Algorithms

AI Applications

Deep Learning

Collaborative Filtering Based on Attention Mechanism

Hongbin Dong$^{(\boxtimes)}$ ⓘ, Lei Yang ⓘ, and Kunming Han

Harbin Engineering University, Harbin 150001, China
donghongbin@hrbeu.edu.cn

Abstract. Each item has different characteristics. Different users place different importance to these aspects which can be thought of as a preference/attention weight vector. Therefore, using the traditional method to predict the unknown ratings ignores this difference in importance, resulting in a false assumption that all users have the same attention to different characteristics of the same item. In this paper, we propose a collaborative filtering system based on attention mechanism and design the feature-topic model to extract the characteristics of the item from review texts. Then, we use an attention network to get the importance of the item's characteristic to the user. Considering the shortcomings of linear interaction features, we adopt the idea of collaboration to predict unknown scores. Extensive experiments on real-world datasets show significant improvements in our proposed CFAM framework over the state-of-the-art methods.

Keywords: Collaborative filtering · Topic model · Attention neural network · Nonlinear feature

1 Introduction

In online shopping, users often leave a lot of historical records such as ratings, clicks, browsing time and reviews. The personalized recommendation analyzes the user's history to mine the user's preferences and make satisfactory recommendations for the user. Matrix Factorization [1] is one of the most popular collaborative filtering techniques, which maps users and items into the same space. It uses latent vectors to denote a user and an item which make an inner product to model the interactive features finally. But the inner product for the latent vector has some disadvantages that can not learn the complex interaction between the user vector and item vector. Neural networks have the excellent ability to fit any function. Salakhutdinov first proposed the use of restricted Boltzmann networks for collaborative filtering. With the great success of neural networks in modeling user preferences and alleviating the data sparse, a deep cooperative neural network model (Deep CoNN) is proposed. The idea uses two parallel neural network models to learn user features and item features. Then an interactive layer is built on the two neural networks to predict unknown rating. Aspect-based model [2, 3] uses external domain knowledge to determine aspect information from the review. The topic-based approach (Latent Dirichlet Allocation [4]) adopts topic models to extract aspect information [5–7] automatically.

© Springer Nature Singapore Pte Ltd. 2019
K. Knight et al. (Eds.): ICAI 2019, CCIS 1001, pp. 3–14, 2019.
https://doi.org/10.1007/978-981-32-9298-7_1

However, a rating only reflects a user's overall satisfaction towards an item without explaining the underlying rationale. For example, a user gives a high score on a mobile phone, because he likes the camera of this mobile phone. But another user gives the same mobile phone a high score just because of its stylish appearance. The mentioned model assumes that all users pay the same attention to the different characteristics of the same item. This assumption is not true in the real scene. The difference in attention can be captured by analyzing the review texts. In order to solve this problem, this paper proposes collaborative filtering based on attention mechanism (CFAM).

2 Related Work

2.1 Topic Model

Early approaches to capturing potential topics from reviews rely on domain knowledge and manually tag the reviews, which requires expensive domain knowledge and labor costs. In the following studies, most methods [8] try to automatically extract potential topics from comments. These methods adopt a topic model [9] or non-negative Matrix Factorization [10], which learn potential factors from comments. The factors can be understood as preference or characteristic extracted from the topic model. HFT [11] and TopicMF [11] use a predefined transformation function to associate the item feature vector with the topic distribution of the item. ITLFM and RBLT [8] assume that potential topics are in the same space as factors, and then use linear functions to combine them to form latent representations of users and items, which can simulate MF ratings. CTR [12] assumes that the feature vector of an item depends on the potential topic distribution of its reviews. When the rating is modeled, the feature vector is used to represent the topic distribution of the item. RMR [13] uses a topic model to learn the features of the item, which applies a mixture of Gaussian distributions rather than Matrix Factorization to predict rating. The topic model is also utilized in the method of our paper. Different from the previous method [5], we adopt a new topic model. This feature topic model not only extracts the user review features but also the item review features from the review, and combines the review features with the rating features to obtain user preferences.

2.2 Neural Networks and Recommendation Systems

Deep learning in the recommendation system has developed a trend. The Generalized Matrix Factorization [2] and Factorization Machine [10] employed for neural collaborative filtering have achieved good results. DeepCoNN and TransNet use deep learning to mine review texts to predict scores. Neural collaborative filtering [18] models linear interaction features and nonlinear interaction features between users and items. If only an element-by-element multiplication operation is performed on the potential representation of users and items, these operations are linear and cannot capture nonlinear interactive features. Neural Network has a greater advantage in fitting nonlinear functions. There are many applications in the field of recommending systems using the neural network to learn interactive features [13, 14]. So we will consider linear features and nonlinear features by inner and MLP, and learn more useful features. This MLP can make up for the loss of a part of the interaction features using the inner product.

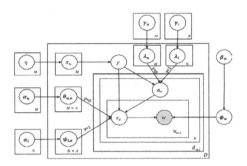

Fig. 1 Feature-topic model structure

3 CFAM Model Framework

We utilize attention mechanisms for collaborative filtering. Firstly, the rating feature vectors of the user and the item are obtained from the rating matrix. A topic model is used to obtain review vector of the user and review vector of the item from the review texts. After obtaining the two vectors, if the latent vectors only make inner operation, there are still many complex nonlinear interactions that cannot be represented. So we merge the basic vector into a multi-layer perceptron to learn nonlinear interaction. Finally, these linear interaction features are connected in parallel with the nonlinear interaction features and then use the attention neural network to predict the unknown rating.

3.1 Feature-Topic Model

Given a document set D containing the user's reviews $\{d_{u,i} \in D, u \in U, i \in I\}$ on the item, we assumed that the hidden topic set (i.e., K topics) covers all the topics of the user in the comment. λ_u is the probability distribution of the topic in the user preference and $\lambda_{u,a}$ represents the importance of the topic a to the user u. λ_i represents the topic probability distribution in the characteristics of the item i, and $\lambda_{i,a}$ represents the importance of the topic a to an item i. The relationship between the topic and user preferences or item characteristics can be expressed as a probability distribution. $\theta_{u,a}$ is the user's interest in the aspect of a, which can be expressed user's Polynomial distribution of topics. Similarly, $\psi_{i,a}$ is the characteristics of an item in the aspect of $a.\theta_{u,a}$ is generated by all user's comments and $\psi_{i,a}$ is learned from all comments on the item i. The feature-based topic model designs that the topic is the polynomial distribution of the words in the comment. To get the parameters $\{\lambda_u, \lambda_i, \theta_{u,a}, \psi_{i,a}, \pi_u\}$, we use Algorithm 1 to simulate the generation of the comment corpus.

Algorithm 1 . Generation Process of Topic

1	for *each topic k = 1, ..., k* do
2	Draw $\phi_{k,w} \sim Dir(. \mid \beta_w)$;
3	for *each user* $u \in U$ do
4	Draw $\lambda_u \sim Dir(. \mid \gamma_u)$;
5	for *each item* $i \in I$ do
6	Draw $\lambda_i \sim Dir(. \mid \gamma_i)$;
7	for *each user* $u \in U$,*each aspect* $a \in A$ do
8	Draw $\theta_{u,a} \sim Dir(. \mid \alpha_u)$;
9	for *each item* $i \in I$,*each aspect* $a \in A$ do
10	Draw $\psi_{i,a} \sim Dir(. \mid \alpha_{i,a})$;
11	for *each review* $d_{u,i}$,$u \in U, i \in I$ do
12	for *each sentence* s $\in d_{u,i}$ do
13	Draw y $\sim Bernoulli(. \mid \pi_u)$;
14	if $y_s == 0$ then
15	Draw $a_s \sim Mult(\lambda_u)$;and then $z_s \sim Mult(\theta_u, a_s)$;
16	if $y_s == 1$;then
17	Draw $a_s \sim Mult(\lambda_i)$; and then;$z_s \sim Mult(\psi_i, a_s)$
18	for *each word* $w \in S$ do
19	Draw $w \sim Mult(\phi_{z_s}, w)$

The structure of the topic model is shown in Fig. 1. The shaded circles represent observed variables and the others represent latent variables. The model simulates the idea of writing comments sentence by sentence where one sentence usually discusses the same topic. This topic comes from user preferences or item characteristics. In order to determine whether the topic comes from user preferences or item characteristics, we use an indicator variable $y \in \{0, 1\}$ which is the Bernoulli distribution with the parameter π_u. If $y = 0$, the item characteristics. The parameter π_u is user-independent, which indicates a trend to comment based on personal preferences or item characteristics. a_s is the topic of the sentence s. When $y = 0$, a_s comes from λ_u, which represents an aspect of user preference, z_s is decided by preferences θ_{u,a_s}. On the contrary, a_s comes from λ_i, which represents an aspect of item characters, z_s is decided by preferences ψ_{i,a_s}. The words w of a sentence s are word distribution $\varphi_{z_s,w}$. The topic comes from the user's preference and whereas from the item's characteristics.

In the topic model, $\alpha_u, \alpha_i, \gamma_u, \gamma_i, \beta, \eta$ is pre-defined, except for these parameters, $\lambda_i, \lambda_u, \theta_{u,a}, \psi_{i,a}, \pi_u$, which also need to be estimated. A lot of approximate reasoning methods are proposed for the parameter estimation of the topic model, such as variational recommendation, Gibbs sampling. Since Gibbs sampling has been successfully applied to many topic models [7]. We use it to estimate the model parameter.

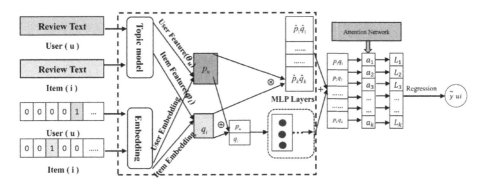

Fig. 2 The structure of CFAM model

3.2   Neural Collaborative Filtering

The model structure is shown in Fig. 2 The input module consists of two parts. The first part is obtaining users and items potential factor from the rating matrix. We convert the ID of the user and the item into a 0–1 sparse vector by a one-hot encoding. And we use a layer of the fully connected network, which maps the sparse vector into a dense vector. The other part is to extract the user's review features by the feature-topic model. After extracting the feature θ_u of the user u and the feature φ_i of the item i, we fuse the potential vector and the review feature to form a vector representation of the user u and the item i. We have adopted the fusion method in [7] and ITLFM [8].

After we get the potential feature of the user u and the potential feature of i by fusion. The process is similar to Matrix Factorization. We use a deformed Matrix Factorization operation. The first step is to make an element-wise product operation for p_u and q_i to get $\hat{p}_i\hat{q}_i$, where $p_u \in \mathbb{R}^K$ and $q_i \in \mathbb{R}^K$. We know Matrix Factorization is a linear operation, so the element-wise product has nothing to learn about nonlinear interaction features. We connect p_u and q_i in series and use a multi-layer perceptron to learn complex interaction features. W_L, b_L is the weight matrix and offset vector of the L-layer of the multi-layer perceptron respectively. a_L is the activation function of the L-th layer. According to experimental [17] about the neural network activation function, the use of ReLU can significantly improve the performance of the model. From the NCF [18], it is concluded that the structure of the MLP adopts the tower structure. The lowest layer is the widest, and the number of neurons in the next layer is reduced to half of the upper layer. Multi-layer perceptron finally outputs a vector of the same dimension as $\hat{p}_k\hat{q}_k$. Combining these two vectors using (2) to get p_kq_k, the fusion vector first passes through a fully connected layer. ReLU is used as the activation function in this layer. The experiment [16] can clearly see the layer significantly improved the effectiveness of the experiment.

$$z_1 = \phi_1(p_u, q_i) = \begin{bmatrix} p_u \\ q_i \end{bmatrix}$$
$$\phi_2(z_1) = a_2(W_2^T z_1 + b_2) \tag{1}$$
$$pq_{MPL} = a_L(W_L^T z_{L-1} + b_L)$$

$$p_k q_k = \hat{p}_k \hat{q}_k \oplus pq_{MPL} \tag{2}$$

3.3   Attention Network

This paper designs a feature-topic model, which extracts user preferences and item characteristics from review texts. When users comment on an item, they are more willing to review the aspects of the items they care about. Hence we represent the preference of user u from the review as θ_u and the characteristics of the item's different aspects of as φ_i. By the interaction between θ_u and φ_i, we can estimate the attention vector $a_{u,i}$ of the user u to the item i. We introduce an attention neural network to model this interaction. The attention network is used to train the user's attention vector on different aspects of the target item. Because the user's preference or the characteristics of the item can be displayed from the reviews, we rely on the review-based feature to capture the user's attention weight $a_{u,i,k}$ on the characteristics k of the item i. At the same time, we combine the previously learned $\hat{p}_k \hat{q}_k$ to represent the attention interaction feature $L = [L_1, L_2, \cdots, L_K]$ of each user-item pair. The attention vector equation is as follows:

$$\hat{a}_{u,i} = v^T \text{ReLu}(W_a[\theta_u; \varphi_i; p_u; q_i] + b_a) \tag{3}$$

$$L = a_{u,i} \odot p_k q_k \tag{4}$$

$[\theta_u; \varphi_i; p_u; q_i]$ means that the input is connected in series. We use W_a and b_a to map the input to the middle layer and then use v^T to map the middle layer to the attention weight vector as the output. Activation function chooses ReLU. The neural attention network is used in many ways. By studying other aspects, a standard $a_{u,i}$ can be obtained by (5).

$$a_{u,i,k} = \frac{\exp(\hat{a}_{u,i,k})}{\sum\limits_{i=1}^{K} \exp(\hat{a}_{u,i,k})} \tag{5}$$

$a_{u,i} \in \mathbb{R}^K$ represents u attention weight to the i. From (5), $a_{u,i,k}$ can be calculated, which indicates the importance of the Kth characteristic of the item to the user. $a_{u,i,k}$ is unique. Focusing on the attention network, we can get the attention weight matrix and use the (6) to learn the attention interaction feature. At the final output layer, we perform a regression operation on the obtained L_K and then get a user's predicted rating for an item. In (7), w and b are the weight matrix and the offset vector, respectively.

$$L_K = a_{u,i,k} \cdot p_{u,k} \cdot q_{i,k} \tag{6}$$

$$\hat{r}_{u,i} = w.L + b \tag{7}$$

3.4 Learning

CFAM predicts the rating of unrated items by extracting user preferences and topic characteristics. This learning process is the same as the general deep neural network. We consider it as a regression task. This paper uses the square loss function of (8) below to train. $r_{u,i}$ represents the rating given in dataset D. $\hat{r}_{u,i}$ represents the predicted rating for the unknown user-item pair and the stochastic gradient descent method is used to optimize the model's parameters.

$$\zeta = \sum_{(r_{u,i}) \in D} \left(\hat{r}_{u,i} - r_{u,i} \right)^2 \tag{8}$$

4 Experiment

We implemented our experiments on Amazon's public dataset, which contains reviews from a wide variety of products. It has a total of 24 product categories. Our experiments have selected the five products that people use the most. The data sizes and sparse of these five types of products are different. We have made some restrictions: (1) Each user and item has at least five evaluations. We remove the user item review that does not meet the requirement. (2) Removing the copied and pasted review. (3) Extracting the user ID and item id and rating from each review record in the dataset. (4) Deleting punctuation marks, stop words in the review.

4.1 Experimental Settings

Data Set. Table 1 describes our data set details. We divide the data set into a training set, a validation set and a test set according to 0.8:0.1:0.1 [5, 6]. Each user has at least five reviews and we take three comments from it for training. We guarantee that each record at least has one scoring in the verification set and test set. Because the actual scene of the user review is not emulated, we can only use the review information on the training set. We can't get the data in the real scene for the verification set and the test set.

Baselines. The approach we proposed predicts unknown ratings based on ratings and reviews, so we compared the baselines that combine ratings with the reviews. To ensure fairness in the experiment, the hyperparameters we used in the comparative experiments were adjusted to the suitable values.

LDA. [11] As a baseline that combines text with product features, we consider Latent Dirichlet Allocation. By itself, LDA learns a set of topics and topic proportions (stochastic vectors) for each document.

LFM. [19] It uses known scoring matrices to estimate unknown rating.

HFT. [4] This is the state-of-the-art method that combines reviews with ratings. HFT models the ratings using a matrix factorization model with an exponential transformation function to link the stochastic topic distribution in modeling the review text and the latent vector in modeling the ratings. The topic distribution can be modeled on either users or items. On most datasets, the item specific topic distribution produces more accurate predictions. We report the results whichever are more accurate.

RMR. [13] This method also uses both ratings and reviews. Different from HFT and CTR, which use MF to model rating, it uses a mixture of Gaussian to model the ratings.

RBLT. [8] This method is the most recent method, which also uses MF to model ratings and LDA to model review texts. Instead of using an exponential transformation function to link the latent topics and latent factors (as in HFT), this method linearly combines the latent factors and latent topics to represent users and items, with the assumption that the dimensions of topics and latent factors are equal and in the same latent space. The same strategy is also adopted by ITLFM.

Evaluation Indicators and Parameter Settings. In this paper, root mean square error (RMSE) is used as an indicator to evaluate the performance of the experiment. The smaller the value, the better the experimental results. The number of the multi-layer perceptron layers is set to 2 [18]. Paper adopts the dropout technique [14] to prevent overfitting. The dropout factor is set to 0.5 and the learning rate is 0.001 [18]. For all of our experiments, the Adam [15] optimization method was used.

4.2   Analysis of Results

Performance Comparison. Table 2 is the MSE results. The best MSE of each dataset is in bold. We listed the performance of various models on the datasets. The compared methods of Table 2 are arranged in the order of time from the left to the right. In order to take into account the fairness of the comparison, we adjust the parameters to make the optimal each method to achieve the best performance. we find that the dimension K of the prediction vector affects the prediction effect to a certain extent. The experimental data of these five items shows that our experiment works best when K is 25.

Table 1. Data set details

Datasets	users	items	ratings	Sparsity
Baby	17,177	7,047	158,311	0.9987
Shoes	73,590	48,410	389,877	0.9988
H&K[a]	58,901	28,231	544,239	0.9997
Watches	62,041	10,318	68,356	0.9996
Sports	31,176	18,355	293,306	0.9995

[a] H&K is the abbreviation of Home&Kitchen

Table 2. Experimental results of rating prediction

Dataset	LDA	LFM	HFT	RMR	RBLT	CFAM
Baby	1.599	1.596	1.161	1.152	1.139	**1.112**
Shoes	0.279	0.293	0.226	0.251	0.209	**0.181**
H&K[a]	1.201	1.182	1.151	1.143	1.125	**1.106**
Watches	1.517	1.497	1.486	1.458	1.487	**1.445**
Sports	1.087	1.071	1.049	1.026	1.017	**0.980**

[a].H&K is the abbreviation of Home&Kitchen

Table 3. Experimental results under different activation functions

Datasets	Tanh	Sigmod	ReLu
Baby	1.119	1.143	**1.112**
Shoes	0.193	0.201	**0.181**
H&K	1.175	1.137	**1.106**
Watches	1.457	1.520	**1.445**
Sports	0.985	1.085	**0.980**

The first three methods in Table 2 only use the scoring information to predict the unknown rating. The latter two methods also use the probability distribution model of the potential topics, which extracts the potential topics from the review texts to obtain user preference and item characteristics. So RMR, RBLT are more effective than the previous three methods. In the HFT paper, the author is divided into User-HFT and Item-HFT. We quote an average of their two best effects.

Effects of Attention. From Figs. 3, 4 and 5, it can be found that when the size of the prediction vector changes, the value of the RMSE on each kind of prediction also changes. A common law can be derived from the three graphs. When K is less than 25, K becomes bigger and the RMSE of the experiment is smaller. When K is greater than 25, the performance of the experiment will decrease. Though RMR and RBLT incorporate reviews, we employ attention mechanism and remove the false assumption.

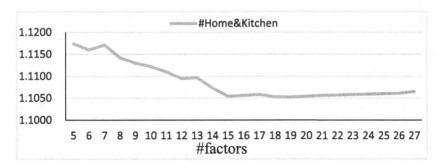

Fig. 3 Relationship between predictive performance of class Home&Kitchen commodity and size of prediction vector

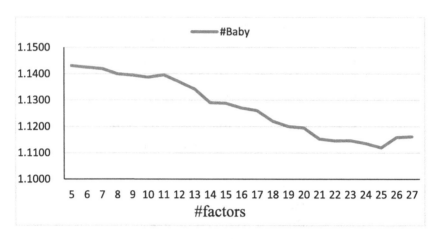

Fig. 4 Relationship between predictive performance and the size of prediction vector

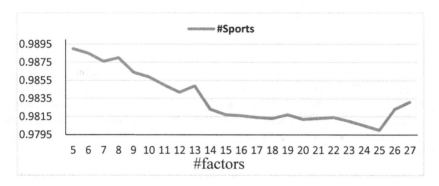

Fig. 5 Relationship between predictive performance and the size of prediction vector

CFAM is 3.4% higher than RMR while increasing by 2.3% over RBLT. It can be seen that adding the attention mechanism can greatly improve the accuracy of prediction.

In the experiment, we make the dimension of the review feature equal the size of the prediction vector. When extracting review features, items with more attributes will naturally extract more features. It can be seen from the Fig. 3 that with the increases the size of the prediction vector. Home&Kitchen first reaches the experimental optimal value. For both Sports and Baby items, when the prediction vectors size close to 25, they reach the best value.

Activation Function. Each layer of the neural network needs to pass a nonlinear transformation at the output layer, which is called activation. If not activated, all layers in the network are linearly transformed. As the nonlinear transformation can greatly increase the information stored in the network. In Table 3, we use three commonly use activation functions to test and found that in deep neural networks ReLU can better propagate the gradient back, so when the activation function for ReLu, the neural network can show the best learning ability and achieve the best results.

5 Conclusion and Discussion

This paper proposes a method which extracts latent factors from scores, learns potential features from reviews, and merges latent factors with potential features to achieve prediction of unknown rating. We design a new topic model to extract user preferences and item characteristics. This paper uses attention mechanisms to capture the different user's attention weights for an attribute of an item. Experiments were carried out on the Amazon five items dataset. The experimental results show that our proposed method significantly improves the accuracy of the prediction In the course of the experiment, we find that there are false comments in the review texts. False reviews will have a negative effect on the final recommendation, so our future research will be developed in the direction of how to identify false comments.

Acknowledgment. We would like to acknowledge the support from the National Science Foundation of China (61472095).

References

1. He, X., Zhang, H., Kan, M.-Y., Chua, T.-S.: Fast matrixfactorization for online recommendation with implicit feedback. In: SIGIR, pp. 549–558 (2016)
2. He, X., Chen, T., Kan, M.-Y., Chen, X.: Trirank: review-aware explainable recommendation by modeling aspects. In: ACM CIKM, pp. 1661–1670 (2015)
3. Blei, D.M., Ng, A.Y., Jordan, M.I.: Latent dirichlet allocation. J. Mach. Learn. Res. **3**(Jan), 993–1022 (2003)
4. McAuley, J., Leskovec, J.: Hidden factors and hidden topics: understanding rating dimensions with review text. In: ACM RecSys, pp. 165–172 (2013)
5. Ling, G., Lyu, M.R., King, I.: Ratings meet reviews, a combined approach to recommend. In: ACM RecSys, pp. 105–112 (2014)
6. Tan, Y., Zhang, M., Liu, Y., Ma, S.: Rating-boosted latent topics: understanding users and items with ratings and reviews. In: IJCAI (2016)
7. Cheng, Z., Ding, Y., Zhu, L., Kankanhalli, M.: Aspect-aware latent factor model: rating prediction with ratings and reviews. In: WWW, pp. 639–648 (2018)
8. Tan, Y., Zhang, M., Liu, Y., Ma, S.: Rating-boosted latent topics: understanding users and items with ratings and reviews. In: Proceedings of the 25th International Joint Conference on Artificial Intelligence, pp. 2640–2646 (2016)
9. Qiu, L., Gao, S., Cheng, W., Guo, J.: Aspect-based latent factor model by integrating ratings and reviews for recommender system. Knowl.-Based Syst. **110**, 233–243 (2016)
10. Bao, Y., Fang, H., Zhang, J.: TopicMF: simultaneously exploiting ratings and reviews for recommendation. In: Proceedings of the 28th AAAI Conference on Artificial Intelligence. pp. 2–8 (2014)
11. Zhang, W., Wang, J.: Integrating topic and latent factors for scalable personalized review based rating prediction. IEEE Trans. Knowl. Data Eng. **28**(11), pp. 3013–3027 (2016)
12. Wang, C., Blei, D.: Collaborative topic modeling for recommending scientific articles. In: Proceedings of the 17th ACM SIGKDD International conference on Knowledge Discovery and Data Mining, pp. 448–456. ACM (2011)

13. Ling, G., Lyu, M., King, I.: Ratings meet reviews, a combined approach to recommend. In: Proceedings of the 8th ACM Conference on Recommender systems, pp. 105–112. ACM (2014)
14. Srivastava, N., Hinton, G., Krizhevsky, A., Sutskever, I., Salakhutdinov, R.: Dropout: a simple way to prevent neural networks from overfitting. J. Mach. Learn. Res. **15**(1), 1929–1958 (2014)
15. Kingma, D.P., Ba, J.: Adam.: a method for stochastic optimization. arXiv preprint arXiv: 1412.6980 (2014)
16. Maas, A.L., Hannun, A.Y., Ng, A.Y.: Rectifier nonlinearities improve neural network acoustic models. In: Proceedings ICML (2013)
17. Cheng, Z., Ding, Y., He, X., Zhu, L., Song, X., Kankanhalli, M.: A3NCF: an adaptive aspect attention model for rating prediction. In: IJCAI (2018)
18. He, X., Liao, L., Zhang, H., et al.: Proceedings of the 26th International Conference on World Wide Web, Neural Collaborative Filtering, pp. 173–182 (2017)
19. Shen, Y., Jin, R.: Proceedings of the 18th ACM SIGKDD International Conference on Knowledge Discovery and Data Mining Social Latent Factor Model for Social Recommendation, p. 1303 (2012)

Sub-pixel Upsampling Decode Network for Semantic Segmentation

Guanhong Zhang⑩, Muyi Sun⑩, and Xiaoguang Zhou⁽✉⁾

Automation School, Beijing University of Posts and Telecommunications,
Beijing, China
{zghzgh1779,sunmuyi,zxg}@bupt.edu.cn

Abstract. Many recent semantic segmentation frameworks use encoder-decoder design strategy. Most of them employ a relatively simple encoder with a complex stratified decoder. In this work, we explore the potential of combining a robust encoder with a simple one-layer but finely designed decoder. We refine the structure of the decoder module with larger effective receptive field and residual refinement connection. We also explore different strategies for fusing multi-stage contextual feature, to improve the effectiveness of the proposed framework. Specifically, we introduce a Sub-pixel Upsampling Decoder module to decode more detailed information by learning spacial downsampled label in different decoder channel groups. We test the proposed model in the dataset of PASCAL VOC 2012 segmentation task, and the results show the improvement of our method under our computational limitation.

Keywords: Semantic segmentation · Convolutional neural networks · Computer vision

1 Introduction

Semantic segmentation is a task of assigning category labels such as "cat" or "dog" to every pixel in the image. It is a challenging task and a fundamental technique in computer vision, which could facilitate numerous applications such as automatic driving, robot sensing, and augmented reality.

Recent semantic segmentation frameworks [3,17,20,21,29,36] are mostly developed from the Fully Convolutional Network (FCN) [18]. FCN is a typical segmentation framework with encoder-decoder structure. The encoder part can embedded contextual information into the encoded feature maps, and a layer-by-layer decoder is employed to generate dense segmentation predictions. Although the encoded feature maps are usually highly semantic, the precision of prediction masks is limited due to the downsampled resolution, which is caused by common backbone models [12,13,16,24,26].

To address this problem, many decoder-based methods [10,17,21] fuse high semantic level but low-resolution features and high-resolution but not highly semantic features to refine details, having obtained impressive results.

© Springer Nature Singapore Pte Ltd. 2019
K. Knight et al. (Eds.): ICAI 2019, CCIS 1001, pp. 15–28, 2019.
https://doi.org/10.1007/978-981-32-9298-7_2

Usually these models use pre-trained convolutional model as encoder without obvious changes, and are composed of stratified decoder modules to utilize multi-stage contexts with different semantic levels. However, Another style of semantic segmentation frameworks such as [3–5, 36] follow a backbone-based design strategy, which introduce powerful context modules to embed different scale context information, predicting segmentation mask without complex layer-by-layer decoder.

Motivated by the above observation, we explore the potential of combining a robust encoder with a simple one-layer but finely designed decoder, trying to combine the advantages of both encoder-decoder based method and backbone-based method. We select a widely used semantic segmentation network [5] as our encoder. We propose a decoder module named Sub-pixel Upsampling Decoder, which can take full advantage of highly semantic encoded feature by utilizing high-resolution features.

Our design strategy have the following two motivations: (1) to finely decode the encoded feature maps with low computational cost, we design a decoder with separable factorized convolution and residual refinement connection; (2) to combine the upsampling process with decoding structure explicitly, we propose a novel upsampling decode module: *Sub-pixel Upsampling Decoder*. We further explore the performance of different feature fusion strategies, and choose the best performing approach on our model. We test the efficiency of our proposed framework on PASCAL VOC 2012 dataset.

Our contributions can be summarized as:

- We propose an efficient semantic segmentation network with a robust encoder and a simple one-layer but finely designed decoder.
- We develop a novel upsampling method which can learn spacial downsampled label in different decoder channel groups.
- In our proposed single-layer decoder structure, we study different feature fusion strategies and decoder structures to find the most efficient design.

2   Related Work

Atrous Convolution. The motivation of atrous convolution is to increase feature map resolution and to enlarge the effective receptive field of kernels by inserting zeros between each pixel in convolutional kernels, which can produce more dense prediction in semantic segmentation networks. Yu and Koltun [32] used stacked atrous convolution layers with increasing rates of dilation to learn better contextual representations with larger receptive fields. Similar strategies are adopted in [29, 33].

Contextual feature fusion. Most networks designed for semantic segmentation employ contextual feature fusion with different purposes and design strategies. RefineNet [17], and [21, 31, 35] have demonstrated the effectiveness of segmentation decoders based on stratified structure. Besides, the backbone-based methods with extra context module have also showed impressive performance. ASPP module is employed in DeepLab [3–5] to capture objects of different size.

Pyramid pooling module in PSPNet [36] serves the same purpose through different implementation.

Depthwise separable convolution. Depthwise separable convolution [23,28], or group convolution [16,30], is an operation that can significantly reduce the computation complexity and number of parameters, while maintaining similar performance. Many recent model designs adopt this strategy [14,34,39].

Sub-pixel convolution. Sub-pixel convolution is an approach of upsampling, which comes from [1,22] in super-resolution task. [29] have employed Sub-pixel convolution to replace the common upsampling method such as bilinear interpolation or transposed convolution, which can recover fine-detailed information that is generally missing in the upsampling operation.

3   Approach

In this section, we briefly review the atrous convolution and Atrous Spatial Pyramid Pooling module [5] which is used as our encoder module, then introduce the proposed decoder module. We also discuss the feature fusion problem and the motivation of upsampling directly by the whole decoder. Finally, we describe our complete network structure and its details.

3.1   Atrous Convolution as Encoder

Atrous Convolution. Atrous convolution, or dilated convolution is a powerful and widely used operation in semantic segmentation task. The operation can change filter's field of view explicitly, while maintaining the effective receptive field of pre-trained classification framework to generate dense prediction mask. Atrous convolution can be defined as:

$$y[i] = \sum_{\omega} x[i + r \cdot \omega] f[\omega]. \tag{1}$$

where $x[i]$ is the input feature map, $y[i]$ is the output feature map, $f[\omega]$ denotes a filter of width ω with atrous rate r which is used to sample the input feature map $x[i]$. A semantic segmentation network with atrous convolution convolves the input x with upsampled filters constructed by inserting $r - 1$ zeros between each pixel in the convolutional kernel. This operation can be applied iteratively to replace all downsampling operations for maintaining the same resolution.

Atrous Spatial Pyramid Pooling. The encoder which is used as our encoder module employ an efficient context module, called Atrous Spatial Pyramid Pooling module [3–5]. ASPP module captures multi-scale information by applying atrous convolution with different rates parallelly, having shown impressive accuracy with considerable computational costs.

As the frameworks based on atrous convolution and ASPP module have proved their success, we implement an encoder with similar structure of [5], using a pretrained ResNetV2 as backbone network. Due to our limited computational budget, we only apply rate = 2 to the last block for output stride = 16 to avoid memory overflow. We extract the last feature map before logits as the encoder output, and combine it with feature maps before striding operations in each convolution blocks as the decoder input.

3.2   Sub-pixel Upsampling Decoder

Most networks designed for semantic segmentation task employ either backbone-based methods with computational costly context module to embed different scales context information, or decoder-based methods which are usually composed of a considerable complex layer-by-layer decoder modules to utilize multistage context with different semantic level.

To combine the advantages of both approaches within practical computing ranges, the decoder must be efficient but low-cost because the proposed encoder is high-performance but costly in terms of computation. Additionally, the Sub-pixel Upsampling operation have shown its advantage over bilinear upsampling method [29]. We further integrate Sub-pixel Upsampling operation with the decoder structure explicitly. Guided by these two principles, we propose a single-layer and light weight decoder, called *Sub-pixel Upsampling Decoder (SUD)*.

Feature Fusion Strategy. Most decoder-based frameworks have an encoder-decoder structure as shown in Fig. 1. The encoder part is usually a convolutional network pretrained on large-scale classification datasets such as [16], which generates feature maps with different semantic levels and resolutions. Then the decoder part fuse these features to predict final segmentation masks. In general, the feature fusion process can be formulated as:

$$y_l = T_l(x_l + Upsample(y_{l+1})) \tag{2}$$

where y_l is the fused feature of l-th level, and x_l corresponds to the same semantic level feature generated by the encoder. T_l is a non-linear transformation for decoding the combined feature, which is usually a complex convolution module. In contrast, we fuse features of different semantic levels simultaneously, and use a single-layer decoder to generate prediction results. Our feature fusion strategy can be defined as:

$$y_l = T_l(x_l + Upsample(x_{l+1}) + Upsample(x_{l+2}) + \ldots) \tag{3}$$

In SUD, 4 different semantic levels of feature maps are extracted from our encoder framework, whose spatial resolutions are $\{1/4, 1/8, 1/16\}$ of the input size. We extract feature maps before and after the ASPP module of encoder, which are same at resolution but different at semantic level. As [6] have mentioned, the coarser feature channels should be about 1/5 of finer features in channel number. We select several subsets of these features to generate prediction

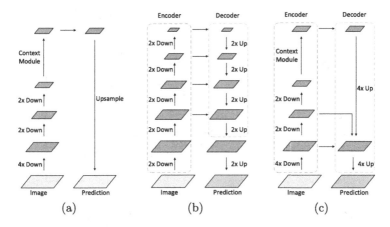

Fig. 1. Different feature fusion strategies. (a) Backbone based model without fusing with low-level features. (b) Encoder-decoder based model with a complex stratified decoder. (c) Our proposed SUD, which fuse low-level and high-level features directly and decode it with a simple one-layer decoder.

results, and use them to retrain the whole system. In Sect. 4 we will introduce these experiments detailedly. We find that combining the encoded feature with 1/4 and 1/8 downsampled features would produce the best performance.

Studies [37] have shown that even the receptive field of a deep convolutional network can be very large, the network tends to gather information from a smaller region in the receptive field rather than receive information equally, and the valid receptive field becomes much smaller. To refine a convolutional layer for boosting its performance, It is naturally to enlarge the convolution kernel size for a larger valid receptive field.

However, enlarging the kernel size directly could be costly in computation, which violates the design principle mentioned above. Consider a decoder architecture as in Fig. 2(a), if we replace the 3×3 convolution with a 5×5 convolution, the computational cost will be $25/9 = 2.78$ times more expensive of the original operation.

Inspired by [27], we factorize the $n \times n$ convolution to a two-stream combination of $1 \times n + n \times 1$ and $n \times 1 + 1 \times n$ convolutions with $2/k$ multiply-add times. This factorization could access a larger receptive field without excessive cost increasing. In practice, we have found that enlarge the convolution kernel size extremely would not lead to better performance, so our SUD module employs two convolution paths of $1 \times 5 + 5 \times 1$ and $5 \times 1 + 1 \times 5$ convolutions. Using nonlinearity after these convolutions would decrease accuracy, we therefore apply linear activation here.

Decoder Structure. Depthwise separable convolution, or group convolution, is a form of factorized convolutions that widely used in compressed model

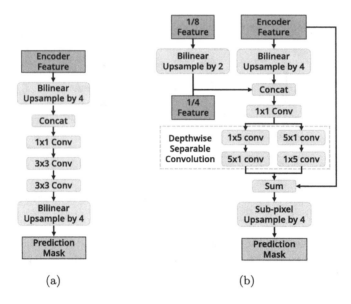

(a)   (b)

Fig. 2. Decoder structure. (a) The baseline decoder with 2 layers of 3×3 convolution. (b) Our proposed Sub-pixel Upsampling Decoder.

(e.g. [14,34]). Depthwise separable convolution factorize a standard convolution into a depthwise convolution across channels and a 1×1 convolution called pointwise convolution, with no obvious performance loss [7]. By splitting convolution as two steps of convolution, we can reduce computation cost of $1/n + 1/\omega_f^2$, where n is the number of output channel and ω_f is the width of filters. We apply depthwise separable convolution on every but 1×1 convolution layer in decoder.

Additionally, we propose a residual connection shown in Fig. 3(b) to refine boundary. Since the robust and high-performance encoder can encode high-semantic and precise prediction, The boundary information restoration can be modeled as a residual branch. The residual branch concat encoded feature and high-resolution features to generate boundary feature, and the decoder combine boundary feature with upsampled encoded feature by element-wise addition, then generate segmentation prediction.

Sub-pixel Upsampling. Sub-pixel upsampling module is used in Super resolution [1,22], Low-light image processing [2], and semantic segmentation [29,35] for different purposes. Consider an input image with size $H \times W$, and the semantic segmentation network would generate prediction logits with dimension $H \times W \times C$, where C is the number of categories. However in general the final feature map before making predictions has dimension $h \times w \times c$, where $h = H/f$, $w = W/f$, f is the downsample factor, and c is the channel number of final feature map that usually greater than C. The key operation of sub-pixel upsampling is to make sure the size of final feature map has size $h \times w \times (f^2 \times C)$, and divide the whole final feature map into equal f^2 subsets with definitely same size. Then the feature map is interleavedly reshaped to $H \times W \times C$ with a softmax layer and an argmax operator to generate final prediction mask.

[29] have shown that the combination of sub-pixel upsampling and a single-layer convolution has capacity to replace bilinear upsampling in segmentation task. Our proposed decoder is not a complex stratified decoder, on the other hand, the decoder part of an encoder-decoder framework designed for semantic segmentation task should be able to upsample feature during processing procedure intrinsically.

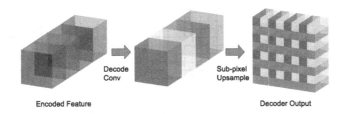

Fig. 3. Illustration of the design of Sub-pixel Upsampling Decoder.

Motivated by the above observations, we design the proposed Sub-pixel Upsampling Decoder module that combines sup-pixel upsampling operation with decoder structure explicitly, for direct upsampling in the decoder part without bilinear upsampling operation. The SUD module receive encoded feature with $f^2 \times C$ channels, where f is the downsample factor and C is the number of categories. The whole label mask is transformed into f^2 small mask with downsample rate of $1/f$, each of them have C channels. These small label masks can be learned with different channel groups in decoder separately. In practice, the channel depth of final output feature map is mapped to spatial grid, as illustrated in Fig. 3. The SUD module can decode features at the original resolution by embedding upsampling operation in decoder channel groups. It is capable of recovering fine-detailed information compared with common upsampling operation.

3.3 Overall Framework

Our overall model are shown in Fig. 4. We use pretrained ResNetV2 [12,13] with atrous convolution and ASPP module [5] as the segmentation encoder. Multi-semantic level feature maps with different downsample rates are extracted from different stages in the encoder network. Sub-pixel Upsampling Decoder fuses multi-scale semantic feature generated by encoder part and low-semantic level features to predict detailed segmentation mask. Score maps of lower resolution will be upsampled directly by the SUD module, and is used to output the dense predict result. The details can be referred to Fig. 4.

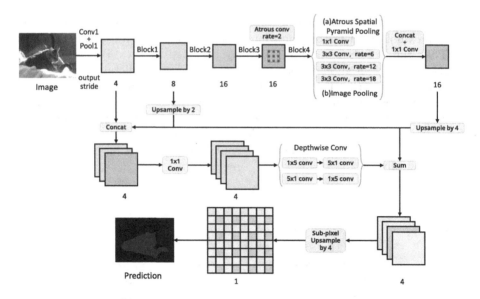

Fig. 4. Overall architecture of our approach. The SUD module fuses encoded feature and high-resolution features, then decodes and upsamples the fused feature simultaneously to generate dense prediction mask.

4   Experiment

We evaluate our approach on the standard benchmark of PASCAL VOC 2012 [8,9], which contains 1464 training images, 1449 validation images, and 1456 test images. All these images are labeled to 20 object categories and one background class with pixel-level annotation. We also employ the Semantic Boundaries Dataset [11] as extra annotation dataset, in which the training dataset is augmented to have 10582 images. The performance is measured in standard terms of mIOU (mean intersection-over-union) across all 21 classes. We use a ResNetV2 101 [12,13] pretrained on ImageNet as our base model, and applying atrous convolution and ASPP module [5] to build our encoder network. We employ atrous convolution with atrous rate = 2 to the last block for encoder output stride = 16, considering our limited computation budget. All our experiments are implemented with Tensorflow [19] and running on a single NVIDIA TITAN X GPU. We use mean subtraction and randomly lift-right flip in training. Randomly scaling the input images from 0.5 to 2.0 is also applied during training. Image patches are cropped to 513 during training. The pretrained ResNetV2 and the encoder context module all include batch normalization [15]. Since large batch size is required to train batch normalization parameters, we magnify the batch size as far as possible on our single GPU. We compute the batch normalization statistics with a batch size of 8. However, we observe that applying batch normalization on factorized $1 \times n + n \times 1$ convolution would degenerate performance as shown in Table 3, so we do not employ batch normalization parameters to these convolution layers.

4.1 Ablation Study

Feature Fusion Strategy. We extract the low semantic level but high-resolution features with downsample factor of 1/4 and 1/8 before striding operation in Block1 and Block2 respectively. We also extract features before and after the ASPP module in encoder, which have same resolution but different semantic level. The feature map before ASPP is coarser and without multi-scales information.

Table 1. Effects of different feature fusion strategies. Extend channel means the input channels of decoder is extended to quadruple (1216 in total). The decoder structure of all experiments in this subsection are same as Fig. 2 (a).

1/4 feature	1/8 feature	1/16 feature	extend channel	mIOU(%)
				73.47
✓				74.23
✓	✓			74.52
✓	✓	✓		73.52
✓			✓	74.23
✓	✓		✓	74.42
✓	✓	✓	✓	73.94

We implement a decoder as shown in Fig. 2(a) as our baseline. The output of encoder have a depth of 256 channels, and the depth would remain unchanged until generating the final logits. Then we fuse different subset of these feature maps to find the most efficient feature fusion strategy. Recent study [6] have shown that features with low semantic level should be about 1/5 of encoded feature in channel number, so the dimensions of coarser features are reduced equally by a pointwise convolution, to 48 channels in total. Then all features are concated to produce the prediction result of 1/4 resolution (129×129 in training), and using bilinear interpolation to upsample to the original size of input.

From Table 1 we can clearly see that fusing low semantic level but high-resolution features with multi-scales and high-semantic encoded feature can significantly improve the performance. However the coarser feature which have same resolution with encoded feature should not be involved in fused feature as it doesn't contain more detailed information than encoded feature.

The best strategy to fuse features is to extract the features with higher resolution than encoded feature, which means the 1/4 and 1/8 downsampled features in our framework.

Table 2. Ablation experiments of the decoder structure. The first line is the baseline decoder same as Fig. 2(a). 5×5 means factorized convolution layer with $1 \times 5 + 5 \times 1$ convolution. The factorized convolution may be applied once or twice. All the factorized convolutions are depthwise separable. Nolinear means that if we employ Nolinear activation after factorized convolution operations. Refinement indicate if the residual refinement connection is applied.

5×5, 1	7×7, 1	5×5, 2	7×7, 2	Nolinear	Refinement	mIOU(%)
						74.23
✓						73.22
	✓					74.17
		✓				74.50
			✓			74.40
		✓		✓		73.11
			✓	✓		73.96
					✓	74.39

Decoder Design. In this subsection we explore the structure design of the proposed decoder. We examine the effect of different decoder architectures by decoding concated feature of encoded feature with size $33 \times 33 \times 256$ and 1/4 downsampled feature with size $129 \times 129 \times 48$. The encoder part of proposed framework maintains unchanged. The factorized convolution layer contains two convolution paths of $1 \times n + n \times 1$ and $n \times 1 + 1 \times n$ convolutions. All the factorized convolutions are depthwise separable. We also test the decoder that repeat the factorized convolution twice. The residual refinement connection adds encoded feature with the output feature of decoder element-by-element to enhance the intra-class consistency. The comparison of these decoder structures is shown in Table 2.

From Table 2, we can see that repeating $1 \times 5 + 5 \times 1$ convolution twice would generate the best performance, and residual refinement connection can also improve the accuracy. Applying nonlinear activation would degenerate performance badly. Our final proposed decoder is a combination of the best branches above, as illustrated in Fig. 2(b).

Sub-pixel Upsampling Decoder. In our proposed decoder structure, the channel number of encoder output and decoder output are both $f^2 \times C$, where $C = 21$ and $f = 4$ in our experiments. Then the summarized feature with size $129 \times 129 \times 336$ is upsampled to $513 \times 513 \times 21$ with Sub-pixel Upsampling method. As a contrast, we test a decoder with 336 channels and bilinear interpolation upsample, and a decoder of original depth 256 which raising the dimension of output feature with a pointwise convolution to 336 channels and then upsampling by Sub-pixel Upsampling operation. The other operations remain unchanged. The comparison of upsampling methods mentioned above is listed in Table 3.

Table 3. Ablation study on the effect of Sub-pixel Upsampling Decoder (SUD). The architecture of SUD is in Fig. 2(b). The base-decoder model is a SUD of 256 channels without Sub-pixel upsampling operation. All experiments in this subsection are modified from the base-decoder.

Method	mIOU(%)
No-decoder	73.47
Base-decoder	74.69
Base+BN	74.46
336+Upsample	74.77
256+1 × 1+SU	74.01
SUD	74.96

4.2 Results and Discussion

Results. Figure 5 visualizes some representative segmentation results of our proposed model and the baseline model without decoder. It is clear that our method have distinct advantage over the baseline model in details.

Limitations. Our proposed model and experiments have some limitations. Due to our limited computational budget, we set the batch size $= 8$, which cannot take full advantage of batch normalization. Large batch size is usually essential for backbone-based models, our reimplementation can hardly achieve the ceiling with extremely large batch size (e.g. batch size 32 or even larger [6, 31]). Besides, We run all our experiments on a single NVIDIA TITAN X GPU, the performance of the proposed model on multi-GPUs remains to be tested.

Another limitation is the output channels of SUD module. The semantic segmentation task on PASCAL VOC 2012 dataset have only 21 classes, however some recent large scale segmentation datasets provide far more classes [38] on which the output channel number of our proposed model would be dramatically large. Inspired by [25], a possible solution is to introduce super-classes for generating segmentation mask, and then perform fine-grained classification without position-sensitive filters.

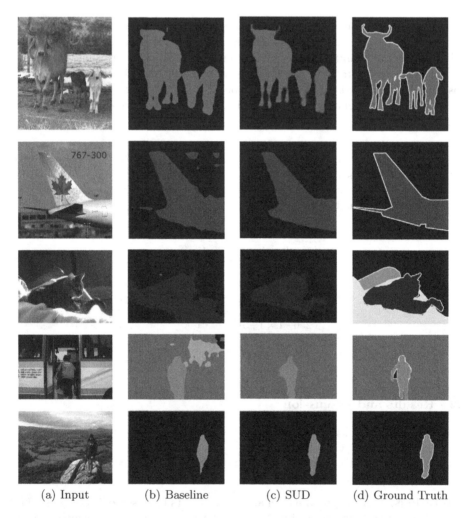

| (a) Input | (b) Baseline | (c) SUD | (d) Ground Truth |

Fig. 5. Representative examples of segmentation results on PASCAL VOC 2012 validation set.

5 Conclusions

In this work, we discuss the design strategy of encoder-decoder based segmentation frameworks, and propose a frameworks with a robust backbone-based encoder and an efficient and low-weight decoder. We experiment the most efficient decoder design and feature fusion method. Eventually, We present a novel Sub-pixel Upsampling Decoder module, which integrate sup-pixel upsampling operation with decoder structure. Our proposed model shows its efficiency on PASCAL VOC 2012 benchmark and have potential to be employed in other segmentation frameworks.

References

1. Aitken, A., Ledig, C., Theis, L., Caballero, J., Wang, Z., Shi, W.: Checkerboard artifact free sub-pixel convolution: a note on sub-pixel convolution, resize convolution and convolution resize. arXiv preprint arXiv:1707.02937 (2017)
2. Chen, C., Chen, Q., Xu, J., Koltun, V.: Learning to See in the Dark. arXiv preprint arXiv:1805.01934 (2018)
3. Chen, L.C., Papandreou, G., Kokkinos, I., Murphy, K., Yuille, A.L.: Semantic image segmentation with deep convolutional nets and fully connected crfs. arXiv preprint arXiv:1412.7062 (2014)
4. Chen, L.C., Papandreou, G., Kokkinos, I., Murphy, K., Yuille, A.L.: Deeplab: semantic image segmentation with deep convolutional nets, atrous convolution, and fully connected crfs. arXiv preprint arXiv:1606.00915 (2016)
5. Chen, L.C., Papandreou, G., Schroff, F., Adam, H.: Rethinking atrous convolution for semantic image segmentation. arXiv preprint arXiv:1706.05587 (2017)
6. Chen, L.C., Zhu, Y., Papandreou, G., Schroff, F., Adam, H.: Encoder-decoder with atrous separable convolution for semantic image segmentation. arXiv preprint arXiv:1802.02611 (2018)
7. Chollet, F.: Xception: deep learning with depthwise separable convolutions. arXiv preprint arXiv:1610.02357 (2017)
8. Everingham, M., Eslami, S.A., Van Gool, L., Williams, C.K., Winn, J., Zisserman, A.: The pascal visual object classes challenge: a retrospective. Int. J. Comput. Vision 111(1), 98–136 (2015)
9. Everingham, M., Van Gool, L., Williams, C.K., Winn, J., Zisserman, A.: The pascal visual object classes (voc) challenge. Int. J. Comput. Vision 88(2), 303–338 (2010)
10. Ghiasi, G., Fowlkes, C.C.: Laplacian pyramid reconstruction and refinement for semantic segmentation. In: ECCV (2016)
11. Hariharan, B., Arbeláez, P., Bourdev, L., Maji, S., Malik, J.: Semantic contours from inverse detectors. In: ICCV (2011)
12. He, K., Zhang, X., Ren, S., Sun, J.: Deep residual learning for image recognition. In: CVPR (2016)
13. He, K., Zhang, X., Ren, S., Sun, J.: Identity mappings in deep residual networks. In: ECCV (2016)
14. Howard, A.G., et al.: Mobilenets: efficient convolutional neural networks for mobile vision applications. arXiv preprint arXiv:1704.04861 (2017)
15. Ioffe, S., Szegedy, C.: Batch normalization: accelerating deep network training by reducing internal covariate shift. arXiv preprint arXiv:1502.03167 (2015)
16. Krizhevsky, A., Sutskever, I., Hinton, G.E.: Imagenet classification with deep convolutional neural networks. In: NIPS (2012)
17. Lin, G., Milan, A., Shen, C., Reid, I.: Refinenet: Multi-path refinement networks for high-resolution semantic segmentation. In: CVPR (2017)
18. Long, J., Shelhamer, E., Darrell, T.: Fully convolutional networks for semantic segmentation. In: CVPR (2015)
19. Abadi, M., Agarwal, A., Barham, P., et al.: Very deep convolutional networks for large-scale image recognition. arXiv preprint arXiv:1603.04467 (2016)
20. Noh, H., Hong, S., Han, B.: Learning deconvolution network for semantic segmentation. In: CVPR (2015)
21. Peng, C., Zhang, X., Yu, G., Luo, G., Sun, J.: Large kernel matters–improve semantic segmentation by global convolutional network (2017)

22. Shi, W., et al.: Real-time single image and video super-resolution using an efficient sub-pixel convolutional neural network. In: CVPR (2016)
23. Sifre, L., Mallat, S.: Rigid-motion scattering for image classification. Ph.D. thesis, Citeseer (2014)
24. Simonyan, K., Zisserman, A.: Very deep convolutional networks for large-scale image recognition. arXiv preprint arXiv:1409.1556 (2014)
25. Singh, B., Li, H., Sharma, A., Davis, L.S.: R-FCN-3000 at 30fps: decoupling detection and classification. arXiv preprint arXiv:1712.01802 (2017)
26. Szegedy, C., et al.: Going deeper with convolutions. In: CVPR (2015)
27. Szegedy, C., Vanhoucke, V., Ioffe, S., Shlens, J., Wojna, Z.: Rethinking the inception architecture for computer vision. In: CVPR (2016)
28. Vanhoucke, V.: Learning visual representations at scale (2014)
29. Wang, P., et al.: Understanding convolution for semantic segmentation. arXiv preprint arXiv:1702.08502 (2017)
30. Xie, S., Girshick, R., Dollár, P., Tu, Z., He, K.: Aggregated residual transformations for deep neural networks. In: CVPR (2017)
31. Yu, C., Wang, J., Peng, C., Gao, C., Yu, G., Sang, N.: Learning a discriminative feature network for semantic segmentation. arXiv preprint arXiv:1804.09337 (2018)
32. Yu, F., Koltun, V.: Multi-scale context aggregation by dilated convolutions. In: ICLR (2016)
33. Yu, F., Koltun, V., Funkhouser, T.: Dilated residual networks. In: CVPR (2017)
34. Zhang, X., Zhou, X., Lin, M., Sun, J.: Shufflenet: An extremely efficient convolutional neural network for mobile devices (2017)
35. Zhang, Z., Zhang, X., Peng, C., Cheng, D., Sun, J.: Exfuse: Enhancing feature fusion for semantic segmentation. arXiv preprint arXiv:1804.03821 (2018)
36. Zhao, H., Shi, J., Qi, X., Wang, X., Jia, J.: Pyramid scene parsing network. In: CVPR (2017)
37. Zhou, B., Khosla, A., Lapedriza, A., Oliva, A., Torralba, A.: Object detectors emerge in deep scene CNNs. arXiv preprint arXiv:1412.6856 (2014)
38. Zhou, B., Zhao, H., Puig, X., Fidler, S., Barriuso, A., Torralba, A.: Scene parsing through ade20k dataset. In: ICCV (2017)
39. Zoph, B., Vasudevan, V., Shlens, J., Le, Q.V.: Learning transferable architectures for scalable image recognition. arXiv preprint arXiv:1707.07012 (2017)

Tracking Algorithm Based on Dual Residual Network and Kernel Correlation Filters

Xiaolin Tian, Fang Li[✉], and Licheng Jiao

Key Laboratory of Intelligent Perception and Image Understanding of Ministry of Education, International Research Center of Intelligent Perception and Computation, International Collaboration Joint Lab in Intelligent Perception and Computation, Xidian University, Xian 710071, Shaanxi Province, China
1517637063100163.com

Abstract. Visual target tracking is a target detection task for a period of time. During this period, the tracking target will undergo significant appearance changes due to deformation, sudden movement, complex background and occlusion. These changes make visual target tracking challenging. In this paper, a target tracking algorithm based on dual residual neural network and kernel correlation filters is proposed, which mainly solves the problems of inaccurate tracking. This method combines depth residual neural network with kernel correlation filter tracking algorithm. The template matches the depth residual feature to determine the location of the target and the kernel correlation filter based on residual feature is designed to detect the target, finally, the template matching result and kernel correlation filtering result are fused to determine the location of the target. The experimental results on a large-scale standard data set show that the proposed algorithm has the advantages of high accuracy. Compared with the previous algorithm, the proposed algorithm has good performance in tracking visual deformation, occlusion and fuzzy video objects.

Keywords: Visual object tracking · Dual residual network · Kernel correlation filters · Template matching

1 Introduction

Visual object tracking is an important part of computer vision research. It combines computer image processing, pattern recognition, artificial intelligence and other related fields to form a kind of automatic track target. Given the initialized state of a target object in a frame of a video, the goal of tracking is to estimate the states of the target in the subsequent frames. An ideal tracker can be used in target recognition, target classification, behavior understanding, intelligent monitoring and it should be accurate an have high durability.

Recently, features based on convolutional neural networks(CNNs) have demonstrated the most advanced results in target recognition and image classification [1,2]. Therefore, people began to use the rich functional hierarchy of CNNs

© Springer Nature Singapore Pte Ltd. 2019
K. Knight et al. (Eds.): ICAI 2019, CCIS 1001, pp. 29–42, 2019.
https://doi.org/10.1007/978-981-32-9298-7_3

to achieve powerful visual tracking [4,5]. These methods pretrained their networks on large scale datasets, and finetuned the networks with the first frame of a specific sequence. In addition, object tracking approaches based on correlation filter(CF) [16] and convolutional neural networks(CNNs) have shown continuous performance improvements in visual object tracking tasks. However, there are several problems in using these two algorithms which remained to be solved. Both HCF(Hierarchical Convolutional Features for Visual Tracking) and HDT(Hedged Deep Tracking) use the rich features extracted from some CNN layers to learn an adaptive correlation filter, and use the correlation response maps to infer the target location. In order to optimize the network during the process of training network, add the pool layer in the network. Because the existence of the pool layer causes loss in the feature extraction process, the features will be degenerated when the network deepens depth. This is a disadvantage of network extracting features. Some algorithms weakened it by resizing the feature map, but no matter which interpolation algorithm used here, resized feature map will be changed and affect the tracking performance. As described in [2,11], many other nontrivial visual recognition tasks also have greatly benefited from very deep models. In the residual convolutional neural network, due to the existence of the residual module, the phenomenon that the network performance decreases rapidly with the increase of the neural network depth disappears.

In this paper, we propose a tracking algorithm based on dual residual network and kernel correlation filters. First, match template based on dual residual network, this dual residual network has a siamese structure, one is named as template network, use it to extract features from the target which to be tracked and use the extracted features as template, the other one is called detection network, for extracting features from the image containing the target of the next frame. By matching the features obtained from the two networks, get the target locations. Second, kernel correlation filter tracking algorithm based on residual network learns adaptive correlation filters on the features from several layers of residual network, these features are often multi-dimensional and contain rich informations of the target image. The kernel filter built with these features is more advantageous in process of visual target localization. After every target position of one frame is determined, the algorithm will update parameters once. There are two innovations in this algorithm. The first is to use the residual network with superior image classification performance for target tracking. The second uses a dual residual network to perform template matching to determine the location, and the template features are updated during the tracking process. Experiments on a large-scale benchmark dataset [3] with 100 challenging sequences show that the proposed algorithm performs favorably against state-of-the-art methods.

2   Related Work

We will make a brief review of tracking methods closely related to this work, refer the readers to a comprehensive review on visual tracking in. [6]

2.1 CF Based Trackers

The most plain modules that CF applies to tracking is the correlation operation which is a measure of the similarity of two signals. The more similar two signals are, the higher correlation value is. In the application of visual tracking algorithm based on CF, it is necessary to design a correlation filter according to the target image and find the max response value of the response map. The highlight of the CF method is its computational efficiency. In the MOSSE [8] algorithm, the FPS achieves a speed of 669 frames per second. This is because CF converts the time domain-dependent convolution operation into the Fourier domain multiplication operation, which saves a lot of time during the training and testing phases. The MOSSE training tracker uses the pixel features of the target itself. Later, the features used in the CF algorithm added grayscale, RGB, HOG, and so on. [16] Although these algorithms have achieved good results, all of these algorithms use only one correlation filter, which limits the power of tracker based on CF. In this work, exploit the computation filters to construct an ensemble tracker where each component filter is based on features extracted from one convolutional layer of a residual networks.

2.2 CNNs Based Trackers

Deep convolutional neural networks have led a series of breakthroughs in image classification and recognition. [9,10] Wang and Yeung proposed a deep learning tracker (DLT) [11] using a multi-layer autoencoder network, this network is pre-trained in an unsupervised manner over a portion of the 80M tiny image dataset. On the other hand, Wang and Liu propose to learn a two-layer neural network in a video repository in which temporary slowness constraints are imposed on feature learning. [12] Li constructed a different instance of the CNNs classifier for the target object to exclude noise samples during model update. [13] The depth feature is combined with the correlation filtering algorithm in HCF, as the features extracted by different layers of the convolutional neural network have different characteristics. The front layer has high resolution, the feature contains more position information, and the back layer contain more semantic information. The authors use a three-layer network to train the correlation filters separately, and weight them to get the final response position of the target. Deep networks naturally integrate low/medium/advanced feature and classifiers in an end-to-end, multi-layered manner and enrich the characteristics of "features". Previously proposed tracking algorithm use the features from CNNs did not considered of the influences of the deepening of the convolutional neural network. Based on these observations, use Resnet50 for the CNNs features extract.

2.3 Template Matching Based Tracker

Template matching is first introduced in visual tracking in [7]. It is extremely simplicity but has achieved state-of-the-art performance in multiple benchmarks.

To find the position of the object in a new frame, this method exhaustively test all possible locations and choose the candidate with the maximum similarity to the past appearance of the object. Pre-trained network before target tracking, SiameseFC trains a Siamese network to locate an exemplar image within a larger search image. Despite this method has superior performance, it is difficult to reproduce and requires a highly equipped hardware environment. In this paper, use the pre-trained ResNet-50 on ImageNet [14] for template matching tracking.

3   Overview

As shown in Fig. 1, the proposed approach consists of four parts: First, the pre-trained ResNet-50 is used to extract feature maps from object regions and use these feature maps to template matching to determine the location of the target. Second, use the extracted feature maps to train the correlation filter and use the filter to complete the target tracking. Finally, use the dual residual network for template matching and weight matching and correlation filtering results. The algorithm implementation steps are as follows: (1) Extracting a depth template network and a depth detection network in a dual residual depth classification network, extracting template image features using a depth template network, extract the detected image features using the depth detection network. Template matching the template features on the detection features to obtain a template matching map to determines the target position. (2) Construct a kernel correlation filter based on the target features extracted by the depth network, uses these kernel correlation filter to determine the target position. (3) Fuses the template matching and the target position determined by the kernel correlation filtering to determines the final target position. Tracking algorithm based on dual residual network and kernel correlation filters can achieve accurate and effective tracking of targets in the case of target deformation, fuzzy or occlusion.

4   Proposed Algorithm

In this section, introduce the technical details of the proposed algorithm, and describe the update scheme of the filter parameters and templates.

4.1   Deep Features Extracted from Residual Network

Convolutional neural networks like VGG-Net, AlexNet etc. have been used in visual tracking and have achieved good tracking results. The proposed method based on ResNet-50 is different from the tracking algorithm which is proposed in the past and based on CNNs. In spite the ResNet-50 and VGG-16 have similar structures, the number of ResNet layers is more than double times of theVGG-16. The two networks both input images of the size of 224×224 pixels, and output the feature maps of the size of 7×7. Know that the residual network performs better than CNNs in image classification and target recognition tasks because of the existence of residual layer in the network.

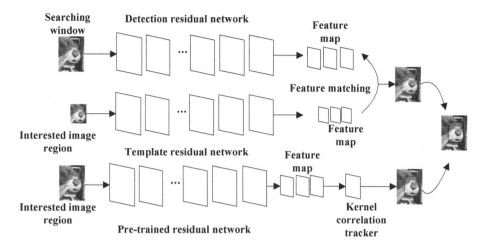

Fig. 1. Thumbnail of proposed algorithm

Introduce the depth residual learning framework to address the degradation problem. These layers are adapted to the residual mapping, rather than having each of the few stacked layers directly fit the required underlying mapping. Formally, denoting the desired underlying mapping as $H(z)$, let the stacked non-linear layers fit another mapping of $F(x) := H(z) - z$. The original mapping is recast into $F(z) + z$. Hypothesize that it is easier to optimize the residual mapping than to optimize the original, unreferenced mapping. To the extreme, if an identity mapping were optimal, it would be easier to push the residual to zero than to fit an identity mapping by a stack of nonlinear layers. The core of extracting depth features using deep residual networks is to solve the side effects (degradation problem) caused by increasing depth, which can improve network performance by simply increasing network depth. [15] The correlated filtering visual target tracking algorithm is similar to image classification because they find the category or position corresponding to the maximum value among all the response values, so they can be applied to video target tracking by using the superior performance of ResNet-50 in image classification. The ResNet-50 used for tracking is pre-trained with the 1.3 million image ImageNet dataset and delivers the most advanced results in the classification challenge.

First, build a dual residual depth classification network model. The input layer parameters of the existing two deep residual neural networks ResNet50 are adjusted, wherein the number of neurons in the first network input layer is set to $224 \times 224 \times 3$, and the neurons of the second network input layer are The number is set to $448 \times 448 \times 3$, and the other layer parameters remain unchanged, and the two depth residual neural networks are used as the front-end network of the dual residual depth classification network model.

Two two-layer fully connected networks are built as the back-end network of the dual residual deep classification network model. The first layer of each fully connected network is the input layer, the second layer is the hidden layer, and the third layer is the output layer. The parameters of the first layer of the two fully connected networks are different, and the parameters of the second layer and the third layer are the same. The parameters of the layers in the two fully connected networks are as follows:

The number of first layer neurons in the first network is set to $1 \times 1 \times 2048$, and the number of first layer neurons in the second network is set to $2 \times 2 \times 2048$; the second layer of the two networks The number of neurons is set to 1024 at the same time, and the activation function is also set to correct the linear unit ReLU function; the number of neurons in the third layer of the two networks is simultaneously set to 1000, and the activation function is simultaneously set to the Softmax function.

The ImageNet classification data set is input into the constructed dual residual classification network model. The stochastic gradient descent method is used to update the weight of each node in the dual residual depth classification network model, and the trained dual residual depth classification network is obtained.

In the obtained trained dual residual depth classification network model, the network layer after the 49th layer of the two deep networks is deleted, and the remaining networks become new networks. The template network model and the detection network model are extracted from the remaining networks according to the input layer parameters, that is, the remaining network with the input layer parameter of $224 \times 224 \times 3$ is used as the template network model, and the remaining network with the input layer parameter of $448 \times 448 \times 3$ is taken as Detect the network model.

Inputting the first frame image in the sequence of video images containing the target to be tracked, and determining a rectangular frame at the center of the initial position of the target to be tracked by one time and width of the target to be tracked; capturing the target image from the rectangular frame, and adjusting the image The size is $224 \times 224 \times 3$ pixels, and the template image is obtained; the template image is input into the template network obtained in step 3 to extract the feature of the template image, and the extracted features are composed into the feature map, and then the last layer of the template network $2048 \times 7 \times 7$ feature maps are output, and these $2048 \times 7 \times 7$ feature maps are used as template feature maps.

The image to be detected containing the target to be tracked is input, and at the center of the initial position of the target to be tracked, a rectangular frame is determined by twice the length and width of the target to be tracked. The target image is intercepted from the rectangular frame, the image size is adjusted to $448 \times 448 \times 3$ pixels, and the detected image is obtained; the detected image is input into the detection network, and the extracted feature is composed into a feature map. Then, $2048 \times 14 \times 14$ feature maps are outputted in the last layer of the detection network, and the $2048 \times 14 \times 14$ feature maps are used as the detection feature map.

Similar to SiameseFC [7], we designed a dual residual network to extract and match the features of the target to be tracked. The difference is that did not pre-train on a large number of video sets. In comparison, our method is easier to reproduce. The algorithm designed is mainly to find the center point of the target, regard the area matching the target center point as the result of the tracking target area. The process of template matching is depicted in Fig. 1. Firstly, crop template image of the target to be tracked by using the groundtruth. The template feature block X is extracted from the template network. Secondly, shear two times area of the template image which is around the estimated target position to obtain the detected image. Input the detected image to the detection network to extract the detected feature block Y. The convolution operation completes the matching of the two feature blocks, and the location of the maximum value indicates the estimated target position. The convolution area of the feature block is one quarter of the detection feature block. This algorithm can be expressed as

$$(x, y) = \arg\max_j(\sum_{i=1}^{n}(X_i * Y_i^j)) \tag{1}$$

4.2 Kernel Correlation Filters

The idea of the kernel-dependented filter algorithm is to learn the filter, which is used to convolve with the image to obtain the associated information map. The position with the largest value in the information graph is the position of the estimated object. Construct a Gaussian shaped label for each dimension feature maps which is obtained from the image, expressed as follow

$$g(x, y) = e^{-\frac{(x-w/2)^2 + (y-h/2)^2}{2\sigma^2}} \tag{2}$$

where w, h is the width and height of the feature map extracted from the resudual networks, σ is the kernel width. In the kernel correlation filter tracking algorithm, use a non-linear mapping Gaussian function $\phi(x)$, so that the mapped samples are linearly separable in the new space. Then use the ridge regression algorithm to find a classifier $\vartheta(x_i) = w^T\phi(x_i)$ in the new space, so the weight coefficient obtained can be formulated as

$$w^* = \arg\min_w ||\phi(x_i)w - \vartheta(x_i)||^2 + \lambda||w||^2 \tag{3}$$

where λ is a regularization parameter ($\lambda \geq 0$). λ can be expressed as a vector of image line vectors $w = \Sigma_i\partial_i\phi(x_i)$, where ∂ is the image expansion factor, so the formula (3) can be expressed as

$$\partial = \min_\partial ||\phi(x_i)\phi(x_i)^T\partial - \vartheta(x_i)||^2 + \lambda||\phi(x_i)^T\partial||^2 \tag{4}$$

The problem is called the dual problem about w. Let the derivative be ∂ 0. ∂ can be computed as

$$\partial = (K + \lambda I)^{-1}y \tag{5}$$

K is a kernel matrix, introducing kernel functions to reduce high-dimensional features to low-dimensional $K_{ij} = k(x_i x_j)$ Convert to the Fourier domain the formula (5) can be rewritten as

$$\hat{\partial} = \frac{\hat{y}}{k^{\hat{x}x'} + \lambda} \tag{6}$$

Finally, determine the position of the current frame target by the following formula

$$\hat{f}(z) = real(\wp^{-1}(k^{\hat{x}z} \bigodot \hat{\partial})) \tag{7}$$

where $k^{\hat{x}x'}$ and $k^{\hat{x}z}$ are converting the filter into a Gaussian kernel auto correlation, transforming the detection target into a cross-correlation value between the region to be detected and the filter Gaussian kernel.

Parameter update is required during the tracking process. The update of the feature map template as

$$H = (1 - \beta)x_{t-1} + \beta x_t \tag{8}$$

After obtaining the optimal location and scale, crop the corresponding region and corresponding search region and extract their feature maps x_t, x_{t-1} is the template feature map of the previous frame. Where β, is the learning rate of the proposed algorithm. The update of the filter parameters as

$$Z = (1 - \mu)z_{t-1} + \mu z_t \tag{9}$$

$$\partial_i^n = (1 - \mu)\frac{y}{k_{i-1}^{x_{t-1}x'_{i-1}} + \lambda} + \mu\frac{y}{k_i^{x_t x'_i} + \lambda} \tag{10}$$

Same with x_t, after obtaining the optimal location and scale, crop the corresponding region and corresponding search region and extract their feature maps z_t and calculate $k_i^{x_t x'_i}$ according to z_t, then combining the template features z_{t-1}, and $k_{i-1}^{x_{t-1}x'_{i-1}}$ where μ is the learning rate of the proposed algorithm.

4.3   Target Location Fusion

The target position determined by the dual residual network and the target position determined by the kernel correlation filter are weighted and fused, and expressed as

$$Q = 0.8Q_1 + 0.2Q_2 \tag{11}$$

where Q is the final target position, Q_1 is the target position determined by the dual residual network, Q_2 is the target position determined by the kernel correlation filter.

5   Experiments

In this section, present the experimental results of the proposed method. First describe the implementation details of the method, and then introduce the comparison between the method proposed in this paper and other methods.

5.1 Implementation Details

For feature extraction, crop the target bounding box and then resize it to 224 * 224 pixels for the ResNet-50 with 49 layers. After forward propagation, use the output from six convolutional layers. Get a template feature maps whose size is one-third of the input image size. Then crop the search box, which is two times size of the bounding box, and resize the search box to the 448 * 448 pixel for detection network input. After forward propagation, get the detection feature block at the output. The template feature block convolutes with the detection feature block to generate a response map, and the location of the maximum matching value is the targets position. In Kernel correlation filter module, select three times the area of the bounding box, resize it to 416 * 416 pixels, get the output feature blocks from 46th layers of the network, and final train filter to complete tracking. Implement our algorithm in Pycharm, and utilize the Keras toolbox in this work. Our implementation runs at 3 frames per second on a computer with an Intel Core i5-6500 CPU 3.20 GHZ * 4, 8 GB RAM, and a GeForce GTX 1080/PCle/SSE2 card.

5.2 Tracking Results

In order to illustrate the feasibility of the proposed algorithm, test our tracker with the others sequences provided in the benchmark. The algorithm is experimentally verified on OTB100.

Figure 2 is the algorithm part tracking effect diagram. The first one in Fig. 2 is the Bolt video sequence, which is a video of a 100-m sprint competition in which the algorithm tracks the 4th player. The first image of the video sequence is the tracking effect of the second frame of the video. The IOU is 0.77. It can be seen from the image that the algorithm designed in this paper can accurately target the player. The second image is the tracking effect IOU of the 34th frame of the video is 0.726. From the image, the tracking target can be seen relative to the initial frame. When the displacement occurs and the deformation occurs, the tracker can still accurately locate the target; the third image is the tracking effect IOU of the video 249th frame is 0.754. It can be seen from the image that the tracking target is compared with the initial frame target. The large displacement and the change from the body to the sideways, the tracker still achieves the accurate positioning of the target in frame 249 by updating; the fourth image is the tracking effect of the last frame of the video IOU is 0.83, as can be seen from the image The tracking target has a large displacement relative to the initial frame target and has changed from the body to the back body. The tracker remains accurate for up to 300 frames while maintaining accurate positioning. Robustness. The second in Fig. 2 is the Woman video sequence, a video of a woman walking on the road, in which the tracking object has a visual appearance and the vehicle blocks the lower body, from the blue line frame and the yellow line in the rendering. The case of the box indicates that the algorithm can still accurately track the target when the target has a large range of occlusion. In the third and fourth video sequences in Fig. 2, the tracking

Fig. 2. The partial tracking effect diagram of OTB100

target also undergoes appearance deformation and occlusion. The algorithm has better performance in different video sequences, indicating that the algorithm is robust.

The first in Fig. 3 is the Jogging video sequence, a video of two women running in the park, in which the algorithm tracks the woman wearing white clothes, and the first image of the video sequence is the second frame of the video. The effect IOU is 0.92. It can be seen from the image that the algorithm designed in this paper is highly accurate. The second image is the tracking effect IOU of the 53rd frame of the video is 0.7. It can be seen from the image that the tracking target is displaced from the initial frame. And the serious occlusion occurs, the tracker can still accurately locate the target. The effect shows that the algorithm can also predict the target motion trend while positioning the image target feature; the third image is the tracking effect IOU of the 64th frame of the video. It is 0.8, which is the image that appears after the target is obscured. From the image, the tracker can still identify and locate the target after several frames. The fourth image is the tracking effect IOU of the last frame of the video is 0.78, and the tracker still performs well. The second one in Fig. 3 is the Girl video sequence, which is a video of a man jumping rope outside. From the video sequence image, it can be seen that the image captured by the tracking target is seriously blurred due to rapid movement, and even a double-target image appears, but the algorithm Effective tracking of the target is still achieved. The third in Fig. 3 is the Jumping video sequence, which is a video taken by a girl indoors. From the third image, the tracking target changes in the video, from the girl's facial features to the girl's back head features. At the same time, another boy interferes with the tracked target. It can be seen from the tracking effect that the tracker can update according to the change of the tracking target during the tracking process. In the fourth video sequence in Fig. 3, the tracking target also undergoes a scale change. Compared with the first image and the last image scale, the algorithm can achieve multi-scale target tracking.

Fig. 3. The partial tracking effect diagram of OTB100

5.3   Comparative Test

We compare our algorithm to 4 recent state-of-the-art trackers: TLD [17], Struck [18], SiameseFC, SiameseRPN [19], and also extracted the features of each layer of Resnet50, and used these features in combination with the algorithm of this paper to detect the video sequence. After a large number of experimental comparisons, selected the characteristics of the 49th layer to complete the target tracking task. In order to show that ResNet-50 is superior to VGG-Net in the field of visual tracking, use the same method to experiment with different network, and make the overlap rates (IOU) which threshold is 0.2 and the accuracy rates results which Threshold is 20 pixels in tabular form. From the Table 1 and the Table 2, we can see that ResNet50 used in this paper is more advantageous than VGG-Net16 in determining the target to be tracked. Similarly, use the features obtained by the two layers Activation49 and Activaton23, using the same method for tracking experiments. There is a residual module between layer Activation49 and Activation23 layer, so the output features of the two layers are

Table 1. IOU results

Sequences	Activation 49	Activation 23	Vgg16	TLD	Struck	SiameseFC	SiameseRPN
Bolt	0.902	0.011	0.014	0.005	0.017	0.034	0.948
David2	0.996	0.069	0.835	1	1	0.998	0.428
Football	0.916	0.160	0.160	0.754	0.660	0.889	0.384
Football1	0.821	0.123	0.424	0.364	0.324	0.891	0.986
Tiger1	0.861	0.946	0.784	0.538	0.042	0.742	0.891
Woman	0.929	0.177	0.161	0.678	0.896	0.668	0.909
Jogging-2	0.983	0.970	0.970	0.918	0.162	0.918	0.960
Girl	0.909	0.114	0.967	0.408	0.970	0.988	0.916
FaceOcc2	0.858	0.106	0.997	0.671	0.987	0.810	0.5
Dudek	0.972	0.192	0.954	0.831	0.851	1	1
Coke	0.906	0.651	0.924	0.525	0.947	0.838	0.955
David3	0.992	0.147	0.462	0.095	0.337	0.531	1

Table 2. Accuracy rates results

Sequences	Activation 49	Activation 23	Vgg16	TLD	Struck	SiameseFC	SiameseRPN
Bolt	0.991	0.022	0.011	0.014	0.020	0.048	1
David2	1	0.083	0.850	1	1	1	1
Football	0.988	0.163	0.163	0.801	0.751	0.942	0.397
Football1	0.917	0.136	0.520	0.527	0.337	0.973	1
Tiger1	0.827	0.934	0.739	0.538	0.042	0.716	0.851
Woman	0.947	0.179	0.189	0.747	1	0.933	0.998
Jogging-2	0.993	0.970	0.977	0.938	0.182	0.944	0.983
Girl	0.961	0.180	1	0.962	1	1	0.928
FaceOcc2	0.855	0.103	0.996	0.718	0.980	0.811	0.942
Dudek	0.832	0.114	0.826	0.473	0.317	0.933	0.996
Coke	0.955	0.689	0.955	0.529	0.949	0.941	0.948
David3	0.996	0.175	0.537	0.111	0.337	0.678	1

different, and prove the practicality of the residual module. Compared with other excellent trackers, we can find that our tracker tracking effect is basically the same. Based on the analysis of the above three parts, the proposed algorithm can accomplish the target tracking task in different environments.

6   Conclusion

In this paper, propose a new tracking algorithm based on dual residual network and kernel correlation filters, which innovatively applies the residual network with superior performance in image classification to visual target tracking, and constructs a dual residual network for template image and detection. Feature extraction of images, using template matching to initially determine the target location. The feature extraction kernel correlation filter extracted by the residual network is used to determine the target position again. The two determined position weights are used to complete the target tracking task. Corresponding to the problem of finding the center point position in the target tracking and the classification problem of the image classification, so that the algorithm with superior performance in the classification field can be applied to the visual target tracking. Using experimental verification to illustrate the proposed algorithm can efficiently perform target tracking. Since the target positioning is directly performed on the feature map, the target jitter is severe between video frames, and it is desirable to consider the relationship with the previous frame target coordinates from the next frame target positioning to shorten the tracking target jitter problem.

Acknowledgements. The work is supported by the National Natural Science Foundation of China (No. 61571342) and Shaanxi Natural Science Basic Research Project (No. 2017JM6032).

References

1. Girshick, R., Donahue, J., Darrelland, T., et al.: Rich feature hierarchies for object detection and semantic segmentation. In: IEEE Conference on Computer Vision and Pattern Recognition. IEEE (2014)
2. Krizhevsky, A., Sutskever, I., Hinton, G.: Imagenet classification with deep convolutional neural networks. In: NIPS (2012)
3. Wu, Y., Lim, J., Yang, M.H.: Object tracking benchmark. IEEE Trans. Pattern Anal. Mach. Intell. **37**(9), 1834–1848 (2015)
4. Wang, L., Ouyang, W., Wang, X., et al.: Visual tracking with fully convolutional networks. In: IEEE International Conference on Computer Vision (ICCV). IEEE (2016)
5. Nam, H., Han, B.: Learning Multi-Domain Convolutional Neural Networks for Visual Tracking (2016)
6. Li, X., Hu, W., Shen, C., Zhang, Z., Dick, A.R., van den Hengel, A.: A survey of appearance models in visual object tracking. ACM TIST **4**(4), 58 (2013)
7. Bertinetto, L., Valmadre, J., Henriques, J.F., et al.: Fully-convolutional siamese networks for object tracking. In: European Conference on Computer Vision (2016)
8. Bolme, D.S., Beveridge, J.R., Draper, B.A., et al.: Visual object tracking using adaptive correlation filters. In: The Twenty-Third IEEE Conference on Computer Vision and Pattern Recognition, CVPR San Francisco, CA, USA. IEEE (2010)
9. Krizhevsky, A., Sutskever, I., Hinton, G.E.: Imagenet classification with deep convolutional neural networks. In: NIPS (2012)
10. He, K., Zhang, X., Ren, S., Sun, J.: Spatial pyramid pooling in deep convolutional networks for visual recognition. In: ECCV (2014)
11. Wang, N., Yeung, D.Y.: Learning a deep compact image representation for visual tracking. In: International Conference on Neural Information Processing Systems. Curran Associates Inc., New york (2013)

12. Wang, L., Liu, T., Wang, G., et al.: Video tracking using learned hierarchical features. IEEE Trans. Image Process. **24**(4), 1424–1435 (2015)
13. Li, H., Li, Y., Porikli, F.: Deeptrack: learning discriminative feature representations online for robust visual tracking. IEEE Trans. Image Process. **25**(4), 1834–1848 (2015)
14. Deng, J., Dong, W., Socher, R., et al.: ImageNet: a large-scale hierarchical image database. In: IEEE Conference on Computer Vision and Pattern Recognition. IEEE (2009)
15. He, K., Zhang, X., Ren, S., et al.: Deep Residual Learning for Image Recognition (2015)
16. Henriques, J.F., Caseiro, R., Martins, P., et al.: High-speed tracking with kernelized correlation filters. IEEE Trans. Pattern Anal. Mach. Intell. **37**(3), 583–596 (2015)
17. Kalal, Z., Matas, J., Mikolajczyk, K.: P-N learning: bootstrapping binary classifiers by structural constraints (2010)
18. Hare, S., Torr, P.H.S.: Struck: Structured Output Tracking with Kernels. In: ICCV (2011)
19. Li, B., Wu, W., Zhu, Z., Yan, J.: High performance visual tracking with siamese region proposal network. In: CVPR (2018)

Image Recognition of Peanut Leaf Diseases Based on Capsule Networks

Mengping Dong, Shaomin Mu$^{(\boxtimes)}$, Tingting Su, and Wenjie Sun

College of Information Science and Engineering,
Shandong Agricultural University, Taian 271018, Shandong, China
dongmengping@126.com, msm@sdau.edu.cn,
ml5065806906@163.com, 766469613@qq.com

Abstract. Peanut is an important kind of economic crop and oil crop, and many diseases are manifested on the leaves. Identifying the leaf disease peanut correctly is of great significance. The convolutional neural networks have great effect on the recognition of peanut leaf disease images. However, the spatial relationships of the images are easy to lost in processing and convolutional neural networks can not handle the problem of rotational invariance properly. Aiming at the problem, Hinton *et al.* propose the capsule networks. A capsule is a group of neurons, which represents the probability of a class. The capsule networks take into account both rotational invariance and spatial relationships with the use of dynamic routing. The essence of dynamic routing is which capsule should be sent to a higher level capsule. Hence, this paper designed and implemented a new method for identifying peanut leaf disease images by taking advantage of the capsule networks. Firstly, constructing the data set of the peanut leaf disease images and data enhancement was used to process the images. Secondly, this paper designed two types of capsule networks: modifying the parameters for the peanut leaf disease images and stacking three convolutional layers to the original model. Finally, the experimental shows that capsule networks are better than the convolutional neural networks.

Keywords: Capsule networks · Convolutional neural networks ·
Peanut leaf disease images · Dynamic routing

1 Introduction

As an important economic crop and oil crop, peanut is planted widely in China [1]. As a peanut producing country, the diseases of peanut are also very serious in China. The most common diseases are rust, black spot and brown spot, which have an effect on the productivity and quality of peanuts severely. Most of the characteristics of peanut diseases are expressed on the leaves; they can be identified by the shape, texture and color of the leaves. At present, the identification of peanut leaf diseases depends on the experience and manpower, which is time-consuming, error-prone and laborious. Therefore, it is essential to recognize the peanut leaf diseases in production. Many machine learning algorithms have been used widely in agricultural disease images recognition, such as Convolutional Neural Networks (CNNs) model and Support Vector Machine (SVM) model.

© Springer Nature Singapore Pte Ltd. 2019
K. Knight et al. (Eds.): ICAI 2019, CCIS 1001, pp. 43–52, 2019.
https://doi.org/10.1007/978-981-32-9298-7_4

The CNNs outperform other method without any pre-processing on image classification [2]. Therefore, they have been used in various fields, besides in the agricultural diseases recognition. Zhang *et al.* [3] presented the model of AlexNet to recognize the citrus canker diseases. Recognition result showed that quantity computed and time expended is reduced obviously using the model in the same recognition accuracy. Long *et al.* [4] studied deep CNNs to learn the characteristics of Camellia oleifera, and transferred the knowledge learned from ImageNet data set by AlexNet model to recognize the Camellia oleifera diseases by the aid of transfer learning. Huang *et al.* [5] used a deep convolutional neural networks model GoogLeNet to detect the panicle blast diseases. Sun *et al.* [6] proposed a model based on the CNNs to recognize the plant leaves diseases, which applied the batch normalization and global pooling method. Ma [1] used the technology of image processing and SVM to study the recognition of brown spot, and segmented the peanut lesion image by algorithm of image segmentation. Srdjan *et al.* [7] proposed a new approach to recognize the plant diseases images by the use of deep CNNs. Zhang *et al.* [8] constructed a recognition model of vegetable leaf diseases images based on the three channel CNNs model, which combines three color components.

The CNNs have played an important role in crop disease images recognition, but they still have some problems. The max-pooling keeps the max feature and throws away other information, and it may contribute to a loss of spatial relationship between object and the part of the object. The lack of rotational invariance could lead to allocating the object another label. In order to address this problem, Sabour *et al.* [9] propose the capsule networks, which replace the scalar output of CNNs with vector output. And then the dynamic routing takes the place of max-pooling. In this paper, we proposed a recognition model of peanut leaf disease images based on capsule networks. Five kinds of peanut leaf disease images were trained and tested, then the paper compared the recognition accuracy with the conventional CNNs model, and we obtained a satisfactory result.

2   Related Works

2.1   Convolutional Neural Networks

The CNNs can infer the feature of the input image without any prior knowledge [10]. And they consist of convolutional layer, pooling layer and full connected layer. Firstly, local features are extracted by the convolutional layer. Secondly, the pooling layers compress the extracted features. Finally, the probability of the prediction is obtained by the fully connected layer. There are three characteristics on the CNNs: local receptive field, weight sharing and sub-sampling.

Although the CNNs have excellent performance on the recognition of leaf disease images, they also have some drawbacks. For one thing, the pooling layer ignores the spatial relationship between two parts of the object. Therefore, the capsule networks replace the pooling layer with dynamic routing. For another, when an object changes the position or orientation, other neurons which recognize the object may not be activated. Data augmentation may solve this problem to some extent, but it can not completely eliminate the problem.

2.2 Capsule Networks

The capsule networks represent the latest breakthrough in neural networks architecture, which overcome the shortcomings of CNNs [11]. A capsule consists of multiple neurons, and the activity vectors get shrunk to a length of 0–1, so that it can represent the probability of a feature extracted by the capsule. And the orientation of the vector represents the instantiation parameters (position, orientation and size) [9]. When changing the characteristics of the image slightly, the length of the activity vector does not change, but the orientation changes. In the capsule networks, the information of lower level capsule will be sent to all capsules in the next higher level layer, but the coupling coefficients between them are different. When the output of the next higher level capsule is satisfied, and the coupling coefficients will increase. Conversely, the coupling coefficients will reduce [12].

Firstly, an affine transformation W_{ij} is performed between the capsule i and the next higher level capsule j, and the output of capsule i is u_i. Therefore, the prediction of the output of the capsule j for the next higher level can be defined by the formula (1).

$$\hat{u}_{j|i} = W_{ij}u_i \tag{1}$$

Secondly, the input vector to next higher level capsule j is obtained by the formula (2), which is the weighted sum of the lower level capsules. Finally, to ensure that the length of the output vector represent the probability that an entity exists, the non-linear squashing function is used to shrink the length of the output vector V_j below 1. And the squashing function is shown in Eq. (3). From the equation, when S_j is large, the first term is equal to 1, and conversely, the first term is equal to 0. And the second term can unitize S_j to the length of 1.

$$s_j = \sum_i c_{ij}\hat{u}_{j|i} \tag{2}$$

$$V_j = \frac{||S_j||^2}{1 + ||S_j||^2} \frac{S_j}{||S_j||} \tag{3}$$

The coupling coefficient c_{ij} is defined to indicate the coupling degree of the adjacent layers and it is calculated by the Softmax function in Eq. (4). And b_{ij} is the log prior probability that capsule i should be coupled to the higher level capsule j. In the dynamic routing iterative algorithm, the initial state of b_{ij} is set to 0. And the process of capsule networks can be seen from Fig. 1.

$$c_{ij} = \frac{\exp(b_{ij})}{\sum_k \exp(b_{ik})} \tag{4}$$

Dynamic routing has been shown to improve the recognition accuracy in capsule networks [9]. In the conventional CNNs, the max-pooling is used to transfer the information to the adjacent layers. However, dynamic routing is used in the capsule networks, and it is used to decide which capsule should be sent to the higher level

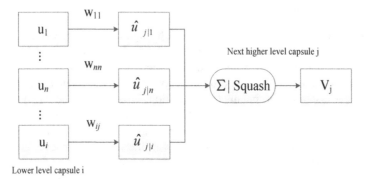

Fig. 1. The process of the lower level capsules send information to the higher level capsules.

capsule. The initial coupling coefficient c_{ij} is set to 0. For lower level capsules, the number of c_{ij} is equal to the number of lower level capsules, and the sum of all coupling coefficients is 1. The scalar product of the output vector V_j and the prediction is defined as agreement a_{ij}, which is used to adjust the initial logit b_{ij}, as is shown in Eq. (5). When the objects are similar, the scalar product is large. When they are different, the scalar product is small.

$$a_{ij} = V_j \widehat{u}_{j|i} \tag{5}$$

The capsule networks allow multiple classes of capsules to exist at the same time, so there is a loss function L_k for each class capsule k, as is shown in Eq. (6). The T_k is set to one when the class capsule k is present; on the contrary, T_k is equal to 0. And $m_+ = 0.9$, $m_- = 0.1$ and the parameters λ is set to 0.5. In addition, additional reconstruction losses are used to encourage the class capsules to encode for the instantiation parameters of the input classes [9], and the class capsules are sent to the decoder which consists of the 3 fully connected layers. The total margin loss is the sum of the losses of all classification error.

$$L_k = T_k \max(0, m^+ - ||v_k||)^2 + \lambda(1 - T_k) \max(0, ||v_k|| - m^-)^2 \tag{6}$$

3 Proposed Method

The capsule networks were used to recognize the images of peanut leaf diseases, and the parameters were modified slightly. The structure is shown in Fig. 2. The capsule networks mainly include a convolutional layer, a Primary Capsule layer and a class capsule layer, and the activation function adopts ReLU function.

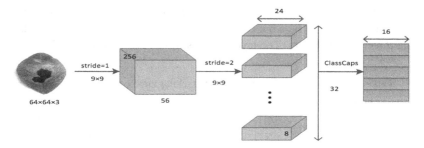

Fig. 2. Proposed model architecture of the capsule networks for peanut leaf diseases.

The size of input images is 64 × 64 with three color channels. And the convolutional kernel size is 9 × 9, the number of convolutional kernels is 256 with the step size of 1. And the output of the image is 56 × 56 × 256 after the first convolutional layer.

The second layer is a Primary layer with a convolutional kernel of 9 × 9 and strides of 2. After the Primary Capsule layer, the model outputs 32 capsule layers, each layer consists of 24 × 24 capsules. And the number of capsules is 24 × 24 × 32, and each capsule is an 8 dimensional vector, which can be used to store information of an entity such as orientation, color and size. However, the class capsule layer outputs one 16 dimensional vector for per capsule, and the 8 dimensional vector should be converted into a 16 dimensional vector by the weight matrix W_{ij}.

The last layer is the class capsule layer, and the class capsule layer outputs five capsules, each capsule represents a kind of peanut leaf disease. The dynamic routing agreement should determine the capsules from the Primary Capsule layer to the class capsule layer. The iteration of routing is 3. And the length of the vector indicates the probability of identifying the kind of peanut leaf disease images.

The decoder is composed of 3 fully connected layers, and the number of neurons is 512, 1024 and 64 × 64 × 3 = 12228. The number of neurons in last fully connected layer is the same as the pixels of the input image, the architecture of the decoder is shown in Fig. 3.

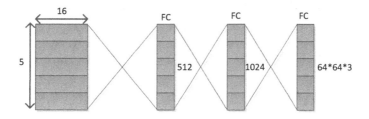

Fig. 3. The decoder.

There is only one convolutional layer in the capsule networks, but the features of peanut leaf disease images are complex. In order to explore more features of the peanut leaf disease images, it is considered to stack more convolutional layers to the model of Hinton's original capsule networks. Therefore, we designed the original model with stacking three capsule layers. The layers of this model are presented in Table 1.

Table 1. The original architecture with stacking three convolutional layers.

Number	The network layer	Kernel size	Kernel number	Input dimension	Output dimension
1	Convolutional layer 1	3×3	16	$64 \times 64 \times 3$	$62 \times 62 \times 16$
2	Convolutional layer 2	3×3	8	$62 \times 62 \times 16$	$30 \times 30 \times 8$
3	Convolutional layer 3	3×3	3	$30 \times 30 \times 8$	$28 \times 28 \times 3$

4   Experiments

4.1   Environment and Data

The experimental environment is a 64-bit system for Windows 7, the Tensorflow 1.9 deep learning framework is used, the programming language is Python, and the NVIDIA GeForce GTX 1080 Ti graphics card accelerates image processing.

There is no standard data set for the recognition of peanut leaf disease images, so this paper constructed the data set of peanut leaf disease images, and the peanut leaf disease images were obtained from a field in Tai'an City, Shandong Province. The images were taken with Canon EOS 80D. To reduce the interference of background, and a single background was used for shooting. The camera is set to the automatic shooting mode with an interval of 2 s, the distance between the camera and the peanut

Table 2. Image samples of peanut leaf diseases.

Classes	Sample 1	Sample 2	Sample 3	Sample 4
Brow Spot				
Black Spot				
Rust				
Anthracnose				
Net Blotch				

leaf disease image is 5–6 cm, the image format is JPG, and the size is 64 × 64. The number of peanut leaf disease images is 6,114, containing five kinds of peanut leaf diseases: brow spot, black spot, rust, anthracnose and net blotch. The samples are shown in Table 2.

There are many parameters of the capsule networks. In order to prevent the data set from over-fitting and increasing the diversity of the training samples, this paper achieved the data augmentation by adding salt and pepper noise to the images and rotating the images.

Salt and pepper noise is added black and white pixels to the images randomly. In order to expand the data set, and not to increase fuzzy of the images, the signal noise ratio of the image is 0.99. Figure 4(b) is an anthracnose image with salt and pepper noise.

(a) Original image   (b) Noise addition

Fig. 4. Salt and pepper noise is added to the original images.

In order to test the rotational invariance of capsule networks, the experiment used the vertical peanut leaf disease images for training data set, and the testing data set was rotated training images 90°, as is shown in Fig. 5. The number and the proportion of the peanut leaf disease images are shown in Table 3.

(a) Original image   (b) Rotate 90 degrees

Fig. 5. Image of anthracnose is rotated 90°.

Table 3. Number and proportion of peanut leaf disease images.

Classes	The number	The proportion
Brow Spot	4028	0.292
Rust	1837	0.138
Black Spot	3449	0.249
Anthracnose	2840	0.207
Net Blotch	1578	0.114

After the data augmentation, the data set are expanded to 13732, the ratio of the training data set and testing data set is 8:2. The number of training data set is 11132, and the rest 2600 images are used as testing data set.

4.2 Experimental Results and Analysis

The first part of our experiment is allocated to training the peanut leaf disease images on the structure of the capsule networks. The batch size is set to 32 and the number of epoch is 100. The loss function is composed of two parts, the margin loss and the decoder loss. The decoder loss is the square error between the 12228 (the input of the image) and the result of the reconstructed image. From the Fig. 6, at the beginning, the value of the loss function appears a rapid dropped trend and it convergences the final results when the training steps are 10000. The training process is depicted in Fig. 7. It can be seen from the figure that the capsule networks begin to converge when it iterates 10000 times.

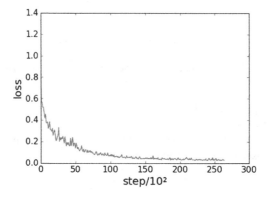

Fig. 6. The loss function.

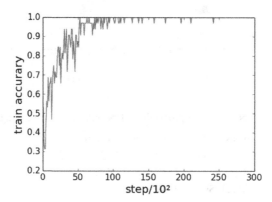

Fig. 7. The process of training the peanut leaf disease images on the capsule networks.

After getting the recognition accuracy of the proposed capsule networks model, then the original model with three convolutional layers was tested. The results are shown in Table 4. From the results, the original model with three convolutional layers achieves higher recognition accuracy compared to the model of modifying the parameters. One reason can be attributed to the fact that the convolutional layers are good at extracting low level features of the peanut leaf disease images.

Table 4. Comparison of recognition accuracy between the proposed capsule networks and CNNs.

Model	Recognition accuracy
Capsule networks	82.17%
Add convolution layers to the capsule networks	83.58%
CNNs	81.14%

Finally, we chose the best model of the capsule networks, and then compared the capsule networks with the conventional CNNs for the same data set of peanut leaf disease images. The model is constructed as Table 5.

Table 5. The architecture of the CNNs.

The network layer	Kernel size	Kernel number	Input dimension	Output dimension
Convolutional layer 1	5×5	32	$64 \times 64 \times 3$	$60 \times 60 \times 32$
Pooling layer 1	2×2		$60 \times 60 \times 32$	$30 \times 30 \times 32$
Convolutional layer 2	5×5	64	$30 \times 30 \times 32$	$26 \times 26 \times 64$
Pooling layer 2	2×2		$26 \times 26 \times 64$	$13 \times 13 \times 64$
Convolutional layer 3	5×5	128	$13 \times 13 \times 64$	$9 \times 9 \times 128$
Pooling layer 3	2×2		$9 \times 9 \times 128$	$4 \times 4 \times 128$
Fully connected layer 1				1024
Fully connected layer 1				512
Fully connected layer 1				5

The test prediction accuracy results are presented in Table 4. And the result of capsule networks is superior to the CNNs, considering the fact that peanut leaf disease images are rotated 90° at the time of testing, the CNNs are not easy to identify the diseases images in different directions, but the capsule networks model are rotation invariant. When the leaf disease images were rotated a certain angle, the CNNs may recognize one disease image as another which may be account to the poorer performance. And the max-pooling layer will lose a lot of rotational information. However, the rotational information can be saved in the capsule networks.

5   Conclusion

This paper studied two types of capsule networks for the recognition of the peanut leaf disease images. And the experimental results demonstrate that the original model with three convolutional layers outperforms than other models. Furthermore, not only do the capsule networks solve the problems in convolutional neural networks, but also improve the recognition accuracy of the peanut leaf disease images. The acquisition of peanut leaf disease images is difficult; most of the diseased leaves are overlapping. In the future, we will explore the multiple overlapping leaves with capsule networks.

Acknowledgement. This work is supported by First Class Discipline Funding of Shandong Agricultural University.

References

1. Ma, X.S.: Image recognition of peanut brow spot based on support vector machine. Anhui University (2018)
2. Afshar, P., Mohammadi, A., Plataniotis, K.N.: Brain tumor type classification via capsule networks. In: 2018 25th IEEE International Conference on Image Processing (ICIP), pp. 3129–3133. IEEE (2018)
3. Zhang, M., Liu, J., Cai, G.Y.: Recognition method of citrus canker disease Based on convolutional neural network. J. Comput. Appl. **38**(S1), 48–52+76 (2018)
4. Long, M.S., Ouyang, C.J., Liu, H.: Image recognition of Camellia oleifera diseases based on convolutional neural network & transfer learning. Trans. Chin. Soc. Agric. Eng. (Trans. CSAE) **34**(18), 194–201 (2018)
5. Huang, S.P., Sun, C., Qi, L., Wang, W.J.: Rice panicle blast identification method based on deep convolutional neural network. Trans. Chin. Soc. Agric. Eng. (Trans. CSAE) **33**(20), 169–176 (2017)
6. Sun, J., Tan, W.J., Mao, H.P., Wu, X.H., Chen, Y., Wang, L.: Recognition of multiple plant leaf diseases based on improved convolutional neural network. Trans. Chin. Soc. Agric. Eng. (Trans. CSAE) **33**(19), 209–215 (2018)
7. Srdjan, S., Marko, A., Andras, A., Culibrk, D.: Deep neural networks based recognition of plant diseases by leaf image classification. Comput. Intell. Neurosci. **2016**(6), 1–11 (2016)
8. Zhang, S., Huang, W., Zhang, C.: Three-channel convolutional neural networks for vegetable leaf disease recognition. Cogn. Syst. Res. **53**, 31–41 (2019)
9. Sabour, S., Frosst, N., Hinton, G.E.: Dynamic routing between capsules. In: Advances in Neural Information Processing Systems, pp. 3859–3869 (2017)
10. Krizhevsky, A., Sutskever, I., Hinton, G.E.: ImageNet classification with deep convolutional neural networks. In: Advances in Neural Information Processing Systems, pp. 1097–1105 (2012)
11. Edgar, X., Selina, B., Yang, J.: Capsule network performance on complex data (2017). arXiv preprint: arXiv:1712.03480
12. Mobiny, A., Van Nguyen, H.: Fast CapsNet for lung cancer screening. In: Frangi, A.F., Schnabel, J.A., Davatzikos, C., Alberola-López, C., Fichtinger, G. (eds.) MICCAI 2018, Part II. LNCS, vol. 11071, pp. 741–749. Springer, Cham (2018). https://doi.org/10.1007/978-3-030-00934-2_82

Multi-scale Convolutional Neural Network for Road Extraction in Remote Sensing Imagery

Jun Li[1,2(✉)], Xue Gong[1], and Yiming Zhang[1]

[1] Key Laboratory of Intelligent Computing and Information Processing, Ministry of Education, Xiangtan University, Xiangtan 411105, China
201610202014@smail.xtu.edu.cn
[2] College of Civil Engineering and Mechanics, Xiangtan University, Xiangtan 411105, China

Abstract. Road is an important semantic region in remote sensing imagery and plays an important role in many applications. Deep learning has obtained a great success in image classification, since it can directly learn from labeled training samples and extract different level image features to encode the input image. In this paper, we propose a multi-scale convolutional neural network (MSCNN) for extracting road from high-resolution remote sensing image, in which road detection can be seen as a regional classification. This core trainable detection engine consists of an encoder-decoder network and a fusion model. Firstly, image was encoded as a feature representation with several stacked convolutional layers. And then guided by the different scale road training data, the pre-trained decoder networks output a series of classification maps. Finally, we investigate the fusion model utilizing different scale classification maps and obtain a final road decision map. To validate the performance of the proposed method, we test our MSCNN based method and other state-of-the-art approaches on two challenging datasets of high-resolution images. Experiments show our method gets the best results both in quantitative and qualitative evaluation.

Keywords: Remote sensing imagery · Road extraction ·
Convolutional neural network · Multi-scale features

1 Introduction

With the rapid development of remote sensing technology, extracting information from remote sensing imagery become one of the most promising research trends in big Earth data analysis. In addition, road-related semantic data can be used in several applications, including urban planning, forest management, climate modeling and intelligent transportation systems. And machine learning based road extraction from remote sensing is usually formulated as a pixel labeling task. The goal is to classify the pixels in each remote sensing view which belongs

© Springer Nature Singapore Pte Ltd. 2019
K. Knight et al. (Eds.): ICAI 2019, CCIS 1001, pp. 53–65, 2019.
https://doi.org/10.1007/978-981-32-9298-7_5

to the road area, in particular, to determine for each pixel whether or not it is a part of road.

Figure 1 shows some remote sensing images with the variety of roads. For example, as shown in Fig. 1(a), concrete roads, asphalt roads and other materials represent different colors in remote sensing images, resulting in different road features in an image. Figure 1(b) shows roads in a dense city area, in which roads may be confused with buildings. While in the mountain area, shows in Fig. 1(c), road is relatively a small target and normally suffers from complex background clutter. Figure 1(d) show that the texture of roads in remote sensing images is easily obscured by trees or confused with rivers, cause feature variation in different imaging conditions. At the same time, the quality of remote sensing image is easily affected by different factors such as the sensor type, spectral and spatial resolution, weather, light variation, and ground characteristic, etc. As a result, different road in an image have different characteristics and different sizes, which makes it difficult to extract road from a remote sensing (RS) image.

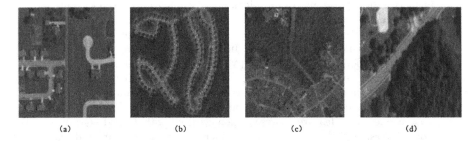

(a)   (b)   (c)   (d)

Fig. 1. The difficulties in road extraction. (a) concrete roads, asphalt roads and other kinds of roads. (b) dense vegetation and buildings. (c) sparse and narrow roads. (d) the disturbances of trees on both sides of roads.

A number of methods have been proposed to extract roads from remote sensing images in recent years, and the features are one of most important factors for road recognition. In recent years, deep learning approaches achieved great achievements for high-level feature representation from labeled data. It is evident that the use of deep learning offers tremendous opportunities to support remote sensing image analysis. Convolutional neural networks (CNNs) in deep learning field are well-known for feature learning [1]. Based on CNNs, many patch-classification methods are proposed to achieve semantic labeling [2–4]. These approaches determine pixel's label by using CNNs to classify a small patch around the target pixel. However, they ignore the inherent relationship between patches and time consumption [5]. It remains an interesting problem to apply deep learning to larger datasets and more challenging tasks.

In this paper, we propose a new DCNN architecture which effectively exploits multi-scale feature information for road extraction. The method employs the receptive field of different sizes to learn image characteristic at different scales through stacked convolutional layers and pooling layers. Moreover, these features are upsampled to obtain classification maps of different scales. Finally, classification maps will be fused which combines high-level coarse semantic information and low-level location information. These classification maps capture road feature at different scales so that the road extraction of different sizes will produce good results, especially for sparse and slender road. Additionally, our method achieves great performance on two challenging datasets, outperforms some state-of-the-art approach.

2   Related Works

Road detection methods primarily depend on pixel-level classification, this accuracy is far from satisfactory because of the misclassification between road and other spectrally similar object such as building block, water areas and parking lots, etc [6]. Classification methods are usually based on support vector machine (SVM), deep learning and Markov random fields (MRFs) classifier. For instance, Song et al. [7] exploited SVM classifier by using spatial information such as texture and shape, classify the image into two groups of categories: a road group and a non-road group. However, SVM methods are difficulties to choice the dimensional space and training samples. To the best of our knowledge, patch-based CNN is the first work on using deep learning techniques for road detection [2]. The basic idea of patch-based CNN is: separate the image into small patches, and apply the CNN model on each patch to predict the class label. This model utilized several stacking CNN and integrated the output with Conditional Random Fields (CRFs) to connect disjointed road regions and complete irregular boundaries of building regions. Panboonyuen et al. [8] employed an improved DCNN framework to extract road objects from aerial images. Landscape metrics (LMs) is added to this DCNN framework to reduce falsely classified road objects.

In order to achieve better classification results, many studies use hierarchical classification technique in DCNN to acquire multi-scale features. Li et al. [9] proposed three CNN submodules to extract features from three nested and increasingly larger rectangular windows. And then, with the same pre-trained DCNN, three saliency maps were estimated corresponding to features with different receptive fields. The boost network aggregated these multiple saliency maps and outputs the final saliency map. Lin et al. [10] introduced the inherent multi-scale and pyramidal hierarchy of deep convolutional networks to construct feature pyramids. A top-down architecture with lateral connection is developed for building high-level semantic feature maps at all scales. After that, this network is able to better image learning by incorporating these feature maps. Hu et al. [11] proposed a multiscale CNN architecture with an improved cross-entropy loss function to produce the segmentation map. They construct the multi-scale network by combining the feature map of each middle layer to learn more detail information of the retinal vessels.

3   Method

In this section, we design a multi-scale CNN architecture to learn probability map and our testing details are described. As shown in Fig. 2, MSCNN is made up of two parts: an encoder-decoder network and a multi-scale feature fusion model. The encoder-decoder model combines the thinking of fully convolutional network (FCN) and deeply supervised nets to extract features. A deconvolution layer is used for getting classification map rm_i ($i = 1, 2, 3, 4$) which in different receptive filed size. Meanwhile, we concatenate rm_i in the fusion model which can effectively integrate multi-level features. This model is capable of utilizing hierarchical features and multi-scale contextual simultaneously to extract roads.

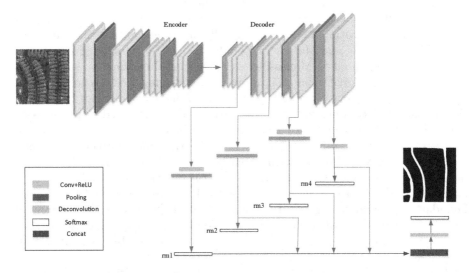

Fig. 2. The architecture of MSCNN. We first use an encoder network to get abundant semantic features, then decoder network is applied to generate multi-scale classification maps. Finally, a concatenation layer is used to fuse these maps.

3.1   Encoder-Decoder Network

Encoder-Decoder network consists of two components: the encoder network and decoder network. The encoder network contains ten convolutional layers and four pooling operations to achieve high-level global feature extraction, as shown in Fig. 2, is configured in the same way as in FCN. It contains the following three modules. For each convolutional layer, the kernel size is 3×3 and stride is 1, rectified linear units (ReLU) layer is attached to each convolution layer. A max-pooling layer with 2×2 kernel and stride 2 is added after each group of convolutional layers, meanwhile we remove the pool5 layer which could reduce the precision of edge pixel extraction [12].

We take advantage of the idea of the decoder network in SegNet, our network stacks a decoder network on top of the encoder part, the decoder network which consists of a hierarchy of decoders one corresponding to each encoder so that restore finer spatial layout of objects. The decoder network consists of four deconvolution layers, corresponding convolutional layers and ReLU layers, and produces sparse classification maps.

3.2 Multi-scale Feature Fusion Model

In this model, different classification maps rm_i $(i = 1, 2, 3, 4)$ are generated. In order to produce a range of rm_i and ensure the same spatial size of rm_i with the original image, we apply an upsample layer. Besides, we add a cross-entropy loss layer connects to upsample layer as a network optimization strategy. Lastly, rm_i are concatenated into a 8-channel image, then a 1×1 convolutional layer is applied to fuse classification maps, and a cross-entropy loss layer is followed to gain the final road extraction map. The cross-entropy loss [13] is defined as:

$$L_{seg}(y, f(x), \theta_1) = \sum_{m=1}^{M} \sum_{n=1}^{N} \sum_{k=1}^{K} -l(y_m^n = k) \log p_k(x_m^n) \qquad (1)$$

Where y_m^n is the ground-truth label of x_m^n; $f(x_m^n)$ is the output of the last convolutional layer at pixel x_m^n; x_m^n is the nth pixel in the mth patch; θ_1 represents the parameters of road extraction network; M is the minibatch size; N is the number of pixels in each patch; K is the number of classes (here, $K = 2$); $l(y = k)$ is an indicator function, it takes 1 when $y = k$, and 0 otherwise; and $p_k(x_m^n)$ is the probability of pixel x_m^n [13] being the kth class, which is defined as:

$$p_k(x_m^n) = \frac{exp(f_k(x_m^n))}{\sum_{l=1}^{K} exp(f_l(x_m^n))} \qquad (2)$$

3.3 Testing Details

After training, we utilize a method to improve road extraction results in the test phase. This test results enhancement strategy includes four steps. And we named the result of this method as MSCNN+voting in Table 2.

Step 1. For each test image T, we can get four different test images named as T_i $(i = 1, 2, 3, 4)$ by rotating it 90°, 180° and flipping it in horizontal and vertical reflections.

Step 2. Testing T and T_i, we can obtain five different road extraction maps M_j $(j = 1, 2, 3, 4, 5)$ by the proposed road extraction framework.

Step 3. M_j is recovered to the original position by rotating and mirroring, and the resulting images are represented by I_k $(k = 1, 2, 3, 4, 5)$.

Step 4. A majority voting strategy is adopted to fuse I_k and obtain the final road extraction map I. It is defined as follows:

$$I(i, j) = \begin{cases} 1, & if \sum_{k=1}^{5} I_k(i, j) \geq 3 \\ 0, & if \sum_{k=1}^{5} I_k(i, j) < 3 \end{cases} \qquad (3)$$

where $I_k(i,j)$ represents the pixel(i,j) in I_k, $I_k(i,j) \in \{0,1\}$; $I(i,j)$ is pixel(i,j) on the final road extraction map I.

4 Experimental Setup and Discussion

To verify the effectiveness of the proposed method, extensive experiments on road extraction from remote sensing images are carried out on two different datasets, the Massachusetts dataset and Cheng's dataset. We compare the results of the proposed approach with a series of state-of-the-art deep learning methods. Lastly, we present the experimental results, which include quantitative comparisons and visual road extraction results.

4.1 Datasets Description

Cheng's Dataset. The dataset is collected by Cheng et al. [13] were used in this work, which consists of 224 high-resolution remote sensing imagery taken from Google Earth. The ground truth was manually labeled by the reference maps, and the dataset is publicly available. We randomly split the data into a training set of 180 images, a validation set of 14 images, and a test set of 30 images (see Table 1). The original images in this dataset are with a spatial resolution of 1.2 m per pixel. There are at least 600×600 pixels in each image, and the road width is about 12–15 pixels.

Massachusetts Dataset. The Massachusetts dataset was built by Mihn [2]. This dataset consists of 1171 aerial images, and includes 1108 training, 14 validation, and 49 testing images (see Table 1). The size of each image is 1500×1500 pixel with the spatial resolution of 1 m per pixel. The dataset roughly covers 2600 km^2 crossing from urban, suburban, and rural regions which is by far the largest and most challenging aerial image labeling dataset.

Table 1. An overview of the datasets.

Dataset	Training	Validation	Testing
Cheng's dataset	180	14	30
Massachusetts dataset	1108	14	49

4.2 Experimental Setting

We use data augmentation and regularization techniques to increase the number of training data and avoid over-fitting in the training stage. In order to ensure that there are enough training images, we first randomly cut out 20 patches with sizes of 300×300 (Massachusetts dataset is 320×320) from the original images and extract 5 fixed position patches (i.e., the 4 corner patches and the center patch). In addition, we use an interpolation method to rotate each patch at the step of 90° and then flip these rotated patches in horizontal and vertical reflections [13]. For each patch, seven another different patches are produced by rotations and reflections in the second stage.

In the experiments, we implement the MSCNN based on the Caffe framework [14]. Additionally, a dropout of 0.5 was added to deeper convolutional layers, which are utilized to overcome the over-fitting problem. For other hyperparameters, the global learning rate is set to 1e−8. The momentum and weight decay is set to 0.99 and 0.0005 respectively.

4.3   Evaluation Metrics

The most common metrics for evaluation for extracting roads are precision, recall, and F-measure (F1) [4,13,15]. Precision means the percentage of matched roads area in the extraction maps. Recall represents the proportion of matched areas in the reference maps. F1 is an overall metric, which combines precision and recall. They are defined as follows:

$$\text{Precision} = \frac{TP}{TP + FP} \tag{4}$$

$$\text{Recall} = \frac{TP}{TP + \text{FN}} \tag{5}$$

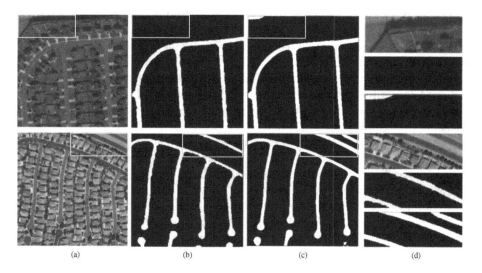

(a)   (b)   (c)   (d)

Fig. 3. Illustration of the ground truth and our results. (a) original image, (b) ground truth, (c) the result obtained by MSCNN, (d) close-ups. Each image in the fourth column illustrates the corresponding closed-ups of the red rectangles in the first three columns. (Color figure online)

$$F1 = \frac{2 \times \text{Precision} \times \text{Recall}}{\text{Precision} + \text{Recall}} \tag{6}$$

Road extraction task can be considered as binary classification, where road pixels are positives and the remaining non-road pixels are negatives. Let TP is defined as the number of correctly classified road pixels, FP is defined as the number of mistakenly classified road pixels, TN is the number of correctly background pixels, FN is the number of mistakenly classified background pixels.

It should be noted that the above evaluation metrics can assess the extraction results to some extent, but not an absolute indicator. Our evaluation metrics are depending on the ground truth, and there are some deviation between the ground truth and the real road. Therefore, the extraction results need to be verified by the original image. It can be seen from each group of pictures in Fig. 3(d), the second ground truth is different from the first original image, while the third extraction result is more similar to the original image.

4.4   Comparing Algorithms

Our MSCNN is compared with the other state-of-the-art methods, including FCN, SegNet, CasNet and MFPN.

(1) FCN [16]: Long et al. introduced FCN for the semantic segmentation, which had achieved great performance. This network is modified by VGG by removing fully connected layer and adding a deconvolution layer. In our method, we modify the architecture of FCN with less convolution layers and less pooling layers.

(2) SegNet [17]: Badrinarayanan et al. proposed a new network architecture, which is an encoder-decoder model. This network stored the max-pooling indices of the feature maps and used them in its decoder network to achieve good performance.

(3) CasNet [13]: The method by Cheng et al. is designed for road detection and road centerline detection. This network is a cascaded end-to-end convolutional neural network. It consists of two encoder-decoder networks which adopted the same configuration as SegNet.

(4) MFPN [15]: In MFPN, an effective feature pyramid and a tailored pyramid pooling module are used for getting multilevel semantic features. This method is verified on Massachusetts Dataset but this network is complex and requires a high hardware environment.

In order to compare our results with those of the above four methods, we use the same data, the same size of input patches, same output patches and same evaluation metrics for the prediction road extraction maps.

4.5   Results

Overall Performance of the Proposed Method. Figure 4 presents the result of road extraction from Cheng's dataset. As we can see, the images of

MSCNN are very similar to the ground truth, which includes more detail for both dense roads in urban areas and sparse roads in the countryside. And as Fig. 5 shows, our method is superior to other methods in different difficult scenes and different scale road extraction, the application of multi-scale features helps to smoothen the edge of roads.

Table 2 illustrates the quantitative performance of different comparing methods on road extraction in Cheng's dataset. From Table 2, MSCNN achieves the best results for each indicator. Especially, the quality value of MSCNN is more than 5% higher than FCN and CasNet. Compare to other methods, our final result MSCNN+voting also achieves better performance and demonstrate the effectiveness of the test results enhancement strategy.

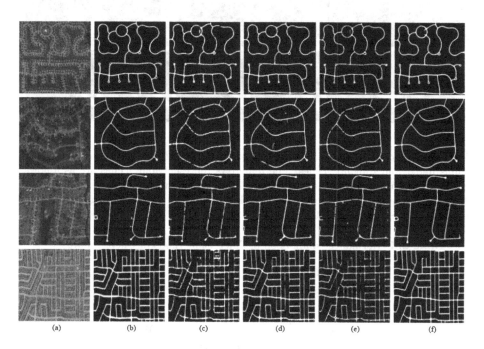

Fig. 4. Visual comparison of road extraction results with different comparing algorithms. There are four rows and six columns of subfigures. (a) original image. (b) ground truth. (c) result of FCN. (d) result of SegNet. (e) result of CasNet. (f) result of MSCNN.

The performance of road extraction in Massachusetts dataset is shown in Table 3. We can see that the improvement of our method in road extraction is also obvious.

Effectiveness of Multi-scale CNN Architecture. In this section, we evaluated the performance of multi-scale classification maps. As discussed in Sect. 3.2, our multi-scale CNN architecture consists of four units, rm_1, rm_2, rm_3 and rm_4.

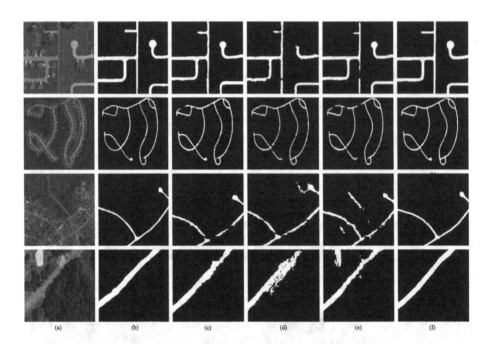

Fig. 5. Visual comparison of road extraction results with different comparing algorithms under four difficult scenes. The four rows represent four scenes are as follows. Different materials road; sparse and narrow road; Road detection under the occlusions of trees. (a) original image. (b) ground truth. (c) result of FCN. (d) result of SegNet. (e) result of CasNet. (f) result of MSDNN.

Table 2. Performance comparison of different methods in Cheng's dataset.

Method	Recall	Precision	F1
FCN [13]	0.931	0.848	0.887
SegNet	0.913	0.891	0.901
CasNet [13]	0.941	0.921	0.931
MSCNN	0.936	0.971	0.953
MSCNN+voting	0.946	0.969	0.957

Table 3. Performance comparison of different methods in Massachusetts dataset.

Method	Recall	Precision	F1
FCN [18]	0.762	0.762	0.762
SegNet [8]	0.765	0.773	0.768
MFPN [15]	0.742	0.851	0.793
MSCNN	0.819	0.932	0.871
MSCNN+voting	0.826	0.930	0.873

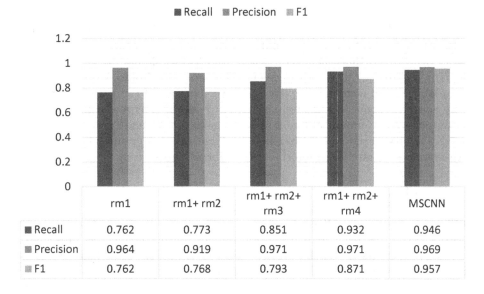

Fig. 6. Investigation of MSCNN with different settings on the Cheng's dataset.

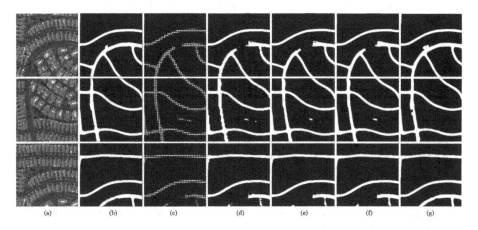

Fig. 7. Visualization of the multi-scale classification maps. (a) original image, (b) ground truth, (c) rm_1, (d) rm_2, (e) rm_3, (f) rm_4, (g) road extraction results.

To prove the effectiveness of our MSCNN architecture, we train four another experiments for comparisons, which respectively take rm_1 only, concatenated rm_1 and rm_2, concatenated rm_1, rm_2 and rm_3, and concatenated rm_1, rm_2 and rm_4. Quantitative results are obtained on Cheng's dataset. As shown in Fig. 6, we can see our method is improved F1 from 0.916 to 0.953 using the multi-scale feature fusion method. This performance demonstrates the validity of multi-scale feature application.

Given an input image, the proposed method efficiently generates four different scale classification maps rm_i, a fusion model is exploited the concatenated multi-scale classification maps to enhance the final map accuracy. As show in Fig. 7, we can see that the fusion result Fig. 7(g) is similar to ground truth Fig. 7(b). Figure 7(c) to (f) give the classification maps from coarsest to finest scale. We can find although Fig. 7(c) gives the contour of the road, it contains many noise. Compare to Fig. 7(c), the road in Fig. 7(f) is more detailed, but the contour is worse. Combine different scale of classifications maps can improve the DCNN's ability to successfully handle both large and small roads, therefore it is necessary to integrate these information to obtain better results.

5   Conclusion

In this paper, a novel multi-scale convolutional neural network to perform road network extracting from high resolution remote sensing imagery is presented. The major contribution of this work is the design of multi-scale feature fusion model, which aims to combine spatial information and multi-scale contextual feature for detecting more details. The experimental results indicate the effectiveness and feasibility of the proposed framework for improving the performance of road extraction of remote sensing imagery in two challenging datasets.

References

1. Mas, J.F., Flores, J.J.: The application of artificial neural networks to the analysis of remotely sensed data. Int. J. Remote Sens. **29**(3), 617–663 (2008)
2. Mnih, V.: Machine Learning for Aerial Image Labeling. University of Toronto, Toronto (2013)
3. Sharma, A., Liu, X., Yang, X.: A patch-based convolutional neural network for remote sensing image classification. Neural Netw. **95**, 19–28 (2017)
4. Alshehhi, R., Marpu, P.R., Woon, W.L.: Simultaneous extraction of roads and buildings in remote sensing imagery with convolutional neural networks. ISPRS J. Photogrammetry Remote Sens. **130**, 139–149 (2017)
5. Maggiori, E., Tarabalka, Y., Charpiat, G., Alliez, P.: Convolutional neural networks for large-scale remote-sensing image classification. IEEE Trans. Geosci. Remote Sens. **55**(2), 645–657 (2017)
6. Wang, W., Yang, N., Zhang, Y., et al.: A review of road extraction from remote sensing images. J. Traffic Transp. Eng. (English edition) **3**(3), 271–282 (2016)
7. Song, M., Civco, D.: Road extraction using SVM and image segmentation. Photogrammetric Eng. Remote Sens. **70**(12), 1365–1371 (2004)
8. Panboonyuen, T., Jitkajornwanich, K., Lawawirojwong, S., et al.: Road segmentation of remotely-sensed images using deep convolutional neural networks with landscape metrics and conditional random fields. Remote Sens. **9**(7), 680 (2017)
9. Li, G., Yu, Y.: Visual saliency based on multiscale deep features. In: Proceedings of the IEEE Conference on Computer Vision and Pattern Recognition (CVPR), pp. 5455–5463 (2015)
10. Lin, T.Y., Dollár, P., Girshick, R.: Feature pyramid networks for object detection. In: Proceedings of the IEEE Conference on Computer Vision and Pattern Recognition (CVPR), vol. 1, no. 2, p. 4 (2017)

11. Hu, K., Zhang, Z., Niu, X., et al.: Retinal vessel segmentation of color fundus images using multi-scale convolutional neural network with an improved cross-entropy loss function. Neurocomputing **309**, 179–191 (2018)
12. Liu, Y., Cheng, M.M., Hu, X., et al.: Richer convolutional features for edge detection. In: Proceedings of the IEEE Conference on Computer Vision and Pattern Recognition (CVPR), pp. 5872–5881 (2017)
13. Cheng, G., Wang, Y., Xu, S., et al.: Automatic road detection and centerline extraction via cascaded end-to-end convolutional neural network. IEEE Trans. Geosci. Remote Sens. **55**(6), 3322–3337 (2017)
14. Jia, Y., Shelhamer, E., Donahue, J., et al.: Caffe: convolutional architecture for fast feature embedding. In: Proceedings of the 22nd ACM International Conference on Multimedia, pp. 675–678. ACM (2014)
15. Gao, X., Sun, X., Zhang, Y., et al.: An end-to-end neural network for road extraction from remote sensing imagery by multiple feature pyramid network. IEEE Access **6**, 39401–39414 (2018)
16. Long, J., Shelhamer, E., Darrell, T.: Fully convolutional networks for semantic segmentation. In: Proceedings of the IEEE Conference on Computer Vision and Pattern Recognition (CVPR), pp. 3431–3440 (2015)
17. Badrinarayanan, V., Kendall, A., Cipolla, R.: SegNet: a deep convolutional encoder-decoder architecture for image segmentation. IEEE Trans. Pattern Anal. Mach. Intell. **39**(12), 2481–2495 (2017)
18. Muruganandham, S.: Semantic segmentation of satellite images using deep learning. Master Thesis, Lulea University of Technology, Lulea (2016)

TCPModel: A Short-Term Traffic Congestion Prediction Model Based on Deep Learning

Xiujuan Xu[1,2], Xiaobo Gao[1,2], Zhenzhen Xu[1,2(✉)], Xiaowei Zhao[1,2],
Wei Pang[3], and Hongmei Zhou[4]

[1] School of Software, Dalian University of Technology, Dalian 116620, China
{xjxu,xzz}@dlut.edu.cn
[2] Key Laboratory for Ubiquitous Network and Service Software of Liaoning Province,
Dalian 116620, China
[3] School of Natural and Computing Sciences, University of Aberdeen,
Aberdeen AB24 3UE, UK
pang.wei@abdn.ac.uk
[4] School of Transportation and Logistics, Dalian University of Technology,
Dalian 116024, China

Abstract. With the progress of the urbanization, a series of traffic problems have occurred because of the growing urban population and the far lower growth rate of roads than that of cars. One of the most prominent problems is traffic congestion problem. The prediction of traffic congestion is the key to alleviate traffic congestion. To ensure the real-time performance and accuracy of the traffic congestion prediction, we propose a short-term traffic congestion prediction model called *TCPModel* based on deep learning. By processing a massive amount of urban taxi transportation data, we extract the traffic volume and average speed of taxis which are the most important parameters for assessing of traffic flow prediction. After analyzing the temporal and spatial distribution characteristics of the traffic flow and average speed, we present a short-term traffic volume prediction model called *TVPModel*, and a short-term traffic speed prediction model called *TSPModel*. Both models are based on a deep learning method Stacked Auto Encoder (*SAE*). By comparing the other traffic flow forecasting methods and average speed forecasting methods, the methods proposed by this paper have improved the accuracy rate. For traffic congestion recognition, we use a novel model called *TCPModel* based on three traffic parameters (average speed, traffic flow and density), which uses standard function method to standardize the parameters and calculate the congestion comprehensive threshold to determine the congestion level by thresholds. According to the experiments, *TVPModel* and *TSPModel* in this paper got satisfied accuracy compared with other prediction models.

Keywords: Short-term traffic congestion prediction · Traffic data · Deep learning · Stacked autoencoder

© Springer Nature Singapore Pte Ltd. 2019
K. Knight et al. (Eds.): ICAI 2019, CCIS 1001, pp. 66–79, 2019.
https://doi.org/10.1007/978-981-32-9298-7_6

1   Introduction

With the progress of the urbanization process, a series of traffic problems have occurred because of the growing urban population and the far lower growth rate of roads than the growth rate of cars. In 2016, Holland navigation technology released the world's most congested city. Among the top 10 cities, China has 4th, namely, Chongqing, Chengdu, Tainan, and Beijing. One of the most prominent problems is traffic congestion, especially in big cities such as Beijing, and Shanghai, and so on. It has been acknowledged that the prediction of traffic congestion is the key to alleviate it.

In this paper, we use the traffic data of urban taxis to study it in Beijing, where traffic congestion frequently occurs. First of all, to predict the two important parameters of traffic flow: traffic volume and average travel speed, we build a deep learning model based on stack autocoder. For the congestion identification, three parameters of traffic flow are extracted from the traffic data of urban taxis in this research: traffic volume, average travel speed and traffic density of road sections. By establishing the correlation among these parameters we generate synthetic judgment formula to identify the traffic congestion patterns. For the prediction of traffic congestions, we use the predicted traffic flow parameters to perform congestion identification, and compared with real traffic congestions in order to assess the prediction performance.

The rest of this paper is organized as follows: Sect. 2 gives a brief review of the related literature, including traffic flow parameters and several common deep learning approaches. Section 3 gives the problem definitions of our research. Section 4 presents an overall framework for our work. Section 5 proposes a novel model called TFPModel to predict traffic flow based on *SAE*. Section 6 presents a model called TCPModel to predict traffic congestion based on traffic congestion level. Section 7 reports the experimental results. Finally, we draw our conclusions in Sect. 8.

2   Related Work

In recent years, traffic congestion prediction has attracted much attention by both researchers and practitioners. Among many models developed, predicting traffic congestion based on traffic flow parameters is particularly interesting.

Nowadays, more and more models are used to predict traffic flow parameters. We divide these prediction methods into the following three categories: parametric [4], nonparametric [2], and simulation-based methods [1]. Examples of parameter models include time series model, Kalman filter model and so on. The nonparametric models include k-nearest neighbor (K-NN) method [2], artificial neural network [3] and so on. The simulation-based method mainly uses simulation tools to predict traffic flow parameters, such as automatic encode methods.

For the prediction of traffic flow parameters, Hamed *et al.* [4] proposed the autoregressive integrated moving average model (ARIMA), which can predict the traffic volume of the main roads city, from which many time series models are derived to forecast the traffic volume. However, due to the nonlinear and stochastic nature of traffic flow parameters, researchers began to pay attention to the nonparametric methods. Davis *et al.* [2] used the K-NN method for short-term freeway traffic forecasting, but they concluded that the K-NN method was less effective than the linear time series method. Chang *et al.* [1] proposed a dynamic multi-interval traffic forecasting model based on K-NN nonparametric regression. Sun *et al.* [10] uses a local linear regression model for short-term traffic forecasting. Jeong *et al.* [5] takes into account the relative importance of the time difference between traffic flow data, and proposes an online learning weighted support vector regression (SVR) for short-term traffic flow forecasting. Gilmore et al. [3] proposed the use of back propagation (BP) neural network model to predict traffic congestion. Zhou *et al.* [13] proposed a traffic flow prediction model based on recurrent neural network. The model can capture the correlation of traffic volume in time series, so its performance is better than that of BP neural network. Zheng *et al.* [12] proposes a Markov chain based traffic status prediction model by defining the average speed corresponding to the states. The results show that the model achieves good performance for short-term traffic status prediction. Because of the spatial and temporal characteristics of traffic, Lv *et al.* [9] learn a traffic flow prediction model from a large amount of traffic data using automatic encoder, and according to their experiments, the method is better than both BP neural network and of SVR. Luo *et al.* [8] proposed the deep belief network for traffic flow prediction. This model compresses the road network data and uses a deep structure to learn the prediction model. They claimed that the prediction performance is better than that of SVR and the shallow layer neural network model.

In addition, some congestion prediction models consider weather and other matters. William *et al.* [11] build individual traffic status prediction models for different seasons. The model does not depend on real-time data, and rather predicts the traffic status through the analysis of historical trends. Li *et al.* [7] proposed a congestion prediction model based on the combination of multiple classifiers. The model not only analyses the traffic flow parameters, but also considers the city environment and other factors, and the prediction performance was higher than the model without considering the environmental factors. Tan *et al.* [6] proposed a congestion prediction model based on auto encoder and Softmax classifier. The model can classify congestions by combining different environmental and human factor data and learning the deep structure from such data, which makes the classification results better than those without feature learning.

3   Problem Definition

3.1   Basic Definitions of Traffic Flow

Traffic flow prediction includes traffic volume prediction and traffic speed prediction.

Problem Definition 1. Traffic volume prediction. We use tf_{t_j} to represent the traffic volume at time t_j on a road section p. We set the sampling interval as β. Suppose we have a set of traffic flow $tf_{t_j-h\beta}, \ldots, tf_{t_j-\beta}, tf_{t_j}$ from time $(t_j - h\beta_t)$ to time t_j on the road section p.

Given large-scale GPS data of a city, we need to forecast the traffic volume $\{tf_{t_j+\beta}\}$ at time β on the road section p:

$$tf_{t_j+\beta} = f(tf_{t_j-h\beta}, \ldots, tf_{t_j-\beta}, tf_{t_j}) \tag{1}$$

Problem Definition 2. Traffic speed prediction. We use ts_{t_j} to represent the traffic speed at time t_j on a road section p. Similarity, we set the sampling interval β. Suppose we have a set of traffic speeds $ts_{t_j-h\beta}, \ldots, ts_{t_j-\beta}, ts_{t_j}$ from time $(t_j - h\beta_t)$ to time t_j on the road section p. Given large-scale GPS data of a city, we need to forecast the traffic speed $\{ts_{t_j+\beta}\}$ at time $t_j + \beta$ on the road section p:

$$ts_{t_j+\beta} = f(ts_{t_j-h\beta}, \ldots, ts_{t_j-\beta}, ts_{t_j}) \tag{2}$$

As a whole, the traffic congestion prediction problem concerned in this research is a typical prediction problem, which is to predicate the traffic congestion level at a certain time in a certain place. The formal definition of the problem is shown in Problem Definition 3.

Problem Definition 3. The level of the traffic congestion. Given large-scale GPS data of a city, find out predict the level of the traffic congestion $ConLevel_{t_j+\beta}$ at time $t_j + \beta$ on the road section p, where β represents the sampling interval.

4   The Overview of Traffic Congestion Prediction Model

First, we introduce the city taxi traffic data and the processing of these data, including data screening and parameter extraction. Second, we introduce the traffic congestion assessment method used in this research. By extracting the parameters of traffic flow, the comprehensive measure formula of parameters is defined according to the relationship among parameters, and finally a traffic congestion level threshold is obtained to carry out the traffic congestion. In this section, we present the model of traffic congestion prediction called TCPModel. Finally, in the experimentation, we present the respective prediction models based on *SAE*.

Figure 1 shows the framework of our approach, which consists of three major parts: data preprocessing, traffic volume forecasting, and traffic congestion prediction.

Data Preprocessing: The traffic volume and average speed of every 5 min are extracted by means of data preprocessing, and the average density is calculated.

Traffic Volume Prediction: Traffic volume includes traffic flow and average speed. We forecast the traffic flow and average speed. Through the analysis on

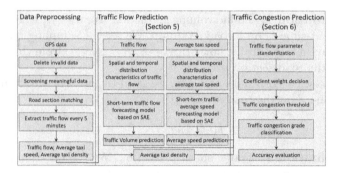

Fig. 1. The models of traffic flow prediction and traffic congestion prediction

spatial temporal distribution of traffic flow, a short-term traffic volume prediction model is established based on *SAE*; through the analysis of temporal and spatial distribution characteristics of average speed, we establish *SAE* prediction model based on short-time average speed. The technical details are given in Sect. 5.

Traffic Congestion Prediction: According to the congestion identification method, we calculate the road traffic congestion threshold and the corresponding level of judgment, as traffic congestion prediction. By comparing the real traffic volume, mean speed, and mean density of the traffic congestion level, we can obtain the accurate rate of traffic congestion prediction. The details are given in Sect. 6.

5   Traffic Flow Prediction Based on *SAE*

In this section, we present a novel model called TFPModel which uses *SAE* to learn the traffic parameters of traffic volume, and use the multiple classifier Softmax in the logic regression predictor at the top of the *SAE* model. We focus on establishment process of traffic flow prediction and traffic speed prediction.

5.1   Input Vectors Selection

First, we define the input feature vectors related to traffic volume. Because of the temporal and spatial characteristics of traffic, it can be found that the amount of traffic between adjacent links is interrelated and interactive. Therefore, the input eigenvector X of the model is as in Eqs. 3 and 4.

$$X_t = \{x_1, x_2, \ldots, x_p\} \qquad (3)$$

$$x_i = (x_{i,t}, x_{i,t-\Delta t}, \ldots, x_{i,t-m\Delta t}) \qquad (4)$$

where $i = 1, 2, \ldots, p$. P is the number of sections, $x_{i,t}$ represents the traffic of the i section at time t. The next time the traffic volume $x_{i,t+\Delta t}$ of any section is predicted by several sections of the adjacent time from $t - \Delta mt$ to t.

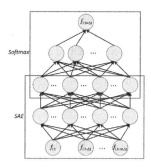

Fig. 2. The structure of traffic volume prediction model

5.2 TVPModel – Traffic Volume Prediction Model

The overall structure of the model is shown in Fig. 2. We use two *AE* to train our model by iteration with two layers. It is trained by minimizing the error between the original input and the reconstructed output, and then the output of the layer is used as the input of the next *AE*. According to this method, we train each layer in sequence. Finally, we obtain a learning model of *SAE*, and removed the output layer of *SAE*. The feature is input into the logical regression layer Softmax, and the traffic volume at the next time is used as a label. The logical regression layer Softmax ensures that the output value of each unit in output layer is 1. Therefore, the output is a set of conditional probabilities, which are the probabilities that the input belongs to every category. Assuming that the input vector of a given logical regression classifier is R, the probability of the input belongs to class i as shown in Eq. 5.

$$P(Y = i | R, W, b) = \frac{e^{W_i R + b_i}}{\sum_j e^{W_j R + b_j}} \tag{5}$$

In Eq. 5, W and b are the weights and biases of the logical regression layer. The sum in the right denominator of Eq. 5 is carried out on all output units. Since the logistic regression is a single-layer neural network, we train the model by the back-propagation algorithm. Therefore, we can tune automatic parameter in front of encoder when the training of logistic regression. Consequently, we integrate the neural network to form a deep and front stack automatic encoder.

The input of the TVPModel are several adjacent sections of the current time and several historical moment. The value of traffic volume is an integer within a certain range, so it can be directly output as a class label this moment. Therefore, we can use the predicted class of input as a value of traffic volume for the next moment.

The traffic flow prediction (*Traffic Volume Prediction*) algorithm based on SAE is summarized as Algorithm 1.

Algorithm 1. TrafficFlowPrediction algorithm

Input: The characteristic vector X_t of traffic volume, the pre-training period Pt, the pre-training learning rate PL, the period of WT, the learning rate of fine tuning WL, implicit layer number L, the number of units of each hidden layer $n[l]$, the sparsity parameter p, input the data dimension D.

Output: The total value of the traffic at the next time C.

1. The training set is used as the input of the automatic encoder. The first layer is trained by minimizing the target function, and the first layer is used as an automatic encoder.
2. The second layer is used as an automatic encoder, and the output of the first layer is used as the input of the automatic encoder.
3. The number of layers required for the iteration in step 2).
4. The output of the last layer is used as the input of the prediction layer, and its parameters are initialized randomly or by supervision training.
5. The BP method is used to adjust the parameters of all layers in a supervised manner.

5.3 TSPModel – Traffic Speed Prediction Model

In order to predict the average speed of short time traffic, we present a novel traffic speed prediction model based on SAE called $TSPModel$. The overall structure of the average speed prediction model is similar with TFPModel.

6 Traffic Congestion Prediction

Generally speaking, traffic volume, average speed and density of traffic flow will change abnormally to affect traffic congestion in the morning, noon and afternoon. Therefore, three traffic flow parameters, including traffic volume, average speed, and average density, are used as indicators to establish a recognition model called $TraConModel$ so as to evaluate traffic congestion. $TraConModel$ is very favorable for the decision of road congestion. When describing the road traffic condition, the weight of the influence parameters is analyzed according to the data analysis. We choose speed, density and flow of traffic as features to estimate road traffic level. These features show how many cars are running within unit section, how many cars are running in unit mileage, and how many cars are in unit mileage within unit time. We identify congested roads when the road with low speed, high density and small flow.

We should normalize three features because three features belong to different dimensional indexes. We use standard function method to normalize them. In the evaluation of road traffic, the greater the average speed of vehicles, the smoother the roads are. The greater the density of the traffic flow, the more congestion the roads are. Meanwhile, there is no direct linear relationship between the traffic flow index and the degree of road congestion. Therefore, we use the average speed as a benefit type parameter, and the traffic flow as a moderate parameter, and the flow density as a cost type parameter. We get the index of traffic congestion as shown in Eq. 6.

$$F = \{(v_1, f_1, k_1), (v_2, f_2, k_2), \ldots, (v_n, f_n, k_n)\} \tag{6}$$

where n represents the number of sections, v_i represents the speed of the section I, f_i represents the traffic of the section I, k_i represents the density of the section I. Consequently, we standardize three parameters as shown in Eqs. 7 to 9.

$$V_j^1 = \frac{v_j - v_{min}}{v_{max} - v_{min}} \tag{7}$$

$$f_j^1 = \frac{f_j - E(f)}{f_{max} - f_{min}} \tag{8}$$

$$k_j^1 = \frac{k_{max} - k_j}{k_{max} - k_{min}} \tag{9}$$

Among them, $E(f)$ is the expected traffic. $v_{min}(f_{min}, k_{min})$ is the minimum value of the field, that is, the lower limit. $v_{max}(f_{max}, k_{max})$ is the maximum value of the field (the upper limit). The standardized congestion index vector, as shown in Eq. 10.

$$F^1 = (v_1^1, f_1^1, k_1^1), (v_2^1, f_2^1, k_2^1), \ldots, (v_n^1, f_n^1, k_n^1) \tag{10}$$

After the standardization of three traffic flow parameters, it is necessary to determine the weight coefficients of evaluation indexes. It includes two factors: the weight coefficient of road section based on road distance and the determination of the importance weight coefficient of three traffic flow parameters.

Traffic congestion often occurs on a section of a road track. For example, the location of traffic congestion is close to the upstream intersection because the queuing behavior caused by congestion has a great influence on the upstream intersection. Meanwhile, queuing can easily lead to congestion in the upstream intersection, and also affect the other road tracks through the intersection. Therefore, the degree of traffic congestion impacts the road from upstream to longer distance. Road weight coefficient based on distance is shown in Eq. 11.

$$W = [W_1, W_2, \ldots, W_n]$$
$$W_i = \frac{\frac{1}{s_i}}{\sum_{i=1}^{n} \frac{1}{s_i}} \tag{11}$$

Among them, n is the number of road sections, and s_i is the distance from I upstream to the upstream intersection.

We use the largest eigenvalue in the analytic hierarchy process to identify the parameter weight coefficient. Since three parameter indexes are important to traffic congestion judgment, traffic density and speed have more important influence on traffic congestion. The traffic volume is smaller than the other two traffic flow parameters. Consequently, the final weight coefficient vector is shown as Eq. 12.

$$B = [b_1, b_2, b_3] = [0.45, 0.10, 0.45] \tag{12}$$

Among them, b_1 denotes the importance weight of average speed. b_2 represents the importance weight of traffic volume, and b_3 means the importance weight of average density.

According to the index vector, the weight coefficient of the section and the weight coefficient of the characteristic parameters, we set the synthetic measure parameter C as Eqs. 13 and 14.

$$C = W \bullet F^1 \bullet B^T \tag{13}$$

$$C = [W_1, W_2, \ldots, W_n] \bullet \begin{bmatrix} v_1^1 \ f_1^1 \ k_1^1 \\ \vdots \ \vdots \ \vdots \\ v_n^1 \ f_n^1 \ k_n^1 \end{bmatrix} \bullet \begin{bmatrix} b_1 \\ b_2 \\ b_3 \end{bmatrix} \tag{14}$$
$$= b_1 \sum_{i=1}^n W_i v_i^1 + a_2 \sum_{i=1}^n W_i f_i^1 + b_3 \sum_{i=1}^n W_i k_i^1$$

Finally, we identify the congestion level according to the comprehensive measure value. The comprehensive measure is shown in Table 1, and the road congestion state is shown in Table 2.

Table 1. Traffic congestion threshold

Threshold	City road	Freeway
Smooth	0.80–1	0.76–1
General smooth	0.70–0.80	0.67–0.76
A little congestion	0.56–0.70	0.59–0.67
Middle congestion	0.40–0.56	0.42–0.59
Serious congestion	0–0.40	0–0.42

Table 2. The table of traffic congestion comprehensive threshold

Congestion	Smoothly	Generally	Mild	Moderate	Serious
City	0.80–1	0.70–0.80	0.56–0.70	0.40–0.56	0–0.40
Country	0.76–1	0.67–0.76	0.59–0.67	0.42–0.59	0–0.42

7   Experiments

In this paper, the performance of the prediction model is mainly based on the performance measurement of the regression task based on the traffic volume and the traffic average speed prediction model. The common several performance indexes are mean square error (MSE), mean absolute error (MAE) and square root error ($RMSE$).

We build our system based on a real taxi dataset generated by over 30,000 taxis on a month in Beijing City. There are totally 200 GB of data for a month, composed of 6 GB data each day. The dataset recorded the taxi trajectories.

7.1 Experimental Results for TFPModel

We need to identify some parameters of input and output for *TFPModel*. In the part of *SAE* learning feature, we train these parameters based on experience. Therefore, we determine parameters' value according to the influence of several important parameters on prediction error. We analyze the following parameters: the number of hidden layers, the number of nodes in the hidden layer, and the influence of the dimension of input vectors on the prediction results.

We compare TFPModel with several traffic forecasting models that can deal with GPS data, such as *SVR* [5], *BP* algorithm [3], recurrent neural network (*RNN*) [13] and long and short term memory network (*LSTM*) [13]. These algorithms predict the traffic volume of weekdays and weekends, respectively.

Table 3. The performance of different traffic flow prediction model for weekdays

Second ring		TFPModel	SVR	BP	RNN	LSTM
West	MAE	6.80	15.30	9.30	7.20	6.90
	MRE(%)	13.86	30.46	18.91	15.01	12.87
	RMSE	9.10	19.20	10.60	9.90	9.40
North	MAE	7.70	16.50	9.60	7.90	7.60
	MRE(%)	16.83	31.21	19.17	16.96	16.33
	RMSE	10.20	18.40	11.20	9.90	8.50
East	MAE	8.60	14.70	9.40	8.70	8.30
	MRE(%)	9.51	29.82	19.01	12.34	10.98
	RMSE	11.20	15.70	12.50	12.30	11.10
South	MAE	6.40	14.90	8.60	6.70	6.40
	MRE(%)	18.71	31.58	16.13	14.77	13.33
	RMSE	8.50	15.04	9.80	8.60	8.20

Table 3 shows the performance of these traffic volume prediction models for working day traffic volume prediction. We forecast the traffic volume of four sections on every 5 min by the same traffic volume dataset. We use as a training dataset which includes working days from 1st December to 21st December, 2014. Meanwhile, the testing dataset uses traffic volume data from December 22nd to December 28th. In terms of these performance indicators, the mean absolute error of this method is less than the other methods of comparison, which is close to the performance of *RNN* and *LSTM* in depth learning.

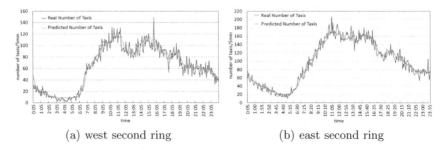

(a) west second ring   (b) east second ring

Fig. 3. Comparison of prediction and real traffic flow of second ring in 22nd Dec, 2014

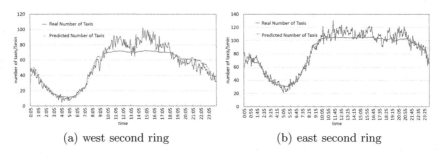

(a) west second ring   (b) east second ring

Fig. 4. Comparison of prediction and real traffic flow of second ring in 22nd Dec, 2014

For weekends, we also choose the volume of traffic per 5 min from 1st December to 21st December, 2014 for weekends as a training set. Meanwhile, we predict the volume of traffic for every 5 min among weekends from December 22nd to Saturday December 28th.

Figure 3(a) presents the results of prediction and real traffic flow of the second ring on west in 22nd (Monday) Dec, 2014. The predicted traffic volume and the actual traffic volume value fit well except a few abnormal points. Meanwhile, the predicted traffic volume is more accurate before noon than afternoon. Figure 3(b) shows the results of prediction and real traffic flow of east second ring in 22nd (workdays) Dec, 2014. Figure 3(b) shows results of traffic volume peak prediction in are somewhat deviant, but for other time traffic prediction results are very close to real values. similarity, Fig. 4(a) and (b) compare the traffic volume in 27th (weekends) Dec, 2014.

7.2   Experimental Results for TSPModel

We predict the average speed of four sections of the second ring road with *TSP-Model* from December 22nd to December 28th, respectively. Figure 5(a) shows the comparison of the real and predicted values of the average traffic speed of the West Second Ring Road section of Beijing City, December 22, 2014 (Monday), using this method. Similarity, Fig. 5(b) compared prediction and real average speed of east second ring in 22 Dec, 2014.

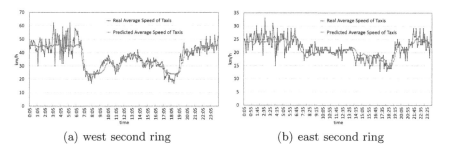

(a) west second ring   (b) east second ring

Fig. 5. Comparison of prediction and real average speed of second ring in 22 Dec, 2014

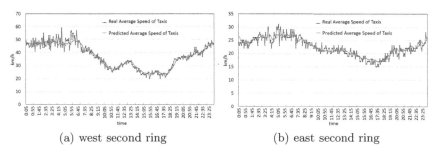

(a) west second ring   (b) east second ring

Fig. 6. Comparison of prediction and real average speed of west second ring in 27 Dec, 2014

Figure 6(a) shows the comparison between the average speed of the West Second Ring Road in Beijing on December 27, 2014 (Saturday) and the predicted value. Figure 6(b) presents the predicted speed with real speed of taxis in the East Second Ring Road in Beijing on 27 Dec, 2014 (Saturday).

7.3 Experimental Results for TCPModel

In this section, we calculate the traffic flow parameters predicted by the traffic congestion identification model mentioned, and get the traffic congestion comprehensive threshold. Consequently, we identify the forecast traffic congestion level through the range of the threshold. According to the actual traffic volume and the average speed to get a real traffic congestion level, we compare the predicted grade with the actual level to get the accuracy of the prediction. Traffic congestion prediction is made for weekdays and weekends respectively.

Figure 7(a) compares the predicted levels of traffic congestion with the actual levels in the West Second Ring Road in Beijing on December 22, 2014. We found that the predicted level of traffic congestion is more accurate in the early and late peak period than the time before 7 a.m. Figure 7(b) shows the predicted levels of traffic congestion compared with the actual levels on the West Second Ring in Beijing on 27 December 2014. From Fig. 7(b), the traffic congestion forecast in this section is well fitted. It is more accurate in the morning and evening

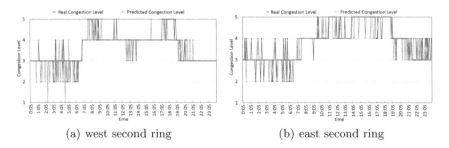

(a) west second ring   (b) east second ring

Fig. 7. Comparison of prediction and real traffic congestion of second ring in 22 Dec, 2014

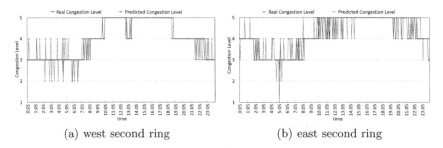

(a) west second ring   (b) east second ring

Fig. 8. Comparison of prediction and real traffic congestion of second ring in 22 Dec, 2014

peak periods than the period before 7 a.m. Similarity, Fig. 8(a) and (b) show the predicted levels of traffic congestion on 27nd (weekends) Dec, 2014.

8   Conclusion

With the growing vehicles in big cities, efficiently and comprehensively predicting traffic congestion in roads of cities has been emerging as an important research topic. In this paper, we have conducted a comprehensive analysis of traffic flow, including traffic volume, traffic speed and vehicles density.

For short-term traffic congestion, we present an auto encoder (SAE) model based on deep learning. We use the Softmax regression according to the temporal and spatial distribution characteristics of traffic flow. We proposed a SAE short-term traffic flow prediction model to predict traffic flow. We analyzed prediction results by several performance indicators. The results show that the performance of the model proposed in this paper is better than that of other models for workday traffic forecasting. For non-workday traffic forecasting, the performance of our model is better than that of the shallow network model, and it is closer to that of *RNN* and *LSTM* in depth learning. Finally, we predicted traffic congestion range according to the traffic congestion identification method. We compares the predicted traffic congestion range with the real traffic flow parameters to determine the accuracy of traffic congestion prediction.

In future, we hope that our work can improve the accuracy rate of traffic congestion in big cities by considering more related information about city traffic, including weather, time. Meanwhile, future research should be done with more small road sections, for example 1 km, and should be done with our presented models in this paper, to forecast traffic status accuracy than now.

Acknowledgment. This work was supported in part by the Natural Science Foundation of China grant 61502069, 61672128, 61702076; the Fundamental Research Funds for the Central Universities DUT18JC39.

References

1. Chang, H., Lee, Y., Yoon, B., Baek, S.: Dynamic near-term traffic flow prediction: system-oriented approach based on past experiences. IET Intel. Transport Syst. **6**(3), 292–305 (2012)
2. Davis, G.A., Nihan, N.L.: Nonparametric regression and short-term freeway traffic forecasting. J. Transp. Eng. **117**(2), 178–188 (1991)
3. Gilmore, J.F., Abe, N.: Neural network models for traffic control and congestion prediction. IVHS J. **2**(3), 231–252 (1995)
4. Hamed, M.M., Al-Masaeid, H.R., Said, Z.M.B.: Short-term prediction of traffic volume in urban arterials. J. Transp. Eng. **121**(3), 249–254 (1995)
5. Jeong, Y.S., Byon, Y.J., Castro-Neto, M.M., Easa, S.M.: Supervised weighting-online learning algorithm for short-term traffic flow prediction. IEEE Trans. Intell. Transp. Syst. **14**(4), 1700–1707 (2013)
6. Tan, J., Wang, S.: Research on prediction model for traffic congestion based on deep learning. Appl. Res. Comput. **32**(10), 2951–2954 (2015)
7. Li, C., Tang, Z., Cao, Y.: Study on traffic congestion prediction model of multiple classifier combination. Comput. Eng. Des. **31**(23), 5088–5091 (2010)
8. Luo, X., Jiao, Q., Niu, L., Sun, Z.: Short-term traffic flow prediction based on deep learning. Appl. Res. Comput. **34**(1), 91–93 (2017)
9. Lv, Y., Duan, Y., Kang, W., Li, Z., Wang, F.Y.: Traffic flow prediction with big data: a deep learning approach. IEEE Trans. Intell. Transp. Syst. **16**(2), 865–873 (2015)
10. Sun, H., Liu, H., Xiao, H., He, R., Ran, B.: Use of local linear regression model for short-term traffic forecasting. Transp. Res. Rec. J. Transp. Res. Board **1836**(1), 143–150 (2003)
11. Williams, B., Durvasula, P., Brown, D.: Urban freeway traffic flow prediction: application of seasonal autoregressive integrated moving average and exponential smoothing models. Transp. Res. Rec. **1644**(1), 132–141 (1998)
12. Zheng, J., Lin, X., Zheng, L., Zhang, S.: Traffic congestion state prediction based on Markov chain model. Traffic Eng. **22**, 76–79 (2012)
13. Zhou, C.: Predicting traffic congestion using recurrent neural networks. In: The World Congress on Intelligence Transport Systems, Chicago, October 2002

Image and Video Processing

3D Level Set Method via Local Structure Similarity Factor for Automatic Neurosensory Retinal Detachment Segmentation in Retinal SD-OCT Images

Yue Sun[ID], Sijie Niu[✉][ID], Jiwen Dong[ID], and Yuehui Chen[ID]

Shandong Provincial Key Laboratory of Network based Intelligent Computing,
School of Information Science and Engineering, University of Jinan, Jinan, China
sjniu@hotmail.com

Abstract. Qualitative assessment of pathological changes is vital for clinicians to determine the degree of neurosensory retinal detachment (NRD). However, accurate segmentation is challenging due to the diversity of NRD size and location. Spectral domain-optical coherence tomography (SD-OCT) imaging technology can yield high-resolution, three-dimensional images of the retinal histopathological structure without invasion and injury, whereas always accompanied by high-level noise and low-contrast intensity. In this paper, a novel automatic segmentation approach was presented based on two-stage clustering and an improved 3D level set method to quantify NRD regions in SD-OCT images. Considering the difference intensity distribution of NRD regions and the background, an unsupervised two-stage clustering method based on k-means was used to get initial surfaces for subsequence segmentation. In order to reduce the effect of noise and utilize spatial constraint information, 3D local structure similarity factor was proposed by combining the characteristic of the intensity of a voxel can be similar to those of its neighbors, which was introduced into level set model. Comparing with six methods, the experiment of 23 volumes from 12 patients demonstrates that the proposed method can improve the accuracy of location and segmentation for NRD regions in SD-OCT images.

Keywords: Neurosensory retinal detachment ·
Spectral domain-optical coherence tomography · Level set ·
3D local structure similarity factor

1 Introduction

Neurosensory retinal detachment (NRD), a prominent feature of central serous chorioretinopathy (CSC) [1,2], is characterized by the accumulation of subretinal fluid at the posterior pole illustrated in Fig. 1. Spectral domain-optical coherence

Granted by No. 61701192, No. ZR2017QF004 and 2016ZDJS01A12.

K. Knight et al. (Eds.): ICAI 2019, CCIS 1001, pp. 83–92, 2019.
https://doi.org/10.1007/978-981-32-9298-7_7

tomography (SD-OCT) imaging technology has become increasingly important in the diagnosis and treatment of eye diseases [3,4], as it can non-invasively provide details of retinal structure, such as NRD disease. At present, quantitative assessment of the distribution and extent of subretinal fluid in SD-OCT images is fundamental to judge the disease severity and progression. However, it is time-consuming and laborious to manually delineate lesion areas, and easily susceptible to subjective factors for users in practical applications. Therefore, the research related in automatic segmentation of lesion regions is accumulating currently.

Some unsupervised [5–8], semi-supervised and supervised [9–12] segmentation methods have been applied for the retinal fluid on OCT scans. For unsupervised methods, they are often based on the grayscale and morphologic features of lesion areas, such as fuzzy level sets with cross-sectional voting [5] and thresholding based approaches [6]. Wu et al. [7] detected and segmented the fluid-associated abnormalities by using thickness map priors and a fuzzy level set method. Ji et al. [8] constructed an aggregate generalized Laplacian of Gaussian method based on Hessian to achieve fluid segmentation into small blob candidate regions. For semi-supervised or supervised methods, they are always applied to generate the segmentation results of higher accuracy by combining labels. Lang et al. [9] adopted a random forest classifier to classify the fluid pixels. Xu et al. [10] proposed a novel classification method with a layer-dependent stratified sampling strategy for fluid detection. Wang et al. [11] proposed an interactive segmentation method based on a conventional Markov random field framework. In recent years, deep learning methods have been developed rapidly and succeed to applied in many fields. Long et al. [12] presented a fully convolutional network (FCN) to the segmentation tasks.

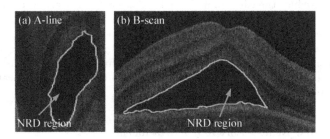

Fig. 1. NRD regions are marked by manual annotations (green curves) in SD-OCT images. (a) A-line image; (b) B-scan image. (Color figure online)

However, the segmentation accuracy in some of these methods would drop-down when dealing with retinal OCT images as its low contrast and high-level speckle noise caused by the imaging principle or others. Besides, for semi-supervised or supervised methods, the higher accuracy results are always contributed by extra expert information or labels, which are easily susceptible to subjective factors. In this paper, a fully automatic unsupervised segmentation method was proposed based on the improved 3D level set model for NRD regions in SD-OCT images. Initial surfaces were obtained by the two-stage clustering based on k-means to divide NRD regions from background roughly without any manual method or training phase. Considering the spatial constraint information and reducing the effect of high-level noise, 3D local structure similarity (3LSS) factor was proposed and introduced into the conventual level set model to obtain the final segmentation results by refining the initial surfaces.

2 Methodology

An overview of the proposed method was illustrated in Fig. 2, which can be divided into two stages: the obtainment of initial surfaces based on clustering and refining NRD region segmentation guided by the improved 3D level set model.

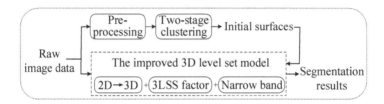

Fig. 2. Flowchart of the proposed method for NRD segmentation in SD-OCT images.

2.1 Initial Surfaces Based on Two-Stage Clustering

Backgrounds and unrelated tissues can cause incorrect segmentation results, so it is necessary to form a restricted region (see Fig. 3) that usually consists of lesions and some layer structure between two surfaces, e.g. ONL-IS/OS surface and RPE-choroid surface obtained by a business software named OCTExplorer (https://www.iibi.uiowa.edu/oct-reference), which need to be corrected manually later.

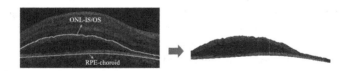

Fig. 3. The area restricted by outer plexiform layer (ONL)-inner segment with outer segment (IS/OS) surface and retinal pigment epithelium (RPE)-choroid surface.

Note that Fig. 4 shows the signal intensity of NRD region is much lower than that of the background in SD-OCT images. Besides, initial surfaces obtained by the manual methods or conventional methods are too coarse to be considered as an adequate initialization for subsequence segmentation due to the randomness and limitation, thus an automatic and robust initialization method based on k-means clustering was proposed to correctly initialize surfaces by utilizing the characteristic of the intensity contrast pattern of the background and NRD regions.

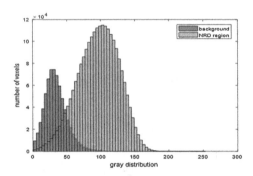

Fig. 4. The gray distribution of NRD regions and the background in the restricted area on one volume.

Firstly, due to the high-level noise caused by the imaging principle of interferometric of coherent optical beams in SD-OCT images and the location-free constraint of k-means method, original images should be pre-processed by filtering operation, such as mean filtering. Then, considering the varied intensity distribution among different samples and the adaptive selection of clustering centers, the two-stage clustering method based on k-means was applied (See Fig. 5). In the first stage, the image data of the whole restricted area was taken as input, and the minimal/maximal value of the intensity domain that usually ranges from 0 to 255 was used as the initial centers, the result was illustrated in Fig. 5(a). We could observe that the incorrect clustering voxels were usually exiting in the background area because the intensity distribution often ranges from 20 to 50 and 80 to 120 in NRD regions and the background respectively, which caused most of the voxels to concentrate upon the class whose initial clustering center was the minimal value. Thus, the second stage clustering was used to make the result more accurate. The class whose initial center was the minimal value at first stage was as input, and the initial clustering centers were assigned by the minimum and maximum values of the input data, the result was shown in Fig. 5(b), which are more consistent with the background and NRD region. Thus, NRD regions and backgrounds can be coarsely divided by the two-stage clustering method.

Finally, the clustering results were transformed into initial surfaces for subsequence segmentation after removing few false objects.

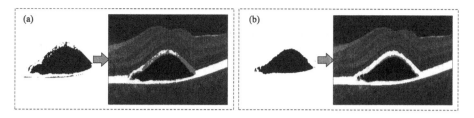

Fig. 5. The results of two-stage clustering based on k-means. (a) The first stage clustering result. (b) The second stage clustering result. The white, black and red area denote the background, NRD region, and incorrect clustering results respectively. (Color figure online)

2.2　The Improved 3D Level Set Model-Based Segmentation

The intensity range may be inconsistent both between and within images to make parameters selection difficult and not conducive to segmentation. To deal with this problem, normalization was applied to uniform the intensity range and enhance the contrast between the background and NRD regions.

In order to fully exploit spatial structure constraints in SD-OCT volumes and inspired by [13], 3D local structure similarity factor was proposed to reform the energy functional of the conventual level set model.

3D Local Structure Similarity Factor. Though some region-based models are less sensitive to Gaussian noise to some extent, they still lack enough robustness on images corrupted by noise of different characters and strength. On the other hand, most methods utilize Gaussian filter to remove noise, based on the assumption that the noise follows a Gaussian distribution, resulting in the boundary leakage on contour evolution.

In order to solve the problem of removing noise and maintaining boundary as discussed above, 3D Local Structure Similarity (3LSS) factor was developed by considering the intensity of a voxel can be similar to those of its neighbors in one volume. For each voxel within a local cube, the contribution of the statistical information lies on the distance from the cube center. The 3LSS factor can be defined as:

$$3LSS(\boldsymbol{x}, lc) = \int\limits_{v_m} \frac{|I(\boldsymbol{y}) - lc|^2}{d(\boldsymbol{x}, \boldsymbol{y})} d\boldsymbol{y} \tag{1}$$

where $I(\boldsymbol{x})$ is the voxel, $v_m(\boldsymbol{y} \epsilon v_m, v_m \epsilon v, m = 1, 2, ..., n, v_1 \bigcup v_2 \bigcup \cdots \bigcup v_n = v)$ is a local cube centered at $\boldsymbol{x}$ and each cubic size is $k_m \times k_m \times k_m$, v denotes the SD-OCT volume. $d(\boldsymbol{x}, \boldsymbol{y})$ is the spatial Euclidean distance between $\boldsymbol{x}$ and $\boldsymbol{y}$, lc is local average intensity defined in a local cube as explained in next subsection.

Improved 3D Level Set Model. In order to fully exploit the spatial structure constraints and come to grips with three-dimensional image data better, the conventual level set model was extended to the higher dimension, e.g., three-dimension. As mentioned above, 3LSS factor was introduced for keeping boundary details based on eliminating the impact of noise. Thus, energy functional can be improved as:

$$E^{improved}(\phi, lc_1, lc_2) = \mu \int_v \delta(\phi(\boldsymbol{x})) |\nabla \phi(\boldsymbol{x})| d\boldsymbol{x}$$

$$+ \lambda_1 \int_v (\int_{v_m} \frac{|I(\boldsymbol{y}) - lc_1(\boldsymbol{x})|^2}{d(\boldsymbol{x}, \boldsymbol{y})} d\boldsymbol{y}) H_\varepsilon(\phi(\boldsymbol{x})) d\boldsymbol{x} \qquad (2)$$

$$+ \lambda_2 \int_v (\int_{v_m} \frac{|I(\boldsymbol{y}) - lc_2(\boldsymbol{x})|^2}{d(\boldsymbol{x}, \boldsymbol{y})} d\boldsymbol{y})(1 - H_\varepsilon(\phi(\boldsymbol{x}))) d\boldsymbol{x}$$

where λ_1, λ_2 and μ are constant parameters, ϕ is the level set function. $H_\varepsilon(x)$ is Heaviside function:

$$H_\varepsilon(x) = \frac{1}{2} \times (1 + \frac{2}{\pi} \times arctan(\frac{x}{\varepsilon})) \qquad (3)$$

$\delta(z)$ is Dirac function:

$$\delta_\varepsilon = \begin{cases} \frac{1}{2\varepsilon}[1 + cos(\frac{\pi z}{\varepsilon})], & |z| \le \varepsilon \\ 0, & |z| > \varepsilon \end{cases} \qquad (4)$$

lc_1 and lc_2 are local average intensity values inside and outside the surface respectively:

$$lc_1(\boldsymbol{x}) = \frac{\int_{v_M} I(\boldsymbol{y}) H_\varepsilon(\phi(\boldsymbol{y})) d\boldsymbol{y}}{\int_{v_M} H_\varepsilon(\phi(\boldsymbol{y})) d\boldsymbol{y}} \qquad (5)$$

$$lc_2(\boldsymbol{x}) = \frac{\int_{v_M} I(\boldsymbol{y})(1 - H_\varepsilon(\phi(\boldsymbol{y}))) d\boldsymbol{y}}{\int_{v_M} (1 - H_\varepsilon(\phi(\boldsymbol{y}))) d\boldsymbol{y}} \qquad (6)$$

where v_M is a local cube centered at $\boldsymbol{x}$ and its side length k_M usually larger than k_m of v_m. lc_1 and lc_2 can be used to avoid small and irregular contours on segmentation results according to combine the spatial constraints.

Therefore, the proposed method can better tackle the fine segmentation of three-dimensional medical images with the characteristic of high-level noise.

Derivation of Energy Functional. In this paper, we denote by $\frac{\partial E}{\partial \phi}$ the first variation of the energy functional E, and the following evolution equation:

$$\frac{\partial \phi}{\partial t} = -\frac{\partial E}{\partial \phi} \qquad (7)$$

is gradient flow to minimize the functional E, then we can infer Eq. 8. Finally, the evolution equation of the level set function is defined in Eq. 9, where $\triangle t$ is the time step.

$$
\begin{aligned}
\frac{\partial \phi}{\partial t} &= \mu \delta(\phi(\boldsymbol{x})) div(\frac{\nabla \phi(\boldsymbol{x})}{|\nabla \phi(\boldsymbol{x})|}) \\
&- \delta(\phi(\boldsymbol{x})) \lambda_1 (\int_{v_m} \frac{|I(\boldsymbol{y}) - lc_1(\boldsymbol{x})|^2}{d(\boldsymbol{x}, \boldsymbol{y})} d\boldsymbol{y}) \\
&+ \delta(\phi(\boldsymbol{x})) \lambda_2 (\int_{v_m} \frac{|I(\boldsymbol{y}) - lc_2(\boldsymbol{x})|^2}{d(\boldsymbol{x}, \boldsymbol{y})} d\boldsymbol{y})
\end{aligned}
\tag{8}
$$

$$
\phi^{i+1} = \phi^i + \triangle t \times \frac{\partial \phi^i}{\partial t}
\tag{9}
$$

Narrow Band Method. The evolution of level set function mainly focuses on the details of boundary due to the initial surface has segmented the background and NRD regions coarsely, thus, narrow band method was applied to reduce the computational complexity. The basic idea is to update intensity values of a zonal region around the surface. First of all, generate the narrow band whose width is k_{nb} centered on surfaces. Secondly, the narrow band need to be re-initialized when surfaces are evolved to its boundary. Thirdly, for non-boundary voxels $\boldsymbol{x}$ in the narrow band, $\phi_{\boldsymbol{x}}^{i+1}$ is obtained according to the evolution equation.

3 Experiment

3.1 Dataset and Parameter Estimation

23 volumes collected by a Cirrus HD-OCT machine of 6.0 version from 12 patients who were diagnosed as NRD disease were used to evaluate the performance of the proposed method. Each volume dimension is $1024 \times 512 \times 128$ pixels with the size of $2 \times 6 \times 6 \, mm^3$. Two experienced physicians manually delineated the NRD-associated subretinal fluid regions in all SD-OCT volumes as ground truth data.

In Dirac function $\delta(z)$, ε was set to 0.1. k_M was set to 5 for lc_1 and lc_2. And in energy functional, the parameters were chosen as: $k_m = 3$, $\lambda_1 = \lambda_2 = 1$, $\triangle t = 4$, and $\mu = 1e - 3$. All of the parameter values mentioned above were fixed by experimental experience.

3.2 Experimental Results

Six state-of-the-art algorithms were used to compare with the proposed method: fully convolutional network (FCN) [12], machine learning approaches using stratified sampling k-nearest neighbor classifier (KNN) [10], random forest (RF) classifier [9], a semi-supervised segmentation algorithm utilizing label propagation and higher-order constraint (LPHC) [11], a fuzzy level set method guided by

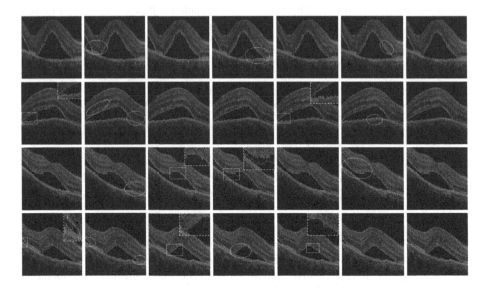

Fig. 6. Examples of NRD segmentation results. From the first column to seventh column, the segmentation results were obtained by FCN [12], KNN [10], RF [9], LPHC [11], WML [7], ALBD [8], and the proposed method respectively. The green curves denote the ground truth marked by expert 1, the yellow curves denote the ground truth marked by expert 2, and the red curves denote automatic segmentation results. (Color figure online)

Fig. 7. Three-dimensional models of NRD segmentation results are corresponded of first row to forth row (the proposed method) in Fig. 6.

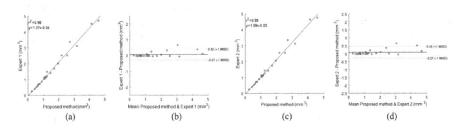

Fig. 8. (a)/(c) Statistical correlation analysis between the proposed method and expert 1/2. (b)/(d) Bland-Altman plots for the proposed method and expert 1/2.

enface fundus imaging (WML) [7] and a large blob detection (ALBD) for sub-retinal fluid segmentation [8].

From Figs. 6 and 7, the obvious inaccuracy of segmentations can be found for other methods. For FCN and ALBD methods, a problem of insufficient evolution could exist. For KNN method, there are some incorrect circular areas. And for LPHC method, the segmentation results were quite inaccurate. Through the qualitative analysis of Figs. 6 and 7, the proposed method can locate and segment NRD regions more accurately by comparing with other methods.

In this paper, the positive volume fraction (TPVF), positive predictive value (PPV) and dice similarity coefficient (DSC) were applied to provide quantitative analysis of the proposed method with the six algorithms mentioned above. Among them, the quantitative analysis results of KNN, RF, LPHC, WML and ALBD came from [7] and [8]. The NRD segmentation results of FCN method were obtained by retraining it with the medical dataset used in this paper and make it can achieve a good performance. Table 1 illustrates the comparison of ground truth and segmented results, the TPVF and DSC values of the proposed method are highest, reaching over 96% and 94% respectively and the PPV values reach more than 92%. Figure 8 shows the comparison of segmentation results obtained by the proposed method between expert 1 and 2 by linear regression analysis and Bland-Altman, which indicated a stable agreement between the automated segmentation results obtained by the proposed method and manual segmentations. From the above experimental results, the proposed method is helpful to improve the accuracy of segmentation results for NRD regions.

Table 1. Comparative analysis of experimental results.

	Expert 1			Expert 2		
	TPVF (%)	PPV (%)	DSC (%)	TPVF (%)	PPV (%)	DSC (%)
FCN [12]	82.6 ± 20.3	$\mathbf{95.0 \pm 5.1}$	86.6 ± 15.6	82.4 ± 20.0	$\mathbf{96.0 \pm 5.1}$	87.0 ± 15.6
KNN [10]	80.9 ± 6.6	91.9 ± 3.8	86.1 ± 4.1	80.3 ± 6.5	91.8 ± 3.8	85.9 ± 4.1
RF [9]	92.6 ± 4.4	92.4 ± 2.0	87.1 ± 4.3	92.5 ± 4.3	91.9 ± 2.2	88.9 ± 4.2
LPHC [11]	81.3 ± 9.4	55.6 ± 13.3	65.3 ± 10.4	81.2 ± 9.3	55.8 ± 12.8	65.7 ± 10.5
WML [7]	94.2 ± 5.2	93.0 ± 4.8	93.7 ± 4.0	94.5 ± 5.1	94.1 ± 5.3	94.6 ± 4.1
ALBD [8]	95.1 ± 2.3	93.1 ± 5.0	94.0 ± 3.1	95.2 ± 2.2	94.2 ± 4.9	94.7 ± 3.0
Proposed	$\mathbf{96.1 \pm 2.6}$	92.2 ± 2.7	$\mathbf{94.0 \pm 1.8}$	$\mathbf{96.3 \pm 2.2}$	93.4 ± 2.9	$\mathbf{94.8 \pm 1.8}$

4 Conclusion

In this paper, a fully automatic unsupervised segmentation method was proposed based on the improved 3D level set model for NRD regions in SD-OCT volumes. Inspired by the contrast pattern of NRD region with the background, and considering the limitation of manual and conventual initialization methods, two-stage clustering method based on k-means was proposed to generate the appropriate initial surfaces for subsequence segmentation. In order to have a balance between eliminating the effect of noise and preserving the details of the boundary, 3LSS factor was proposed by combining the spatial constraint information in volumes, then introduced into the energy functional. Finally, the improved 3D level set model was applied to guide initial surfaces close to the desired boundary. Compared with other methods, an experiment demonstrated the proposed method is helpful to improve the accuracy of automatic segmentation for NRD regions. However, it is sensitive to layer segmentation data. In future work, we need to reduce its dependence on layer segmentation data to improve the performance of segmentation.

References

1. Balaratnasingam, C., et al.: Bullous variant of central serous chorioretinopathy. Ophthalmology **123**(7), 1541–1552 (2016)
2. Daruich, A., et al.: Central serous chorioretinopathy: recent findings and new physiopathology hypothesis. Prog. Retinal Eye Res. **48**, 82–118 (2015)
3. Chen, H.Y., Xia, H.H., Qiu, Z.Q., Chen, W.Q., Chen, X.J.: Correlation of optical intensity on optical coherence tomography and visual outcome in central retinal artery occlusion. Retina **36**(10), 1964–1970 (2016)
4. Gao, E.T., et al.: Comparison of retinal thickness measurements between the topcon algorithm and a graph-based algorithm in normal and glaucoma eyes. PLoS ONE **10**(6), e0128925 (2015)
5. Wang, J., et al.: Automated volumetric segmentation of retinal fluid on optical coherence tomography. Biomed. Opt. Express **7**(4), 1577 (2016)
6. Wilkins, G.R., Houghton, O.M., Oldenburg, A.L.: Automated segmentation of intraretinal cystoid fluid in optical coherence tomography. IEEE Trans. Biomed. Eng. **59**(4), 1109–1114 (2012)
7. Wu, M.L., Chen, Q., He, X.J., Li, P., Fan, W., Yuan, S.T.: Automatic subretinal fluid segmentation of retinal SD-OCT images with neurosensory retinal detachment guided by enface fundus imaging. IEEE Trans. Biometrical Eng. **65**(1), 87–95 (2017)
8. Ji, Z., et al.: Beyond retinal layers: a large blob detection for subretinal fluid segmentation in SD-OCT images. In: Frangi, A.F., Schnabel, J.A., Davatzikos, C., Alberola-López, C., Fichtinger, G. (eds.) MICCAI 2018, Part II. LNCS, vol. 11071, pp. 372–380. Springer, Cham (2018). https://doi.org/10.1007/978-3-030-00934-2_42
9. Lang, A., et al.: Automatic segmentation of microcytic macular edema in OCT. Biomed. Opt. Express **6**, 155–169 (2015)
10. Xu, X., Lee, K., Zhang, L., Sonka, M., Abramoff, M.D.: Stratified sampling voxel classification for segmentation of intrarenal and subretinal fluid in longitudinal OCT data. IEEE Trans. Med. Imaging **34**, 1616–1623 (2015)
11. Wang, T., et al.: Label propagation and higher-order constraint-based segmentation of fluidassociatedregions in retinal SD-OCT images. Inf. Sci. **358–359**, 92–111 (2016)
12. Long, J., Shelhamer, E., Darrell, T.: Fully convolutional networks for semantic segmentation. In: IEEE Conference on Computer Vision and Pattern Recognition, vol. 79, pp. 3431–3440 (2015)
13. Niu, S.J., Chen, Q., de Sisternes, L., Ji, Z.X., Zhou, Z.M., Rubin, D.L.: Robust noise region-based active contour model via local similarity factor for image segmentation. Pattern Recogn. **61**, 104–119 (2017)

Real-Time Scale Adaptive Visual Tracking with Context Information and Correlation Filters

Yue Xu[1,2], Mengru Feng[1,2], Jiezhi Yang[1,2], Mingjie Peng[1,2],
Jiatian Pi[1,2(✉)], and Yong Chen[1,2]

[1] Chongqing Normal University, Chongqing 401331, China
xu.yue.fh@qq.com, pijiatian@cqnu.edu.cn
[2] Chongqing Engineering Research Center for Digital Agricultural Service,
Chongqing, China

Abstract. Visual tracking has been studied for several decades but continues to draw significant attention because of its critical role in many applications. However, there still need to improve the overall tracking capability to counter various tracking issues, including large scale variation, occlusion, and deformation. This paper presents an appealing tracker with robust scale adaptive, which applies the discriminative correlation filter for scale estimation as an independent part after finding the optimal translation based on the kernelized correlation filter and context information. Compared to an exhaustive scale space search scheme, our approach provides improved performance while being computationally efficient. In order to reveal the effectiveness of our approach, we use benchmark sequences annotated with 11 attributes to evaluate how well the tracker handles different attributes. Numerous experiments demonstrate that the proposed algorithm performs favorably against several state-of-the-art algorithms.

Keywords: Correlation filters · Scale estimation · Visual tracking

1 Introduction

Visual object tracking is one of the most fundamental tasks in the field of computer vision and is related to a wide range of important real-world applications, such as surveillance, robotics, autonomous driving, etc. Although great progress has been made in the past decade, it remains a challenging problem due to baffling factors, such as illumination variations, background clutter, shape deformation, etc.

Canonical examples of the tracking-by-detection paradigm include those based on Support Vector Machines (SVM) [1], Random Forest classifiers [2], or boosting variants [3]. Zhang et al. [4] propose a projection to a fixed random basis, to train a Naive Bayes classifier, inspired by compressive sensing theory. Tracking-Learning-Detection (TLD) tracker [5] exploits a set of structural constraints with a sampling strategy using boosting classifier. Recent benchmark [6–8] studies show that the top-performance trackers are usually deep-learning-based trackers [9, 10]. However, in the pursuit of ever increasing tracking performance, their characteristic speed and real-time

© Springer Nature Singapore Pte Ltd. 2019
K. Knight et al. (Eds.): ICAI 2019, CCIS 1001, pp. 93–105, 2019.
https://doi.org/10.1007/978-981-32-9298-7_8

capability have gradually faded. Lately, correlation filter based trackers [11–14] also have achieved appealing performance despite their great simplicity and superior speed. Those trackers train a discriminative filter, where convolution output can indicate the likeness between candidate and target. Because the element-wise operation in Fourier domain is equal to the convolution operation in time domain (spatial domain in tracking), they evaluate the cyclically shifted candidates very efficiently. However, Minimum Output Sum of Squared Error (MOSSE) tracker [11], Circulant Structure Kernels (CSK) tracker [12], and Kernelized Correlation Filter (KCF) tracker [14], are limited to only estimating the target position with the fixed size. Discriminative Scale Space Tracker (DSST) [13] has proposed an efficient method for estimating the target scale by training a classifier on a scale pyramid, which is the best tracker in the competition [15]. However, there is still room for improvement in translation estimation in the DSST.

In this paper, we incorporate the proposed scale estimation approach in the DSST into the KCF tracker without much computational overhead. Compared to the traditional DSST and the KCF tracker, the improved algorithm is more robust and can deal with more challenging scenarios. The key contributions of this work can be summarized as follows. Firstly, we apply the context information for the KCF tracker to obtain an accurate position of the target. Secondly, we extend the KCF tracker with the capability of handling scale changes and verify that the applied scale estimated approach is generic. Finally, we perform extensive experiments on 50 sequences in the recent benchmark evaluation [6]. Experimental results show that the proposed tracker achieves outstanding performance both in accuracy and robustness against the state-of-the-art trackers.

2 The Proposed Tracker

In this section, we give the details of our proposed tracker. In order to incorporate the scale estimation approach into the KCF tracker, we decompose the task into translation and scale estimation of objects.

2.1 Translation Estimation

Recently, the tracking system based on the Kernelized Correlation Filter (KCF) achieves favorable performance with high speed. In that work, Henriques et al. [14] demonstrate that it is possible to analytically model natural image translations, which shows that the resulting data and kernel matrices become circulant under some conditions. The diagonalization by the Discrete Fourier Transform (DFT) provides a general blueprint for creating fast algorithms that deal with translations. By considering correlation filters as classifiers, the goal of training is to find a function $f(\mathbf{z}) = \mathbf{w}^T \mathbf{z}$ that minimizes the squared error over samples x_i and their regression targets y_i according to:

$$\min_{\mathbf{w}} \sum_i (f(x_i) - y_i)^2 + \lambda_1 \|\mathbf{w}\|_2^2 \tag{1}$$

where $\mathbf{w}$ denotes the parameters, and λ_1 is the regularization parameter to prevent over fitting. The Ridge Regression has the close-form solution according to:

$$\mathbf{w} = \left(\mathbf{X}^T\mathbf{X} + \lambda_1\mathbf{I}\right)^{-1}\mathbf{X}^T\mathbf{y} \tag{2}$$

where the data circulant matrix $\mathbf{X}$ has one sample per row x_i and each element of $\mathbf{y}$ is a regression target y_i. $\mathbf{I}$ is an identity matrix.

To introduce the kernel functions for improving the performance, input data x can be mapped to a non-linear-feature space as $\varphi(x)$, and $\mathbf{w} = \sum_i \alpha_i \varphi(x_i)$. Then the solution to the kernelized version of Ridge Regression in the KCF tracker is given by:

$$\alpha = (\mathbf{K} + \lambda_1\mathbf{I})^{-1}\mathbf{y} \tag{3}$$

where $\mathbf{K}$ is the kernel matrix and α is the vector of coefficients α_i. With the help of circulant matrix, all the translated samples around the target can be collected for training with no significant decrease in the speed. Given a base sample $\mathbf{x} = (x_0, \ldots, x_{n-1})$, all the cyclic shift visual samples are concatenated to form the circulant matrix $\mathbf{X} = C(\mathbf{x})$. With the various interesting properties of circulant matrices, the solution of can be expressed as follows:

$$\hat{\alpha} = \frac{\hat{\mathbf{y}}}{\hat{\mathbf{k}} + \lambda_1} \tag{4}$$

where $\hat{\alpha}$, $\hat{\mathbf{y}}$ and $\hat{\mathbf{k}}$ denote the DFT of α, $\mathbf{y}$ and $\mathbf{k}$, respectively. Although dot-product, radial basis kernel and polynomial kernels functions are found to satisfy this condition, we apply the Gaussian kernel which can be expressed as follows:

$$\mathbf{k}^{\mathbf{XX}'} = \exp\left(-\frac{1}{\sigma^2}\left(\| \mathbf{x} \|^2 + \| \mathbf{x}' \|^2 - 2F^{-1}(\hat{\mathbf{x}}^* \odot \hat{\mathbf{x}}')\right)\right) \tag{5}$$

where $\hat{\mathbf{x}}$ denote the DFT of the base sample $\mathbf{x}$, and $\hat{\mathbf{x}}^*$ represents complex conjugation.

The surroundings of the tracked object can have a big impact on tracking performance. Therefore, we apply the context information into the correlation filter as mentioned in [16]. We sample m context patches $\mathbf{a}_i \in \mathbb{R}^n$ around the object of interest $\mathbf{a}_0 \in \mathbb{R}^n$. Their corresponding circulant matrix are $\mathbf{A}_i \in \mathbb{R}^{n \times n}$ and $\mathbf{A}_0 \in \mathbb{R}^{n \times n}$, respectively. The formulation 1 can be expressed as follows:

$$\min_{\mathbf{w}} \| \mathbf{A}_0\mathbf{w} - \mathbf{y} \|_2^2 + \lambda_1 \| \mathbf{w} \|_2^2 + \lambda_2 \sum_{i=1}^{m} \| \mathbf{A}_i\mathbf{w} \|_2^2 \tag{6}$$

where the context patches are regressed to zeros controlled by the parameter λ_2. Then the Ridge Regression with context information has the close-form solution according to:

$$\mathbf{w} = \left(\mathbf{B}^T\mathbf{B} + \lambda_1\mathbf{I}\right)^{-1}\mathbf{B}^T\bar{\mathbf{y}} \tag{7}$$

where $\mathbf{B} = \left[\mathbf{A}_0\sqrt{\lambda_2}\mathbf{A}_1 \cdots \sqrt{\lambda_2}\mathbf{A}_m\right]^T$ and $\bar{\mathbf{y}} = [\mathbf{y}0\cdots 0]^T$.

Then the solution of $\boldsymbol{\alpha}^{trans}$ with context information can be expressed by:

$$\alpha^{trans} = \left(\mathbf{B}^T\mathbf{B} + \lambda_1\mathbf{I}\right)^{-1}\bar{\mathbf{y}} \tag{8}$$

If the new sample is $\mathbf{z}$ and the corresponding circulant matrix is $\mathbf{Z}$, a confidence map $\mathbf{y}^{trans}$ can be obtained by:

$$\mathbf{y}^{trans} = \mathbf{Z} \odot \mathbf{B}^T \odot \alpha^{trans} \tag{9}$$

Using the identity for the circulant matrices, α^{trans} can be decomposed of a concatenation of dual variables $\left\{\alpha_0^{trans}, \alpha_1^{trans}, \cdots, \alpha_m^{trans}\right\}$. Then the confidence map in the Fourier domain can be rewritten as:

$$\hat{\mathbf{y}}^{trans} = \hat{\mathbf{z}} \odot \hat{\mathbf{a}}_0^* \odot \hat{\alpha}_0^{trans} + \sqrt{\lambda_2}\sum_{i=1}^{m} \hat{\mathbf{z}} \odot \hat{\mathbf{a}}_i^* \odot \hat{\alpha}_i^{trans} \tag{10}$$

The position with a maximum value in $\mathbf{y}^{trans}$ can be predicted as new position of the target.

2.2   Scale Estimation

The Kernelized Correlation Filter (KCF) with context information in Sect. 2.1 is used for estimating the translation, then we can find the accurate position of the target without scale change. To handle the challenging problem of scale change, we incorporate separate filters [13] for scale estimation. The discriminative correlation filter is closely related to the MOSSE filter [11], which produces stable correlation filters when trained on a small number of image windows. Consequently, the discriminative correlation filter is an efficient and ideal approach for robust scale estimation. After finding the accurate position with the KCF tracker and context information, we apply the discriminative correlation filter for scale estimation. Firstly, the MOSSE filter need a set of training images f_i, as well as a set of training outputs g_i. Training is conducted in the Fourier domain to take advantage of the simple element-wise relationship between the input and the output. To find a filter that maps training inputs to the desired training outputs, MOSSE finds a filter h that minimizes the sum of squared error. The minimization problem takes the form according to:

$$\min_{\hat{h}^*} \sum_i |\hat{f}_i \odot \hat{h}^* - \hat{g}_i|^2 \tag{11}$$

Where $\hat{f}_i$, $\hat{g}_i$ and the filter $\hat{h}$ are the Fourier transform of f_i, g_i and h, respectively. $\hat{h}_i^*$ represents complex conjugation. By solving for $\hat{h}^*$, a closed form expression for the MOSSE filter is found

$$\hat{h}^* = \frac{\sum_i \hat{g}_i \odot \hat{f}_i^*}{\sum_i \hat{f}_i \odot \hat{f}_i^*} \tag{12}$$

where $\hat{f}_i^*$ represents complex conjugation.

In the DSST, the MOSSE filter has been extended to multidimensional features. Assuming the feature dimension number $l \in \{1, 2, \ldots, d\}$, the solution for the optimal correlation filter $\hat{h}$, which consists of one filter $\hat{h}^l$ per feature, is obtained in the DSST as follows:

$$\hat{h}^l = \frac{\hat{g}^* \odot \hat{f}^l}{\sum_{k=1}^d \hat{f}^k \odot \hat{f}^{k*} + \lambda} \tag{13}$$

where λ is the regularization parameter to prevent over fitting, and $\hat{g}^*$ represents complex conjugation. To reduce the computational cost of the scale estimation with separate filters, we use Principal Component Analysis (PCA) to reduce the feature dimensionality. The projection matrix R_t, is $\hat{d} \times d$, where $\hat{d}$ the dimensionality of the compressed feature representation. We obtain R_t by minimizing the reconstruction error of the target template u_t. Then the compressed numerator $\tilde{A}_t^l$, and denominator $\tilde{B}_t$, of the filter is updated as follows:

$$\begin{aligned} \tilde{A}_t^l &= \eta \hat{g}^* \tilde{U}_t^l \\ \tilde{B}_t &= (1-\eta)\tilde{B}_{t-1} + \eta \sum_{k=1}^{\hat{d}} \tilde{F}_t^{k*} \tilde{F}_t^k \end{aligned} \tag{14}$$

where $\tilde{U}_t = F\{R_t u_t\}$, $u_t = (1-\eta)u_{t-1} + \eta f_t$, $\tilde{F}_t = F\{R_t f_t\}$, $F\{\}$ represents DFT. The correlation scores $\mathbf{y}^{scale}$ are then computed as follows:

$$\mathbf{y}^{scale} = F^{-1}\left\{ \frac{\sum_{l=1}^{\hat{d}} \tilde{A}^{l*} \tilde{Z}^l}{\tilde{B} + \lambda} \right\} \tag{15}$$

The scale with a maximum value in $\mathbf{y}^{scale}$ can be predicted as the new scale of the target.

2.3 Tracking Algorithm

The main steps of our tracker are presented in Algorithm 1. We use two independent correlation filters for translation and scale estimation. The Kernelized Correlation Filter (KCF) with context information is only applied for translation estimation and the discriminative correlation filter cooperates on scale estimation. Unlike our tracker, the DSST [13] uses separate filters for translation and scale estimation, which are all based on discriminative correlation filters. In addition, we extract translation sample with fixed size to find the target position without considering the scale, whereas the DSST extracts translation sample according to the previous scale. Thus, we really separate the translation and scale estimation in a way. Furthermore, the major difference between the KCF tracker and our tracker is that the KCF tracker is unable to deal with the challenge of scale change.

Algorithm 1: Proposed tracking algorithm: iteration at time t
---
Inputs:
- A bounding box with previous target position p_{t-1} and scale s_{t-1} in Image I_t.
- Training sample feature A_{t-1}^{trans} and parameter a_{t-1}^{trans} for translation model.
- Training scale model A and B.

Translation estimation:
- Extract a translation sample z^{trans} with fixed size at p_{t-1} in I_t.
- Compute the translation response y^{trans}, using z^{trans}, A_{t-1}^{trans} and α_{t-1}^{trans}.
- Set p_t to the target position that maximizes the response y^{trans}

Scale estimation:
- Extract a scale sample z^{scale} with scale s_{t-1} at p_t in I_t.
- Compute the translation response y^{trans}, using z^{trans}, A_{t-1}^{trans} and α_{t-1}^{trans}.
- Set p_t to the target scale that maximizes the response y^{scale}.

Model update:
- Extract sample feature with fixed size at p_t in I_t to update A_t^{trans} and α_t^{trans}.
- Extract sample feature with scale s_t at p_t in I_t to update $\tilde{A}_t^{scale}$ and $\tilde{B}_t^{scale}$.

Output:
- Estimated target position p_t and scale s_t.
- Updated the translation model A_t^{trans}, α_t^{trans} and scale model $\tilde{A}_t^{scale}$, $\tilde{B}_t^{scale}$.
---

3 Experiments

In this section, our proposed algorithm is evaluated with other 7 state-of-the-art methods on 50 challenging sequences of the new OTB dataset [6]. The compared trackers include other correlation filter-based trackers, such as DSST [13], CSK tracker [12] and KCF tracker [14]. The other compared trackers are Adaptive Structural Local Appearance (ASLA) tracker [17], Incremental Learning Tracker (IVT) [18], Distribution Fields Tracker (DFT) [19] and Compressive Tracking (CT) tracker [4]. For each tracker, we use the default parameters which are tuned well by the authors. We select 50 difficult and representative ones in the OTB dataset for analysis. The proposed algorithm runs at 40 frame per second (FPS) with a matlab implementation on an Intel Core (TM) i5-4590 3.00 GHz CPU with 4 GB RAM without any optimizing.

3.1 Experiment Setup and Evaluation Criteria

In our experiments, we use a Gaussian function to initialize the desired translation and scale filter output, respectively. The regularization parameter is set to 10^{-4}, the learning rate is set to 0.02. The bandwidth of the Gaussian kernel $\sigma = 0.5$, spatial bandwidth for the desired translation filter output is $\sqrt{mn}/10$ for a $m \times n$ target, and the standard for the desired scale filter output is $1/16$ times the number of scales $S = 33$. We use the Principal Component Analysis Histogram of Gradient (PCA-HOG) [20] for target representation. The cell size of HOG is 4×4 and the orientation bin number of HOG is 9. In order to get fair experimental results, all the parameters are kept constant for the following experiments.

We use two metrics to evaluate the performance. The first one is the precision plot which is based on the location error. The other one is the success plot which is based on the overlap rate. To compare the performance of different trackers, the results at error threshold of 20, as well as in the benchmark [8], are used to ranking in the precision plot. The success plot shows the ratios of successful frames while the overlap thresholds vary from 0 to 1. We use the area under the curve (AUC) of each success plot to rank the tracking algorithms. To analyze the robustness of initialization, each sequence is partitioned into 20 segments. This evaluation metric is referred as temporal robustness evaluation (TRE) in the benchmark [8]. We use both the one-pass evaluation (OPE) and temporal robustness evaluation (TRE) to evaluate our tracker.

3.2 Experiments on the Whole Dataset

We set up a comparison on the whole OTB-50 dataset [6] with one-pass evaluation (OPE). Results are shown in Figs. 1 and 2, where our tracker outperforms other state-of-the-art trackers. Experimental results show that our trackers achieve 68.1% on the precision, which is 7% improvement over the KCF tracker and 5.6% improvement over the DSST tracker. So the context information and the scale estimation can be indeed incorporated into the KCF tracker framework to improve the precision of the target position. Due to the success plot represents the overlap score between the tracked bounding box and the ground truth bounding box, the scale adaptability of trackers can be presented excellently. Experimental results in the Fig. 2 show that our tracker achieves 49.2% on the AUC score, which is 8.9% improvement over the KCF tracker and 3.1% improvement over the DSST tracker. Our tracker performs more favorable than DSST because we apply the KCF tracker with context information to find the optimal translation before scale estimation, which is more accurate than the DSST and can improve the scale estimation to some extent.

To further evaluate the robustness and efficiency of our tracker, we set up a comparison on the whole OTB-50 dataset with temporal robustness evaluation (TRE). Precision plots and success plots with TRE are shown in Figs. 3 and 4, which show our tracker is also superior compared to other trackers. In addition to high accuracy, our tracker runs efficiently at an average speed of 40.0 FPS, which is about more than 1.4

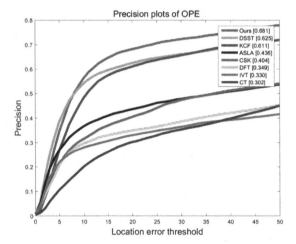

Fig. 1. Precision plots with one-pass evaluation (OPE) over all 50 benchmark sequences.

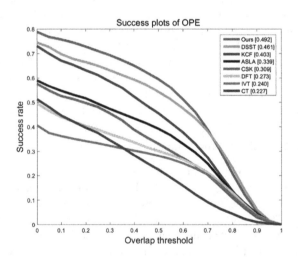

Fig. 2. Success plots with one-pass evaluation (OPE) over all 50 benchmark sequences.

times faster than the DSST. Although the speed of the KCF tracker is 140.0 FPS on average and is faster than ours, it is not able to handle scale changes.

3.3 Experiments with Sequence Attributes

For better analysis of our tracker, we use the sequences annotated with the other 11 attributes in the benchmark [6] to evaluate how well the tracker handles different attributes. The name of the attributes are listed as follows: fast motion (FM), motion blur (MB), deformation (DEF), in-plane rotation (IPR), occlusion (OCC), out-of-plane rotation (OPR), out-of-view (OV), illumination variation (IV), background clutter

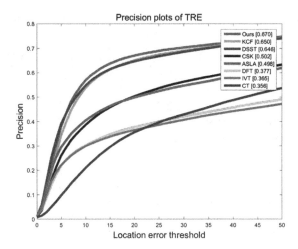

Fig. 3. Precision plots with temporal robustness evaluation (TRE) over all 50 benchmark sequences.

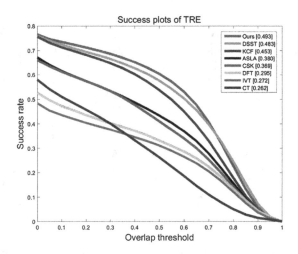

Fig. 4. Success plots with temporal robustness evaluation (TRE) over all 50 benchmark sequences.

(BC), low resolution (LR) and scale variation (SV). The precision rate at error threshold of 20 in precision plots with OPE are presented in Table 1. Our tracker achieves best performance to 10 of the 11 attributes. Moreover, the AUC score of success plots with OPE in each attribute are demonstrated in Table 2.

According to the experimental result, the proposed algorithm achieves best performance to 9 of the 11 attributes. Experimental results show that our tracker is superior compared to the KCF and DSST. Furthermore, our tracker improves significantly for sequences with fast motion and deformation. This is largely due to the fact that adding

Table 1. The precision rate (%) at error threshold of 20 of precision plots in 11 attributes. The best result is highlighted in red color and the second result is highlighted in blue color.

Attris	Ours	KCF	DSST	CSK	ASLA	DFT	IVT	CT
FM	57.6	52.9	48.0	29.9	28.9	26.5	21.6	24.7
MB	59.8	57.3	53.2	31.9	29.5	28.3	23.7	23.9
DEF	67.4	58.2	55.5	33.7	37.9	37.1	25.0	28.2
IPR	65.8	58.7	61.9	38.5	41.5	33.9	31.8	33.0
OCC	63.8	60.4	61.7	37.2	44.1	38.0	39.4	37.4
OPR	65.3	59.8	58.5	36.4	43.7	36.0	32.8	34.1
OV	47.2	44.1	41.1	25.2	40.3	38.9	33.8	42.8
IV	70.4	66.7	69.0	40.5	48.4	40.9	31.5	23.3
BC	67.8	62.3	66.0	46.0	43.8	41.0	28.2	23.4
LR	56.1	54.6	58.1	38.9	53.2	29.7	43.5	39.8
SV	65.5	56.5	59.1	34.9	44.1	29.7	34.5	31.4

Table 2. The AUC scores (%) of success plots in 11 attributes. The best result is highlighted in red color and the second result is highlighted in blue color.

Attris	Ours	KCF	DSST	CSK	ASLA	DFT	IVT	CT
FM	43.6	37.0	38.3	25.5	25.5	21.9	18.0	22.0
MB	44.0	39.5	41.9	27.1	24.4	22.8	18.8	20.5
DEF	48.7	40.0	42.0	26.5	28.7	29.7	17.4	24.0
IPR	48.1	38.9	44.9	29.4	32.9	27.4	22.2	25.1
OCC	46.7	39.6	44.3	27.6	34.4	28.7	27.1	26.2
OPR	47.8	39.6	41.9	27.1	34.5	28.3	22.1	25.3
OV	35.9	32.7	32.4	21.4	31.6	29.5	24.9	32.2
IV	49.3	43.3	51.1	31.8	36.9	32.2	23.1	19.2
BC	50.0	41.7	48.5	34.4	35.0	31.9	20.9	19.9
LR	36.5	30.7	38.9	26.3	37.2	19.8	29.3	20.5
SV	46.4	35.3	42.8	27.0	35.0	22.8	25.3	22.6

the context information allows for a larger search region and obtains an accurate position before the scale estimation.

3.4   Qualitative Analysis

The intuitive illustration on several representative challenging sequences is shown clearly in Fig. 5. When the motion blur occurs in the sequences *BlurBody*, such as frame #42 and frame #214, most algorithms fail to track the target except ours. The *CarScale* sequences contain scale variation, the proposed tracker can locate the target with a changeable rectangle box. In the *Human3* sequences, the man is occluded by the street lamp at frame #47. During this process, most tracker could not capture the target. Only our tracker could accurately locate the target during occlusion. The *Trellis* sequences contain illumination variation, the proposed tracker also performs well.

From the abovementioned various sequences, our experimental demonstrate that the proposed tracker achieves an encouraging performance in terms of accuracy and robustness.

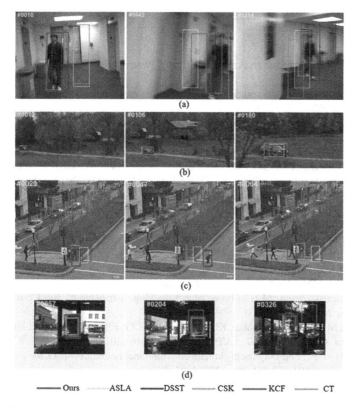

Fig. 5. Performance on (a) 'BlurBody', (b) 'CarScale', (c) 'Human3' and (d) 'Trellis' sequences by 6 trackers. Our algorithm results are marked with red line as shown in the up panel.

4 Conclusion

In this paper, we propose a robust tracking algorithm which combines the method of discriminative correlation filters with the Kernelized Correlation Filter (KCF) tracker. Our tracker handles the problem of fixed template size in KCF tracker without much decrease in the speed. In addition, we apply the context information to the KCF tracker, which leads to accurate and robust tracking results. Experiments on benchmark sequences demonstrated that the proposed algorithm performs favorably in terms of accuracy and robustness. Considering that our tracker uses only the HOG feature, we plan to incorporate more robust features into our tracker in the future.

Acknowledgments. This work is supported by National Natural Science Foundation of China under Grant No. 61603258, Basic and Cutting-edge Research Program of Chongqing under Grant No. cstc2018jcyjAX0470, Scientific and Technological Research Program of Chongqing Municipal Education Commission under Grant No. KJQN201800521 and No. KJQN201800536.

References

1. Avidan, S.: Support vector tracking. IEEE Trans. Pattern Anal. Mach. Intell. **26**, 1064–1072 (2004)
2. Saffari, A., Leistner, C., Santner, J., et al.: On-line random forests. In: Conference 2009, ICCV, pp. 1393–1400 (2009)
3. Grabner, H., Leistner, C., Bischof, H.: Semi-supervised on-line boosting for robust tracking. In: Forsyth, D., Torr, P., Zisserman, A. (eds.) ECCV 2008. LNCS, vol. 5302, pp. 234–247. Springer, Heidelberg (2008). https://doi.org/10.1007/978-3-540-88682-2_19
4. Zhang, K., Zhang, L., Yang, M.-H.: Real-time compressive tracking. In: Fitzgibbon, A., Lazebnik, S., Perona, P., Sato, Y., Schmid, C. (eds.) ECCV 2012. LNCS, vol. 7574, pp. 864–877. Springer, Heidelberg (2012). https://doi.org/10.1007/978-3-642-33712-3_62
5. Kalal, Z., Mikolajczyk, K., Matas, J.: Tracking-learning-detection. In: Conference 2012, IJCV, pp. 1409–1422 (2012)
6. Wu, Y., Lim, J., Yang, M. H.: Object tracking benchmark. In: Conference 2015, IJCV, pp. 1834–1848 (2015)
7. Kristan, M., Leonarids, A., Matas, J.: The visual object tracking VOT2017 challenge results. In: Conference 2017, ICCV, pp. 1949–1972 (2017)
8. Wu, Y., Lim, J., Yang, M.H.: Online object tracking: a benchmark. In: Conference 2013, CVPR, pp. 2411–2418 (2013)
9. Danelljan, M., Bhat, G., Shahbaz Khan, F., et al.: ECO: Efficient convolution operators for tracking. In: Conference 2017, CVPR, pp. 6638–6646 (2017)
10. Nam, H., Han, B.: Learning multi-domain convolutional neural networks for visual tracking. In: Conference 2016, CVPR, pp. 4293–4302 (2016)
11. Bolme, D.S., Beveridge, J.R., Draper, B.A., et al.: Visual object tracking using adaptive correlation filters. In: Conference 2010, CVPR, pp. 2544–2550, June 2010
12. Henriques, J.F., Caseiro, R., Martins, P., Batista, J.: Exploiting the circulant structure of tracking-by-detection with kernels. In: Fitzgibbon, A., Lazebnik, S., Perona, P., Sato, Y., Schmid, C. (eds.) ECCV 2012. LNCS, vol. 7575, pp. 702–715. Springer, Heidelberg (2012). https://doi.org/10.1007/978-3-642-33765-9_50
13. Danelljan, M., Häger, G., Khan, F., et al.: Accurate scale estimation for robust visual tracking. Conference 2014, BMVC. BMVA, Nottingham (2014)
14. Henriques, J.F., Caseiro, R., Martins, P., et al.: High-speed tracking with kernelized correlation filters. In: IJCV, pp. 583–596 (2015)
15. Kristan, M., et al.: The visual object tracking VOT2014 challenge results. In: Agapito, L., Bronstein, Michael M., Rother, C. (eds.) ECCV 2014. LNCS, vol. 8926, pp. 191–217. Springer, Cham (2015). https://doi.org/10.1007/978-3-319-16181-5_14
16. Mueller, M., Smith, N., Ghanem, B.: Context-aware correlation filter tracking. In: Conference 2017, CVPR, pp. 1396–1404 (2017)
17. Jia, X., Lu, H., Yang, M.H.: Visual tracking via adaptive structural local sparse appearance model. In: Conference 2012, CVPR, pp. 1822–1829 (2012)
18. Ross, D.A., Lim, J., Lin, R.S., et al.: Incremental learning for robust visual tracking. Int. J. Comput. Vision **77**(1–3), 125–141 (2008)

19. Sevilla-Lara, L., Learned-Miller, E.: Distribution fields for tracking. In: Conference 2012, CVPR, pp. 1910–1917. (2012)
20. Felzenszwalb, P.F., Girshick, R.B., McAllester, D., et al.: Object detection with discriminatively trained part-based models. IEEE Trans. Softw. Eng. **39**(9), 1627–1645 (2010)

An Improved Word Spotting Method for Printed Uyghur Document Image Retrieval

Eksan Firkat[1], Askar Hamdulla[2(✉)], Palidan Tuerxun[1],
and Abdusalam Dawut[1]

[1] School of Software, Xinjiang University,
Urumqi 830046, People's Republic of China
[2] School of Information Science and Engineering, Xinjiang University,
Urumqi 830046, People's Republic of China
askar@xju.edu.cn

Abstract. Key word spotting plays an important role in the field of Uyghur printed document retrieval. However, the cursive nature of Uyghur script causes some drawbacks for retrieving a word correctly with word spotting approach based on SIFT (Scale Invariant Feature Transform) feature. To overcome this limitation, this paper proposes a new approach by introducing the concept of considering the retrieval result of EDM (Euclidean distance mapping) matching algorithm as a geometry information by homograph matrix and perspective transformation to further improve the accuracy of word spotting. The comparative experiment of the proposed method is evaluated on a dataset of 190 Uyghur printed document images which contain about 17648 words. Experimental result demonstrates that the proposed approach is an effective method of retrieving the word comparing with the previous word spotting method used in Uyghur printed document image.

Keywords: Word spotting · SIFT · Homograph matrix ·
Perspective transformation · Uyghur printed document retrieval

1 Introduction

The document printed image is the most convenient format of the digital libraries, which offers better services for readers [1]. Though the major obstacle these digital libraries need to deal with is providing an effective searching approach which means effectively retrieving content from printed document images [2, 3].

Currently there are two major approaches for printed document retrieval. First is using the Optical Character Recognition (OCR) system which performs at the character level. However, OCR technology is still premature for some circumstances. For instance, if images are degraded, they will have serious effect on recognition and the cursive nature of the Uyghur script causes some drawback for segment the word into characters. And the second approach is word spotting, which has been widely applied in the field of document image retrieval [4, 5]. As an alternative to OCR, word spotting just avoids the aforementioned problems. Most of the word spotting methodologies contain two major steps, first of which is extraction of the features, which include ZOI

© Springer Nature Singapore Pte Ltd. 2019
K. Knight et al. (Eds.): ICAI 2019, CCIS 1001, pp. 106–116, 2019.
https://doi.org/10.1007/978-981-32-9298-7_9

(zone of Interest) [6, 7], BoVW (badge of visual word) [8], and spacial organization of local descriptors [9] such as SIFT. Then is to match the features for retrieval. There are some matching algorithms which were mainly used, such as EDM, DTW (dynamic time warping) and SSD (sum of squared difference) [10]. The previous word spotting method used in Uyghur printed document image adopted SIFT as local descriptors and EDM as matching algorithm [11]. But there is one drawback of this word spotting approach. Because it's based content, the approach has difficulties in thoroughly searching the information especially in Uyghur script. Also, the local descriptor extracted from word image is always mismatched by EDM. This drawback my lead to an unreliable precision problem that similar words are often matched by mistake in word spotting approach.

To overcome above obstacles, this paper proposes a novel approach that based on SIFT and uses EDM retrieval result for further retrieving. This approach extracts the SIFT features and takes advantage of the matching information of the EDM by transforming those matching points into geometrical characteristics with the help of homograph matrix and perspective transformation. Then according to the geometrical characteristics, the model retrieve the matching word.

The proposed method shows competitive performance during the experiments on 190 Uyghur printed document images which are collected from Internet. Rest of the paper is organized as follows. In Sect. 2, some related works about printed document retrieval are listed. In Sect. 3, the author describes the proposed methodology in detail. In Sect. 4, results of the conducted experiment are illustrated. And in Sect. 5, conclusions and summarization of the paper are demonstrated.

2   Related Work

Word spotting has been applied to printed document image for many years and researchers have developed many effective word spotting approaches that based on local descriptors which have been certified very effective for retrieving printed document image, because of its scale and orientational stability. SIFT is the most widely used feature in Uyghur printed document image retrieval.

Yang Al proposed Uyghur character recognition method based on SIFT [12]. They first built the character sample corpus, then detected the correlation of SIFT feature points between the test images and sample images. Finally they generated orientation descriptors to retrieve the character. However, this approach only works at character level.

Batur AI proposed Uyghur printed document image retrieval based on SIFT features. They extracted the SIFT features from Uyghur printed document images [13] at feature extraction stage, then used EDM to filter the mismatched word in the process of matching. However, this approach only works at printed image level which means it can't retrieves a certain word from printed document image.

SIFT is also implied into other printed language Document image retrieval. Lee Al proposed a novel matching algorithm for retrieval by using SIFT as a local descriptor [10]. Their approach works well on multilingual document images. The process of their approach could be described as the following steps. They first detect the key points extracted from SIFT, and then match and cluster those key points for retrieval. Experiments on English and Korean documents shows that this method is promising.

Hussian Al proposed a novel approach for word spotting which takes advantage of the shape information of extracted feature to retrieve the query word [14]. At first the document images are segmented into partial word then clustered according to their features. Based on shape information of the clusters, they query word instance to fulfill printed Urdu document image retrieval.

This paper proposes a novel method considering the retrieval result of EDM as geometry information by homograph matrix and perspective transformation to further improve the accuracy of word spotting.

3   Methodology

3.1   General Presentation of the Method

The processing steps of the proposed approach are describe as follows. As is shown in Fig. 1, the segmentation algorithm segments printed document images to word image corps, the uses SIFT [9] as local descriptor to extract the local features from segmented word image and stores those feature sets for further matching. At the same time, the query word is input and transformed into query image, the feature sets of which is extracted by using SIFT [9] as local descriptor. The local descriptor extracted from segmented word images and local descriptor extracted from query image is matched by EDM matching algorithm. Then the result of EDM matching algorithm is transformed into geometrical characteristics by HMPT (homograph matrix and perspective trans-formation) to retrieve the matched word.

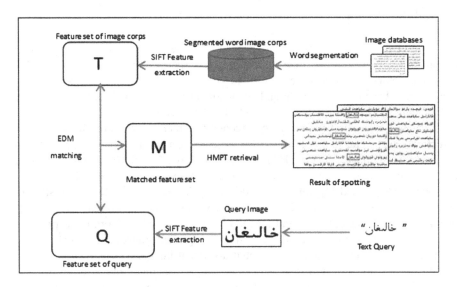

Fig. 1. Framework of the proposed approach

3.2   Word Segmentation

The printed document images need to be segmented into word image corps for further retrieval. The detail of word segmentation is described as follows. At first the printed document images need to be grade out and binarized for segmentation. Then each text line being segmented according to the empty space between each line. Once the lines are segmented, the minimum space between each word in the text line is calculated, which is used to define the convolution kernel. Then the text lines are dilated with the previously defined convolution kernel. A mean filter is used to make each word a connected component. The result after filtering is shown in the Fig. 2.

Fig. 2. The result after filtering.

Finally, the connected component in the line image is segmented into a single potential word by a vertical projection profile to obtain a segmented word image.

3.3   Feature Extraction

Feature extraction is a major step in word spotting, the quality of the extracted feature could influence the effectiveness of word spotting. SIFT is the most used feature in document image retrieval and have been proven to be very effective in Uyghur document retrieval [13]. The generation of SIFT are divided into 3 main steps.

Firstly, pyramid of images blurred by Gaussian filter are constructed as shown in Fig. 3. This step is taken for the purpose of mimic the human interaction with different distance which can be helpful for extract local feature from document image as human perspective.

Fig. 3. The pyramid of Gaussian-blurred images.

Secondly, the adjacent images in each line of pyramid are subtracted by each other to get the Difference of Gaussian space (DOG) image for the purpose of obtain the stable extreme points. The computation of DOG is show below:

$$D(x,y,\sigma) = (G(x,y,k\sigma) - G(x,y,\sigma)) \otimes I(x,y) = L(x,y,k\sigma) - L(x,y,\sigma) \quad (1)$$

Where $L(x, y, k\sigma)$ is the convolution of the original image $I(x, y)$ with the Gaussian filter $G(x, y, k\sigma)$ at scale $k\sigma$. Images in each line of pyramid are decreased by one as show in the Fig. 4 due to the previously applied image subtraction method.

Fig. 4. The DOG image of Uyghur word.

After obtaining the DOG image, some edge response key points of the image needs to be eliminated, because DOG has strong response to those edge response key points. Then the orientation and descriptor vector of remaining key points are calculated. The remaining key points are highly distinctive and partially invariant to image translation, scaling, and rotation. The comparison image of SIFT feature of Uyghur word is shown in the Fig. 5.

Fig. 5. The comparison image of SIFT feature of Uyghur word. (Color figure online)

Key points extracted from image is 3-tuple entity <x, y, H>, where x and y represents positions of feature point, H is a 128-dimensional vector describing gradient distribution. Key points are matched when the value of H are similar. The SIFT features extracted from image is shown as yellow cross in the right part of the Fig. 5. The experiment indicated that the font in the query word must be identical with the font in the printed document image, because the SIFT is very sensitive to the font.

3.4 EDM Retrieval

At the EDM retrieval stage, the Euclidean distance is proposed to match the each feature key points. The matched feature is a 128-dimensional vector which is very time consuming to locate the correct matching pair. To optimized the matching efficiency the FLANN (library for performing fast approximate nearest neighbor searches in high dimensional spaces) is been proposed. Assume the extracted feature key points of query image and test image are A{x1, x2, x3, ..., xN} and B{y1, y2, y3, ..., yN}, where N is the number of features extracted from the image. The kNN (k-Nearest Neighbor) matching algorithm is adopted to match the correct features [16]. The Euclidean distance function is describe as follow:

$$d(A, B) = \sqrt{\sum_{i=1}^{n} (x_i - y_i)^2} \tag{2}$$

For each key point, 2 nearest key points are sorted out, and the matching pairs are found out based on the previous defined threshold [12]. In order to filter some mismatched pairs, the bidirectional matching method is used, to compare each pair with its symmetric matching pair. If the pair $<q_i, t_j>$ is consistent with $<t_i, q_j>$ as shown in Eq. (3), this pair will be put into matching sets.

$$<q_i, t_j> \ = \ <t_i, q_j> \tag{3}$$

After filtering the mismatch pairs, the query images are judged whether they belong to EDM retrieval corps according to result of the EDM Matched.

$$EDM \ Matched \leftrightarrow N_m/N_t \begin{cases} \text{if ratio } > \varepsilon, & \text{YES} \\ \text{else,} & \text{NO} \end{cases} \tag{4}$$

N_m refers to the number of matching points in extracted matching key points where ε refers to the threshold value, N_t refers to the total number of the feature points. The degree of rigor of the match depends on the size of the threshold. However, a strict condition for EDM Matched would cause some serious effects on retrieving the correct word might be filtered out at the EDM retrieval stage. To overcome the problem above, the threshold value is set as 0.6 while the number of matching key points only needs to reach 60% of the total extracted feature key points of the test image. The result of the correct matched word image is shown is the Fig. 6.

Fig. 6. Result of the correct match word through EDM retrieval.

3.5 HMPT Retrieval

After EDM retrieval, lots of mismatched word images have been filtered out and the matching pairs are gotten. However, due to the cursive nature of the Uyghur script and some similar parts of the word images, serious noises may occur for further word retrieval. The result of the mismatched word images due to similar parts is shown in the Fig. 7.

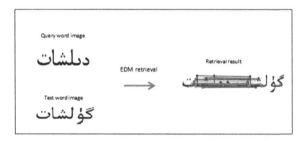

Fig. 7. Mismatched word images due to similar part.

The problem is that we need an approach that can retrieve the word more correctly than EDM matching algorithm.

Homograph Matrix and Perspective Transformation. To deal with the problem above, this paper proposes a whole new method that uses the matching pairs from retrieval result of EDM as geometry information by homograph matrix and perspective transformation to further improve the accuracy of word spotting, in which HMPT is used to filter the mismatched word image. The implementation of homograph matrix and perspective transformation is described as follows. After getting the matching pairs through EDM, the matching key points in query image are mapped to the corresponding points in the test image.

Assuming that the location of one the matching key points in query word image (xi, yi) and the corresponding location of matching key points in test image is (xi, yi), the calculation of a homograph matrix H through those matching key points is shown in the Eq. (5). The accuracy of the homograph matrix depends on the number of matching pairs, of which at least 4 are needed.

$$\begin{bmatrix} x_j \\ y_j \\ 1 \end{bmatrix} = \begin{bmatrix} h_{00} & h_{01} & h_{02} \\ h_{03} & h_{04} & h_{05} \\ h_{06} & h_{07} & h_{08} \end{bmatrix} \begin{bmatrix} x_i \\ y_i \\ 1 \end{bmatrix} \tag{5}$$

Once the homograph matrix is calculated, edge locations of the query images are mapped into the test image through perspective transformation. The mathematical expression is described as follows.

$$[x' \quad y' \quad w'] = [u \quad v \quad w] * \begin{bmatrix} h_{00} & h_{01} & h_{02} \\ h_{03} & h_{04} & h_{05} \\ h_{06} & h_{07} & h_{08} \end{bmatrix}$$
$$y = x'/w'$$
$$y = y'/w' \tag{6}$$

(u, v) is the original edge location of query word image, and (x, y) is the perspective transformed location of (u, v). After getting those 4 mapping points through perspective transformation, a quadrilateral is built to further retrieve the query word. If the calculated quadrilateral belongs to a rectangle and identical with the edge of the test word images the test word images are matched with query word image, as shown in the Fig. 8.

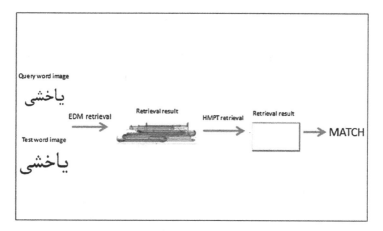

Fig. 8. Retrieving matched word through homograph matrix and perspective transformation.

If the calculated quadrilateral does not belong to rectangle as shown in the Fig. 9, the test image would not be retrieved, which proves that the proposed step could filter some mismatched word images that are missed at the EDM retrieval stage.

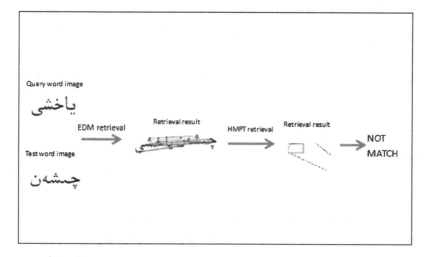

Fig. 9. Filter mismatch word image through homograph matrix and perspective transformation.

4 Experiment and Result

The proposed approach is evaluated on the dataset of 190 Uyghur printed document images. Prior to retrieval, a collection of segmented word images is generated through segmentation algorithm. In the experiment, 17648 word images are produced. For retrieval session, about 30 words are used as query word, which contains a total of 186 instances in 90 printed document. Figure 10 shows the result of proposed approach work on the retrieve session.

Query word	Retrieved Document
تۇرپان	*(Uyghur text document image)*
خالغان	*(Uyghur text document image)*

Fig. 10. Retrieval result of the proposed approach.

In the retrieval session, 178 instances are retrieved correctly, reaching a recall rate of 95.39%, and the precision rate of 98.34% respectively. The result of the experiment are summarized in Table 1 and the result of the retrieval session are illustrated in Fig. 10. To evaluate the effectiveness of the proposed approach, an experiment that retrieves only by EDM matching algorithm (Table 2) is conducted. To evaluate the performance of the proposed approach, the precision and recall are measured. In the analysis of the result of retrieving by EDM matching algorithm, the recall rate is 95.39% and the precision is 90.81% respectively. It could be seen from Table 1 that the proposed method is an effective approach for retrieving Uyghur word in printed document images.

Table 1. Retrieval results of proposed approach.

Query instance	True positive	False positive	False negatives	Recall	Precision
186	178	3	8	95.39%	98.34%

Table 2. Retrieval results of EDM matching algorithm

Query instance	True positive	False positive	False negatives	Recall	Precision
186	178	3	8	95.39%	90.81%

Some of the retrieval words are not correct. After inspection of the wrong retrieved words, the wrong retrieval result caused by some words only different by one character and the shape of the character is very similar, and the extract feature points of the query word and test word image are same those factors cause the wrong retrieval.

5 Conclusion and Future Work

In this paper, the word spotting approach is applied to the Uyghur printed document image retrieval system which uses local feature SIFT as a feature extractor. After the extraction of the local feature in the segmented word images, some mismatched word images are filtered through EDM retrieval. At the final stage, words are retrieved through the homograph matrix and perspective transformation. The proposed approach can filter some mismatch word images through homograph matrix and perspective transformation which is the most astonishing point of this approach.

The major contribution of this study is that it puts forward a flexible way to retrieve word in the Uyghur printed word image and suggests a new matching idea in image retrieval.

In future work, some other feature extracting approaches such as SURF, ORB would be employed on the feature extractor in the purpose of increasing matching efficiency and reducing the time cost at the feature extraction stage. Further study on the key point extraction would enhance the retrieving ability of printed Uyghur document retrieval.

Acknowledgement. This work is supported by Natural Science Foundation of China (61662076).

References

1. Jameson, M.: Promises and challenges of digital libraries and document image analysis: a humanist's perspective. In: International Workshop on Document Image Analysis for Libraries, Proceedings, pp. 54–61. IEEE (2004)
2. Singhai, N., Shandilya, S.K.: A survey on: content based image retrieval systems. Int. J. Comput. Appl. **4**(2), 22–26 (2010)
3. Zheng, Q.C., Zeng, Z.Y., Chi, Y.L., et al.: Advanced image retrieval method based on color and SIFT feature. Comput. Syst. Appl. (2015)
4. Lu, Y., Tan, C.L.: Information retrieval in document image databases. IEEE Trans. Knowl. Data Eng. **16**(11), 1398–1410 (2004)
5. Leydier, Y., Lebourgeois, F., Emptoz, H.: Text search for medieval manuscript images. Pattern Recogn. **40**(12), 3552–3567 (2007)
6. Leydier, Y., Ouji, A., Lebourgeois, F., et al.: Towards an omnilingual word retrieval system for ancient manuscripts. Pattern Recogn. **42**(9), 2089–2105 (2009)
7. Rusinol, M., Aldavert, D., Toledo, R., et al.: Browsing heterogeneous document collections by a segmentation-free word spotting method. In: International Conference on Document Analysis and Recognition (ICDAR). IEEE (2011)
8. Rusi, M., Llad, J.: Word and symbol spotting using spatial organization of local descriptors. In: Eighth IAPR International Workshop on Document Analysis Systems. IEEE (2008)
9. Lowe, D.G.: Object recognition from local scale-invariant features. In: ICCV. IEEE Computer Society (1999)
10. Lee, D.R., Hong, W., Oh, I.S.: Segmentation-free word spotting using SIFT. In: Image Analysis & Interpretation. IEEE (2012)

11. Batur, A.: Research on Uyghur printed complex document image retrieval based on local feature. Xinjiang University (2017)
12. Yang, N.-n., Halidan, A., Yiliyaer, D.: Uyghur character recognition method based on SIFT image registration. Sens. Microsyst. **33**(3), 40–43 (2014)
13. Batur, A., Tursun, G., Mamut, M., et al.: Uyghur printed document image retrieval based on SIFT features. Procedia Comput. Sci. **107**(1), 737–742 (2017)
14. Hussain, R., Khan, H.A., Siddiqi, I., et al.: Keyword based information retrieval system for Urdu document images. In: International Conference on Signal-Image Technology & Internet-Based Systems, pp. 27–33. IEEE (2016)
15. Rath, T.M., Manmatha, R.: Word spotting for historical documents. IJDAR **9**(2), 139–152 (2007)
16. Yi, Z.B.: Research on a SAR image matching method based on improved SIFT algorithm. Comput. Modernization **1**(6), 108–112 (2013)

Video Saliency Detection Based on Eye-Movement Guided Region Matching and LDP Embedded Optical Flow

Zheng Yang, Bing Han[✉], Guowei Wei, and Wenliang Qiu

Xidian University, Xi'an 710071, China
bhan@xidian.edu.cn

Abstract. Video saliency detection has many potential applications in computer vision field and attracts more and more attention recently. In this paper, a video saliency detection method based on eye-movement guided region matching and local directional patterns (LDP) embedded optical flow is proposed. We use the eye fixation of human beings to match the regions of the adjacent frames to overcome the inaccuracy of the dynamic characteristics constructed by the traditional optical flow method when the object displacement is too large or the velocity is too fast. The dynamic LDP embedded optical flow features between two consecutive frames are captured to reduce the influence of illumination. And the final video saliency map to be detected can be generated by the support vector regression (SVR) model. The proposed method achieved better results on several video databases in objective and subjective evaluation compared with stat-of-arts, which demonstrates the effectiveness and robustness of our algorithm.

Keywords: Video saliency · Super-pixel segmentation ·
LDP embedded optical flow · Eye-movement

1 Introduction

The research on cognitive psychology and neuroscience suggest that human vision system (HVS) can effectively focus on the important and conspicuous regions in a complex scene. While saliency detection simulates the human vision system by identifying salient regions from the background in a static image. Saliency detection has attracted considerable attention over past several years because it is closely related to many applications in machine vision, multimedia and entertainment, such as image segmentation [1], image retrieval [2], video surveillance [3], object detection [4, 18] and so on.

This work is supported by the National Natural Science Foundation of China (61572384; 41831072), the National Key Research and Development Program of China under Grant (2016QY01W0200), China's postdoctoral fund first-class funding (2014M560752), Shanxi province postdoctoral science fund Key Research and Development Program of Shaanxi Province (2017KW-017).

© Springer Nature Singapore Pte Ltd. 2019
K. Knight et al. (Eds.): ICAI 2019, CCIS 1001, pp. 117–130, 2019.
https://doi.org/10.1007/978-981-32-9298-7_10

There has significant progress in saliency detection recently. The saliency detection models for image can be divided into task independent model [5, 6] and task dependent model [7, 8], which are called as bottom-up models and top-down models respectively. In addition, many saliency detection models [9–11] based on deep learning have been proposed recently. The above methods would not be introduced in detail because the problem of video saliency detection is focused on in this paper. So far, researchers have paid more and more attention to the video saliency detection. Compared with saliency detection in still images, detecting saliency in videos is a much more challenging problem. It is related to not only spatial features in a single frame but also inherent temporal information. Rahtu et al. [12] applied a salient object segmentation conditional random fields (CRFs) methods, the algorithm uses a motion-first approach to the integration of dynamic and static characteristics, in line with people's perception of visual attention bias. Papadopoulos et al. [13] achieve spatiotemporal consistency optimization by minimizing an energy function, which is solved by max flow algorithm. In the field of deep learning, video saliency detection based on long short-term memory (LSTM) have achieved good performance because LSTM can utilize the temporal information of video sequences. Wang et al. [14] proposed a novel video saliency model that augments the CNN-LSTM network architecture with an attention mechanism to enable fast, end-to-end saliency learning. Besides, optical flow method and its various (3D flow) are widely used to extract the dynamic features [15].

As we know, the optical flow feature is influenced by the illumination easily while the brightness of a point on an object in an image can dramatically change if the object moves into the scene with different illumination. So, the most video saliency models, which integrate the spatial saliency feature with temporal saliency feature obtained from optical flow method would fail when there exist heavily illumination changes in video sequences.

The LDP is proposed by Jabid et al. [16], which is the illumination invariant description of an image. In this paper, we use the LDP features as the spatial features of an image. In addition, the LDP is introduced into the Optical Flow method to improve the traditional Optical Flow method [17]. In [17], LDP coder is used to calculate the data term function, which miss lots of information because LDP is 8-bits descriptor.

The dynamic characteristics constructed by the traditional optical flow method is inaccuracy when the object displacement is too large or the velocity is too fast. The super-pixel optical flow field is considered to be a large displacement optical flow field and region matching is required, but the original LDP optical flow method cannot consider this problem and cannot get the motion fields accurately. To address the problems above, we propose Eye-movement Guided Region Matching and LDP Embedded Optical Flow.

In the following parts of this paper, we firstly introduce the eye fixation information in Sect. 2 and a novel video saliency method is detailed in Sect. 3. Then, we demonstrate the performance of our proposed attention model on 2 video benchmark datasets in Sect. 4. Finally, the paper is concluded in Sect. 5 and references in Sect. 6.

2 Eye Movement Experiment

Ten volunteers (5 male, 5 female) participated in our experiment. The mean age of all observers is 25 (min = 23, max = 27). Observers were undergraduates at our laboratory. Each of these volunteers were unaware of the experiment and were instructed to watch the videos naturally.

In order to ensure the testing accuracy, we conducted the eye-movement experiment in a sound-attenuated room. Subjects were seated 70 cm from a 19-in. monitor with a refresh rate 70 Hz while their eye positions were recorded using an Eyelink 1000 eye tracking system manufactured by SR Research Ltd, which has a sampling rate of 2000 Hz and a typical spatial resolution of 0.25 to 0.5.

The databases of Videoseg [19] and MCL [20] are from the website http://www.brl. ntt.co.jp/, where contains 10 and 8 video clips respectively. These video clips in the datasets, including outdoors scene, natural scene, sports and other content, were presented to eight native subjects. Each clip comprised between 60 and 120 frames, for totaling 8 min.

The eye-movement information of these volunteers was recorded for fourteen video snippets. All the frames from all videos are rescaled to the resolution of the monitor. The eye tracker was calibrated using 9 points calibration at the beginning of each recording session. Participants were offered a five to ten minutes break every video clips, in between each of the videos, a cross screen was displayed for four seconds so that the initial eye-movement during a video are not biased by the previous video. We sampled the ten subjects eye fixation information and took the average position, the last samples are used to match the region features, described in the next section.

Figure 1 shows the eye movement experiment procedure and the eye movement fixation information of subject on a video clips. The red circles indicate the positions of the fixations in the video frames.

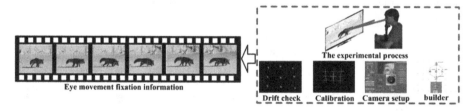

Fig. 1. The examples of the eye fixation information on video frame. (Color figure online)

3 The Proposed Algorithm

The flow chart of our method is shown in Fig. 2, which based on the eye-movement guided region matching and LDP embedded optical flow. The low-level static features of each frame and a novel robust dynamic LDP optical flow features between two consecutive frames are merged to get the final feature of a super-pixel. Then the fixation data was used to match the super-pixel of adjacent frames. The K-means cluster is carried out to divide the training features into K classes. Finally, the relationship between K classes features and corresponding labels is learned to generate the final video saliency by the SVR model.

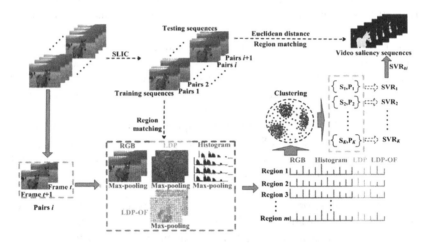

Fig. 2. Flow chart of our proposed method.

3.1 Region Matching Based on Eye Movement Guiding

For an input image X, we first perform over-segmentation to form super-pixels by the simple linear iterative clustering (SLIC) algorithm [21] because of its low computational cost and high performance.

The corresponding super-pixels will be matched between the two consecutive frames and calculate the corresponding optical flow according to eye fixation points. The eye-movement guided region matching scheme is shown in Fig. 3.

Fig. 3. Eye-movement fixation matching framework.

Two consecutive images (I_t and I_{t+1} in Fig. 3) with fixation information will be obtained in the experiment. P_t and Q_t denote the region with fixation points and the region without fixations points in the frame I_t. For P_t. We can map this coordinates of the fixation point to the *ith* super-pixel and obtain the match between P_t and P_{t+1}. The rest regions without objects can be matched according to the existing method [22]. The matching time will be greatly reduced through the above method.

It is worth mentioning that the fixation data is pre-processed to avoid the inaccuracy of human eye-movement information. After obtaining the original fixation data, we filtered the data firstly to remove the fixation points which are meaningless. Then the average position of fixation data is taken to overcome the deviations caused by personal factors.

Then we extract the statistic image features from frame pairs instead of a single image. Three descriptors including RGB, statistic LDP features and histogram of an image are used as the spatial features to describe each region obtained by super-pixel division.

3.2 Dynamic Feature Extraction Based on LDP Embedded Optical Flow

In this section, a method called LDP embedded optical flow in our paper is proposed by combining the TV-L1 optical flow method with LDP descriptors to extract the dynamic feature.

The Eq. (1) is a typical variational formulation of total variation based motion estimation.

$$\min_{u,v} \|\nabla u\|_1 + \|\nabla v\|_1 + \lambda \|\rho(u,v)\|_1 \tag{1}$$

Where $v = (v_1, v_2)^T : \Omega \rightarrow R^2$ is the motion field, and in the Eq. (2),

$$\rho(u,v) = I_t + (\nabla I)^T (v - v^0) + \beta u \tag{2}$$

ρ is the optical flow constraint, I_t is the time derivative of the image intensities, ∇I is the spatial image gradient, and v^0 is the given motion field. The parameter λ is used to define the tradeoff between data fitting and regularization. The function $u : \Omega \rightarrow R$ is expected to be smooth and hence we also regularize u by means of the total variation. The parameter β controls the influence of the interfere term.

Then we introduce LDP coder into Eq. (2) to reconstruct optical flow constraint to overcome the influence of illumination.

$$\rho(u_{i,j}, v_{i,j}) = (I_t)_{i,j} + (\nabla I)_{i,j}^T (M) + \beta u_{i,j} \tag{3}$$

$$M = \begin{cases} \sum_{k=1}^{n} 2^{n-1} \ \left(S_{t+1,k}(v_{i,j}) - S_{t,k}(v_{i,j}) \right)^2 \neq 0 & (k = 1, 2 \cdots n) \\ 0 & else \end{cases} \tag{4}$$

$S_{t+1,k}(v_{i,j})$ and $S_{t,k}(v_{i,j})$ are the two descriptors extracted from the frame pairs $It+1$ and It in the video clips, respectively. (i,j) denote the indices of the discrete locations in the frame domain.

The energy function (1) is divided into two parts, which are then solved iteratively. The final data can be extended in order to be applicable to a multi-channel descriptor as,

$$E_{data} = \sum_{\Omega} \left(\lambda \sum_{i=1}^{n} \rho(u_{i,j}, v_{i,j})^2 + \frac{1}{2\theta}(\omega - \hat{\omega})^2 \right) \tag{5}$$

Where $\omega = (u_{i,j}, v_{i,j})$ and $\hat{\omega} = (\hat{u}_{i,j}, \hat{v}_{i,j})$ is the auxiliary optical flow vector and θ is a threshold. Optimizing the two variables by the alternately method described in Eq. (5), then obtained the final solution by the threshold.

The second part contains the regularization:

$$E_{smooth} = \left(\frac{1}{2\theta}(\omega - \hat{\omega})^2 + \left\| \nabla \hat{u}_{i,j} \right\| + \left\| \nabla \hat{v}_{i,j} \right\| \right) \tag{6}$$

Our optical flow method is based on the region matching, which can match the super-pixel features between adjacent frames quickly and accurately. Then the region-matching term based on optical flow can be expressed as

$$E_{match} = \sum_{p=1}^{P} \sum_{j=1}^{m} \left(I_{t+1,j}^{p}(x+u,y+v) - I_{t,j}^{p}(x,y) \right)^2 + \sum_{q=1}^{Q} \sum_{j=1}^{m} \left(I_{t+1,j}^{q}(x+u,y+v) - I_{t,j}^{q}(x,y) \right)^2$$

$$+ \gamma \sum_{s=1}^{n} \sum_{q=1}^{Q} \sum_{j=1}^{m} C^q(i)\psi\left[(u(x) - us(x))^2 + (v(x) - vs(x))^2 \right]$$

$$+ \partial \sum_{q=1}^{Q} \sum_{j=1}^{m} \psi\left(|\nabla u(x)|^2 + |\nabla v(x)|^2 + g(x)^2 \right)$$

$$\tag{7}$$

P is the regions of the eye-movement matching, Q is the regions non-eye movement matching.

The main requirements for a good descriptor matching method are that the grid is fine enough to capture the motion of smaller structures and that the descriptors are unique enough to limit the number of false watches. So we can get the energy function of the optical flow field as Eq. (8),

$$E = Edata + E_{smooth} + E_{match} \tag{8}$$

3.3 Training the Robust Models of Feature Vector Regression

Then the super-pixel blocks were clustered into K classes by K-means before training. During the training phase, K SVR models are trained respectively by the K classes samples.

During the testing phase, the testing super-pixels feature vectors were detected firstly. Then we calculate saliency value of each testing super-pixel according to the corresponding SVR model and the testing feature vectors. The significant value vector was mapped to each testing super-pixels, so we can get the finally saliency map of the testing sequence.

4 Experiments and Analysis

To demonstrate the effectiveness of the proposed LDP embedded and eye-movement guided region matching method for video saliency detection, various experiments are conducted on the different video public databases, which consist of same videos in Sect. 2. All of the experiments are performed on a PC with an Intel Xeon 2.8-GHz CPU, Nvidia FX580 GPU, and 8-GB memory. The salient objects in these videos were marked in advance as ground truths for convenient quantitative evaluation.

4.1 Parameters Setting Experiments

In our proposed method, there are several parameters to be determined. The important one is the number of the super-pixels. The results of precision and recall (P-R) curve with different number of super-pixels are shown in Fig. 4(a), and Fig. 4(b) gives Precision, Recall, F-measure and time cost among the different number of super-pixels. Figure 4(a) shows the P-R curve covers the larger region when the number of super-pixels N is 600. And we can know from Fig. 4(b) that the F-measures cannot improve with the further increasing of the number of super-pixels at the expense of time cost. So, we set N = 600 in the following experiments.

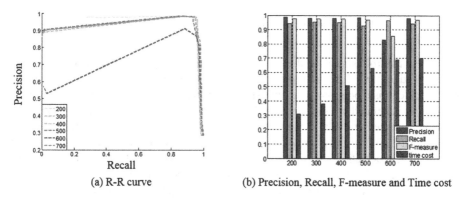

(a) R-R curve   (b) Precision, Recall, F-measure and Time cost

Fig. 4. The results with different number of super-pixels.

4.2   Qualitative Analysis

In order to evaluate the accuracy and the effectiveness of the proposed method on saliency detection, we compared our proposed method with the eleven state-of-the-art video saliency models, including Seg [7], Seo [23], CBSal [24], Rahtu [12], PDA [25], GBVS [6], CE [26], LSD [27], DCMR [28], SLT [29] and STC [30] model. For these eleven models, we used the source codes with default parameter settings or executable provided by the authors. For a fair comparison, all saliency maps generated using different models were normalized into the same range of [0, 255] with the full resolution of original videos.

Figure 5 shows the intuitive qualitative comparison for video saliency detection obtained by the eleven methods and the proposed method on Videoseg datasets. All of methods except ours fail in finding the object under complex background in the fifth video. And it is worth noting that our proposed method performs well for videos containing nothing except objects as shown in the bottom rows of Fig. 5. Due to the contribution of the proposed dynamic LDP optical flow features, our method can accurately locate the salient objects and detect the contour of the object clearly.

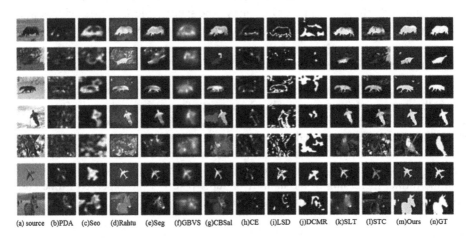

(a) source (b)PDA (c)Seo (d)Rahtu (e)Seg (f)GBVS (g)CBSal (h)CE (i)LSD (j)DCMR (k)SLT (l)STC (m)Ours (n)GT

Fig. 5. Quantitative comparison of different approaches.

To test the robustness of the proposed video saliency detection method against noise, we added five different types of noise to the above video datasets. These noises were speckle noise, poisson noise, gaussian noise, pepper, salt noise and localvar noise. The results generated by the three models, DCMR [28], SLT [29] and Seg [7] were affected by noise seriously and obtained the black values with the increasing of the noisy density. So we only compare our method with 8 top-performance models for video saliency detection mentioned above in Fig. 6. The results show that saliency maps generated using the proposed method has better performance in both highlighting the clear objects and suppressing the complicated background regions noise.

To demonstrate the characteristics and robustness of our model, we also tested our model and other ten video saliency models on MCL dataset. Some challenging videos

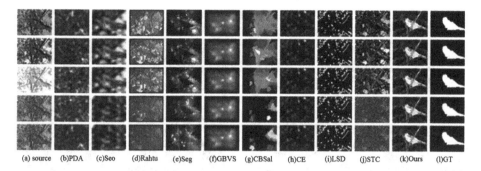

(a) source   (b)PDA   (c)Seo   (d)Rahtu   (e)Seg   (f)GBVS   (g)CBSal   (h)CE   (i)LSD   (j)STC   (k)Ours   (l)GT

Fig. 6. Saliency map under the five noise of different approaches.

contain more than one object, and the object with large motion and complicated background, which were used in this experiment. Figure 7 shows the qualitative comparisons on MCL dataset. We can observe that most video saliency detection models can effectively handle videos with relatively simple background and homogeneous objects, such as the first example shown in Fig. 7 However, for some complicated videos containing the heterogeneous objects, such as vehicles, human beings and multiple objects with different moving directions, Seo [23], CE [26], GBVS [6] and PDA [25] models falsely regard some background regions as saliency area and cannot effectively detect salient objects in these videos. Seo [23] model can detect the moving objects by building the geometric structures which can represent motion orientation of object in video, but it fails in detecting the fast and complicated moving object and highlighting the object boundaries in the background clearly. CE [26] does not take motion information into account. GBVS [6] presents smoothed saliency results, but it fails in predicting salient regions clearly. PDA [25] find the object by suppressing the background regions, which can degrades the saliency detecting performance on videos with moving objects and cluttered background. STC [30], Rahtu [12] and CBSal [24] cannot recognize the foreground and background uniformly. STC [30] and Rahtu [12] have the similar subjective performance and they regard the background as the salient regions. CBSal [24] can cover most salient regions, but it produces the usefulness salient regions randomly. LSD [27] has better subjective evaluation than other previous models, but it cannot give the more complete object boundary. Our proposed method is more effective to highlight multiple salient objects with well-defined boundaries. The salient regions detected by our proposed method can cover most of the information areas of the image.

In addition, our model can detect the objects from the complex background exactly. The saliency detection results of four consecutive frames from Videoseg dataset are given in Fig. 8. During the analysis of all video saliency detection results, we found that our methods have better performance on all of frames in a video sequence than that of other aforementioned models. In this experiment, STC [30], SLT [29] and CBSal [24] methods can generate good saliency maps.

Figure 9 shows a visual comparison between our method and other video saliency models as mentioned above. The detection results of four consecutive frames from

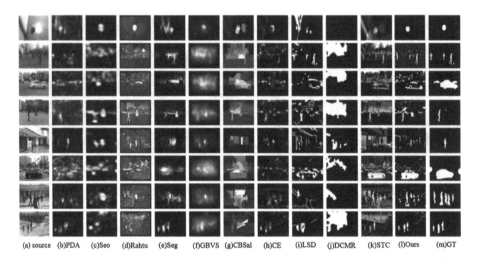

(a) source (b)PDA (c)Seo (d)Rahtu (e)Seg (f)GBVS (g)CBSal (h)CE (i)LSD (j)DCMR (k)STC (l)Ours (m)GT

Fig. 7. Spatiotemporal saliency maps for some video frames in MCL dataset.

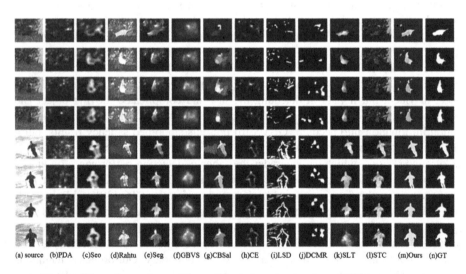

(a) source (b)PDA (c)Seo (d)Rahtu (e)Seg (f)GBVS (g)CBSal (h)CE (i)LSD (j)DCMR (k)SLT (l)STC (m)Ours (n)GT

Fig. 8. Comparison of two video saliency detection results on Videoseg dataset.

MCL dataset show that our model not only separate objects to be detected accurately on all frames, but also locate the multiple objects precisely.

4.3   Quantitative Analysis

For an objective comparison, the P-R curve is generated by using a set of thresholds ranged in [0,255]. Video saliency maps obtained by different stat-of-art video saliency detection models and our proposed method use this threshold to get the binary saliency maps, which is compared with the ground truth. Similarly, we also generate the

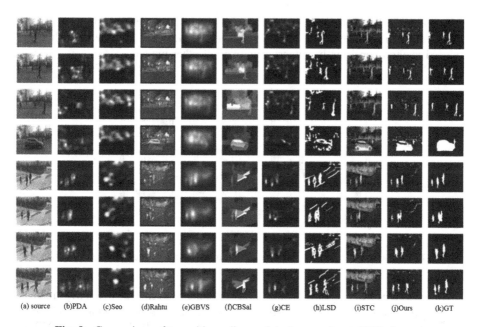

(a) source   (b)PDA   (c)Seo   (d)Rahtu   (e)GBVS   (f)CBSal   (g)CE   (h)LSD   (i)STC   (j)Ours   (k)GT

Fig. 9. Comparison of two video saliency detection results on MCL dataset.

receiver operating characteristic (ROC) curve for each video saliency model on same dataset as the second objective evaluation measure to characterize the performance of video saliency detection methods.

Figure 10(a) and (b) show the P-R curves and ROC curves of different saliency models on VideoSeg dataset. From these 2 figures, we can see that our proposed model outperforms the baseline methods on both P-R curve and ROC curve.

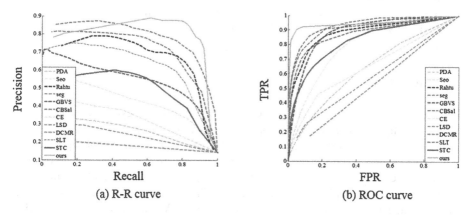

(a) R-R curve   (b) ROC curve

Fig. 10. The evaluation results of different saliency models on VideoSeg dataset.

The objective comparison on F-measure metrics and the superiority of the proposed method under different types of noise are shown in Fig. 11. It can be seen that our proposed method consistently outperforms over all other models.

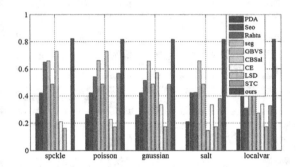

Fig. 11. F-measure of different video saliency models under different types of noise.

We also compared the time cost with all the other eleven video saliency models. Table 1 reports the average running time per frame of each model. Our algorithm is based on the super-pixel level, so the test time is less than the SLT [29] and the DCMR [28] models. Among all models, PDA [25] model has the lowest time complexity, but its accuracy is not higher than our proposed model.

Table 1. Experimental results of compared saliency models running time

Model	PDA [25]	Seo [23]	Rahtu [12]	Seg [7]	GBVS [6]	CBSal [24]
Time(s)	0.037	3.340	5.811	1.056	0.924	1.957
Code	MATLAB	C++	MATLAB	MATLAB	MATLAB	MATLAB
Model	CBSal [24]	CE [26]	LSD [27]	DCMR [28]	STC [30]	Ours
Time(s)	1.957	5.893	1.845	33.071	0.276	3.244
Code	MATLAB	C++	MATLAB	C++	MATLAB	MATLAB

5 Conclusions

We have proposed a video saliency detection method based on eye-movement guided region matching and LDP embedded optical flow in this paper.

Firstly, the low-level static features of each frame and dynamic LDP optical flow features between two consecutive frames are cascaded to get the final feature of a super-pixel. Secondly the fixation data obtained from eye-movement information was used to match the super-pixels of adjust frames. The K-means cluster is carried out to divide the final features into K classes. Then we use SVR models to learn the relationship between features and labels. Finally, the best regression model is selected to reconstruct in the testing process and generate the final saliency map. Extensive experiments on 16 video clips from two challenging datasets demonstrate that proposed spatiotemporal saliency framework outperforms other state-of-the art methods. In future work, we can consider utilizing deep features to further improve the performance of our method.

References

1. Kim, T.H., Lee, K.M., Lee, S.U.: Generative image segmentation using random walks with restart. In: Forsyth, D., Torr, P., Zisserman, A. (eds.) ECCV 2008. LNCS, vol. 5304, pp. 264–275. Springer, Heidelberg (2008). https://doi.org/10.1007/978-3-540-88690-7_20
2. He, J., Li, M., Zhang, H.J., et al.: Manifold-ranking based image retrieval. In: Proceedings of the 12th Annual ACM International Conference on Multimedia, pp. 9–16. ACM, New York (2004)
3. Yubing, T., Cheikh, F.A., Guraya, F.F.E.: A spatiotemporal saliency model for video surveillance. Cogn. Comput. **3**(1), 241–263 (2011)
4. Walther, D., Koch, C.: Modeling attention to salient proto-objects. Neural Netw. **19**(9), 1395–1407 (2006)
5. Itti, L., Koch, C., Niebur, E.: A model of saliency-based visual attention for rapid scene analysis. IEEE Trans. Pattern Anal. Mach. Intell. **1998**(11), 1254–1259 (1998)
6. Harel, J., Koch, C., Perona, P.: Graph-based visual saliency. In: Advances in Neural Information Processing Systems, pp. 545–552. MIT Press, Vancouver (2006)
7. Wang, W., Shen, J., Porikli, F.: Saliency-aware geodesic video object segmentation. In: Proceedings of the IEEE Conference on Computer Vision and Pattern Recognition (CVPR), vol. 40, pp. 3395–3402. IEEE, Boston (2015)
8. Li, N., Sun, B., Yu, J.: A weighted sparse coding framework for saliency detection. In: Proceedings of the IEEE Conference on Computer Vision and Pattern Recognition (CVPR), pp. 5216–5223. IEEE, Boston (2015)
9. Wang, L., Lu, H., Ruan, X., et al.: Deep networks for saliency detection via local estimation and global search. In: Proceedings of the IEEE Conference on Computer Vision and Pattern Recognition (CVPR), pp. 3183–3192. IEEE, Boston (2015)
10. Zhao, R., Ouyang, W., Li, H., et al.: Saliency detection by multi-context deep learning. In: Proceedings of the IEEE Conference on Computer Vision and Pattern Recognition (CVPR), pp. 1265–1274. IEEE, Boston (2015)
11. He, S., Jiao, J., Zhang, X., et al.: Delving into Salient object subitizing and detection. In: Proceedings of the IEEE International Conference on Computer Vision (ICCV), pp. 1059–1067. IEEE, Venice (2017)
12. Rahtu, E., Kannala, J., Salo, M., Heikkilä, J.: Segmenting salient objects from images and videos. In: Daniilidis, K., Maragos, P., Paragios, N. (eds.) ECCV 2010. LNCS, vol. 6315, pp. 366–379. Springer, Heidelberg (2010). https://doi.org/10.1007/978-3-642-15555-0_27
13. Papadopoulos, D.P., Clarke, Alasdair D.F., Keller, F., Ferrari, V.: Training object class detectors from eye tracking data. In: Fleet, D., Pajdla, T., Schiele, B., Tuytelaars, T. (eds.) ECCV 2014. LNCS, vol. 8693, pp. 361–376. Springer, Cham (2014). https://doi.org/10.1007/978-3-319-10602-1_24
14. Wang, W., Shen, J., Guo, F., et al.: Revisiting video saliency: a large-scale benchmark and a new model. In: Proceedings of the IEEE Conference on Computer Vision and Pattern Recognition (CVPR), pp. 4894–4903. IEEE, Salt Lake City (2018)
15. Li, G., Xie, Y., Wei, T., et al.: Flow guided recurrent neural encoder for video Salient object detection. In: Proceedings of the IEEE Conference on Computer Vision and Pattern Recognition (CVPR), pp. 3243–3252. IEEE, Salt Lake City (2018)

16. Jabid, T., Kabir, M.H., Chae, O.: Local directional pattern (LDP) for face recognition. In: 2010 Digest of Technical Papers International Conference on Consumer Electronics (ICCE), pp. 329–330. IEEE, Nha Trang (2010)

17. Rivera, A.R., Castillo, J.R., Chae, O.O.: Local directional number pattern for face analysis: Face and expression recognition. IEEE Trans. Image Process. 22(5), 1740–1752 (2013)

18. Wang, W., Shen, J., Dong, X., et al.: Salient object detection driven by fixation prediction. In: Proceedings of the IEEE Conference on Computer Vision and Pattern Recognition (CVPR), pp. 1711–1720. IEEE, Salt Lake City (2018)

19. Fukuchi, K., Miyazato, K., Kimura, A., et al.: Saliency-based video segmentation with graph cuts and sequentially updated priors. In: 2009 IEEE International Conference on Multimedia and Expo, pp. 638–641. IEEE, New York (2009)

20. MCL Database. http://mcl.korea.ac.kr/database/saliency/. Accessed 24 Aug 2015

21. Yang, C., Zhang, L., Lu, H., et al.: Saliency detection via graph-based manifold ranking. In: Proceedings of the IEEE Conference on Computer Vision and Pattern Recognition (CVPR), pp. 3166–3173. IEEE, Portland (2013)

22. Yous, H., Serir, A.: Blotch detection in archived video based on regions matching. In: 2016 International Symposium on Signal, Image, Video and Communications (ISIVC), pp. 379–383. IEEE, Tunis (2016)

23. Seo, H.J., Milanfar, P.: Static and space-time visual saliency detection by self-resemblance. J. Vis. 9(12), 15 (2009)

24. Jiang, H., Wang, J., Yuan, Z., et al.: Automatic salient object segmentation based on context and shape prior. In: Proceedings of the British Machine Vision Conference (BMVC), vol. 6, no. 7, p. 9 (2011)

25. Zhou, B., Hou, X., Zhang, L.: A phase discrepancy analysis of object motion. In: Kimmel, R., Klette, R., Sugimoto, A. (eds.) ACCV 2010. LNCS, vol. 6494, pp. 225–238. Springer, Heidelberg (2011). https://doi.org/10.1007/978-3-642-19318-7_18

26. Li, Y., Zhou, Y., Yan, J., Niu, Z., Yang, J.: Visual saliency based on conditional entropy. In: Zha, H., Taniguchi, R., Maybank, S. (eds.) ACCV 2009. LNCS, vol. 5994, pp. 246–257. Springer, Heidelberg (2010). https://doi.org/10.1007/978-3-642-12307-8_23

27. Xue, Y., Guo, X., Cao, X.: Motion saliency detection using low-rank and sparse decomposition. In: 2012 IEEE International Conference on Acoustics, Speech and Signal Processing (ICASSP), pp. 1485–1488. IEEE, Kyoto (2012)

28. Huang, C.R., Chang, Y.J., Yang, Z.X.: Video saliency map detection by dominant camera motion removal. IEEE Trans. Circuits Syst. Video Technol. 24(8), 1336–1349 (2014)

29. Li, J., Liu, Z., Zhang, X.: Spatiotemporal saliency detection based on super pixel-level trajectory. Sig. Process. Image Commun. 38, 100–114 (2015)

30. Zhai, Y., Shah, M.: Visual attention detection in video sequences using spatiotemporal cues. In: Proceedings of the 14th ACM International Conference on Multimedia, pp. 815–824. ACM, Santa Barbara (2006)

Performance Evaluation of Visual Object Detection for Moving Vehicle

Zeqi Chen[✉], Yuehu Liu, Shaozhuo Zhai, and Xinzhao Li

Institute of Artificial Intelligence and Robotics, Xi'an Jiaotong University,
Xi'an, China
chenzeqi@stu.xjtu.edu.cn

Abstract. With the rapid development of intelligent vehicle technology, the evaluation of the perception algorithms is crucial for safe driving of intelligent vehicles. This paper proposes a novel evaluation method for visual object detection of intelligent vehicles. The traditional evaluation methods are based on each frame of the video and treat all the objects equally. Distinguished from that, the proposed method applies the length of driving trajectory to evaluate the detection algorithm. Besides, the proposed method brings about the road region constraint, weight of safety and weight of early detection. The road region constraint ensures that only the objects relevant to intelligent vehicles are taken into account. The weight of safety for each object is determined based on its degree of relevance of the safe driving. The weight of early detection is set to evaluate whether the objects are detected in time or not. In experiments, the proposed evaluation method can reflect the relatively true detection ability of intelligent vehicles compared to the traditional methods.

Keywords: Performance evaluation · Visual object detection · Intelligent vehicle

1 Introduction

Intelligent vehicles require extensive testing before they go on the road to ensure that no safety issues arise [1]. However, the test site is limited and cannot cover all the conditions in the real road. In particular, the test of some extreme traffic scene often has risks, since it will consume lots of manpower and material resources and is difficult to reproduce [2]. One of the solutions is to build a test and simulation platform for intelligent vehicles to perform quantitative, repeatable and comparable tests on it [3]. The latest research of intelligent vehicles shows that appropriate testing methods can significantly improve the efficiency of intelligent testing [4]. Therefore, it is important to design evaluation methods according to the practical driving demand of intelligent vehicles.

This work was supported by National Natural Science Foundation of China (Grant No: 91520301).

© Springer Nature Singapore Pte Ltd. 2019
K. Knight et al. (Eds.): ICAI 2019, CCIS 1001, pp. 131–144, 2019.
https://doi.org/10.1007/978-981-32-9298-7_11

In recent years, in order to meet the unmanned demand and better evaluate the environmental cognition and understanding algorithms [5–7], more and more object detection datasets provide data obtained by on-vehicle cameras, such as KITTI [8,9], Caltech-USA [10], RobotCar [11], and Daimler [12]. However, these datasets still use the traditional evaluation method of static images, and there are some shortcomings in evaluating the visual object detection algorithm of intelligent vehicles: (1) For the video captured by on-vehicle cameras, the slower the vehicle passes through the traffic scene, the more frames the object in the video captured persists. And the proportion of the object in the scene among all objects is greater, which is equivalent to the weight of the object increasing when the vehicle speed is low. (2) For intelligent vehicles, it usually does not need to pay attention to all the objects in the field of vision. Over-focusing on the objects unrelated to the safe driving of the intelligent vehicles will waste limited computing resources. (3) Different objects have different correlations with intelligent vehicle driving, but traditional methods do not distinguish them. (4) The traditional evaluation methods only focus on the accuracy and the real-time performance of the algorithm, but do not evaluate whether the algorithm can discover the object in time.

Hence we propose an evaluation method of object detection algorithm based on the "existence length" of object applied to the video captured by on-vehicle cameras. Firstly, the road region is defined, and only the objects in road region are evaluated, then we can get the true positive objects, false positive objects and false negative objects in each frame. Then, the "existence length" of the objects are calculated, and the length is weighted according to the degree of relevance of the objects and safe driving. Finally, precision, recall and f-measure are calculated according to the total lengths of true positive objects, false positive objects and false negative objects.

The contributions of this paper are as follows.

(1) During the evaluation, the number of frames in which the object exists is converted into the "existence length" of the object, which avoids the influence of vehicle speed on the data collected by the on-vehicle cameras.
(2) According to the speed v of the intelligent vehicle and the relative distance d between the intelligent vehicle and the object, different weights of safety are given to the objects. It enhances the weights of the objects which have strong correlation with the safe driving of the intelligent vehicle.
(3) The weight of early detection is set to evaluate whether the intelligent vehicle can find objects in time.
(4) The road region is defined according to the practical demand of intelligent vehicle driving. Only the objects with strong correlation with intelligent vehicle driving are evaluated, which avoids the interference of unimportant data to the evaluation results.

2   Traditional Evaluation Method

The traditional evaluation method of object detection algorithm generally performs single image evaluation on each frame of video sequences or each single image in the entire dataset, and then counts the evaluation result of each image to obtain the final performance of the algorithm. Single image evaluation usually uses the Intersection-over-Union evaluation criteria specified in the PASCAL VOC challenge [13] to evaluate the output of the algorithm and the manually labeled ground-truth.

After performing the single image evaluation on all images, we can get the total number of true positives, false positives and false negatives of the algorithm on the dataset, and calculate precision, recall and f-measure. Generally, we rank the performance of the algorithm according to f-measure.

Different datasets have different characteristics, and their evaluation methods are not the same. For example, the Caltech-USA pedestrian detection dataset provides video obtained by on-vehicle cameras. According to the number of pixels of pedestrian height, the dataset divides the distance between pedestrian and the collected vehicle into three levels. Unfortunately, the dataset is evaluated according to IoU without considering the influence of distance. In addition, the Caltech-USA dataset also sets the *ignore* regions based on the number of pixels and the degree of occlusion of the pedestrian object, but the *ignore* regions is set due to the limited capabilities of the object detection algorithm, rather than for the demand of safe driving of intelligent vehicles.

The KITTI dataset is an internationally renowned dataset for autonomous driving. It provides a variety of sensor data with detailed label, and its video data captured by on-vehicle cameras can be used for the evaluation of various visual tasks. However, the dataset still uses traditional evaluation criteria.

In summary, in order to meet the unmanned demand, the existing datasets use the data obtained by the on-vehicle camera, but they still use the traditional evaluation method to count the evaluation results frame by frame. Although the evaluation method is relatively mature and can correctly evaluate the performance of visual algorithms, it does not really combine the practical demand of unmanned driving. Therefore, we have designed an evaluation method of visual object detection algorithm for intelligent vehicles.

3   Evaluation Based on Existence Length

3.1   The Framework of Our Method

We propose an evaluation method of visual object detection algorithm for intelligent vehicle based on the "existence length" of the objects. The evaluation process is shown in Fig. 1: (1) According to the actual demand of the intelligent vehicle, we evaluate the detection results of objects only in the road region frame by frame in the original video sequence, and get the true positive objects, false positive objects and false negative objects in each frame. (2) Then, we calculate the "existence length" of the objects by the trajectory length of the intelligent vehicles in the process of the objects appearing in the video. (3) Considering the weight of safety and the weight of early detection, the weighted length of true positive, the weighted length of the false positive and the weighted length of the false negative are obtained for each frame. (4) The total length of the true

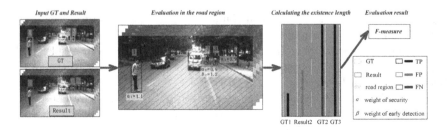

Fig. 1. The framework of our evaluation method.

positive is the sum of the lengths of the true positive in all frames, the total length of the false positive is the sum of the lengths of the false positive in all frames, and the total length of the false negative is the sum of the lengths of the false negative in all frames. (5) Finally, precision, recall and f-measure are calculated according to the total lengths of true positive, false positive and false negative.

3.2 Evaluation in Road Region

General visual detection tasks require that the algorithm can detect all objects in the video images, but in the process of vehicle driving, it is not necessary to detect all objects. In order to avoid the interference of unimportant data on the evaluation results, our method only evaluates the objects that are relevant to the safe driving of intelligent vehicles, that is, only the objects in the road region. The objects outside the road region need not be detected. However, if they are detected, the detections are not considered to be false positives either.

The road region is defined as the region whose boundary is the isolation grassland or road edge which closest to the two sides of the current lane. It is worth noting that the sidewalk belongs to the road region.

Define the IoU of ground truth's bounding box and result's bounding box is above γ for true positive, with $\gamma = 0.5$ in our evaluation, as shown in Eq. (1):

$$IoU = \frac{area(BB_D) \cap area(BB_{GT})}{area(BB_D) \cup area(BB_{GT})} \tag{1}$$

Where, BB_D is the bounding box of the detection result, and BB_{GT} is the bounding box of ground truth.

After the evaluation is completed, we obtain the object detection result of each frame. The true positive objects set of the jth frame of the i-th data group is denoted as $\mathbb{S}_{i,j}^{TP}$, and the false positive objects set is denoted as $\mathbb{S}_{i,j}^{FP}$, and the false negative objects set is denoted as $\mathbb{S}_{i,j}^{FN}$.

3.3 Calculating the Existence Length of the Object

In order to avoid the influence of vehicle speed on the proportion of the video captured by on-vehicle cameras, we convert the number of frames that the objects exist in the video into the "existence length" of the objects. Then, according to the detection result of the objects in each frame, the true positive length, the false positive length and the false negative length of the objects in each frame can be obtained.

The "existence length" of the object is defined the trajectory length of the intelligent vehicle in the process of the object appearing in the video.

According to the on-vehicle GPS, the instantaneous vehicle speed $v_{i,j}$ of the jth frame of the i-th data group is obtained, and the frame rate f of the on-vehicle cameras is constant. The equation for calculating the original trajectory length $l_{i,j}$ of each object in the frame is as follows:

$$l_{i,j} = \frac{v_{i,j}}{f} \tag{2}$$

The design concept of "existing length" is as follows.

In the process of data collection, the frame rate of on-vehicle cameras is usually constant. When the data collection vehicle passes through a certain scene at a slower speed, the object lasts a longer time in the video, which means more frames contain the object. For the traditional evaluation method of frame-by-frame statistics, the data collected at slower speeds account for a higher proportion of the total data, but the frames in the video collected at slower speeds have a high repetition rate, which is obviously unreasonable. The proportion of the objects in a certain test scene to the total objects should be independent of the speed which the intelligent vehicle passes through the scene. It is easy to imagine that in the course of data collection vehicle passing a certain test scene, the trajectory length traveled by the vehicle is independent of the speed of the vehicle. Therefore, we use the trajectory length l that the vehicle is driven in the object's lifetime (t_1, t_2) to replace the number of frames during that time, as shown in the Eq. (3).

$$l = \int_{t_1}^{t_2} v_t dt \approx \sum_{t_1}^{t_2} v_t \Delta t \tag{3}$$

By splitting the length calculation equation shown in Eq. (3) into each frame, we can obtain the calculation equation of the existence length of the single frame shown in Eq. (2).

In particular, when the data collection vehicle is stationary, the "existence length" of the acquired object is 0, which usually corresponds to the case of traffic jam or wait red light, and a special "trajectory length" can be assigned to the object collected in the above situation.

To sum up, we get the calculation equations of the true positive length $l_{i,j}^{TP}$, false positive length $l_{i,j}^{FP}$ and the false negative length $l_{i,j}^{FN}$ of the jth frame of the i-th data group are as follows:.

$$\begin{cases} l_{i,j}^{TP} = \sum^{k \in \mathbb{S}_{i,j}^{TP}} l_{i,j}, \\ l_{i,j}^{FP} = \sum^{k \in \mathbb{S}_{i,j}^{FP}} l_{i,j}, \\ l_{i,j}^{FN} = \sum^{k \in \mathbb{S}_{i,j}^{FN}} l_{i,j}. \end{cases} \tag{4}$$

3.4   The Weight of the Object

Weight of Safety. Our evaluation method combines the actual situation of intelligent vehicle driving and weights the object that is highly relevant to the safe driving of intelligent vehicle.

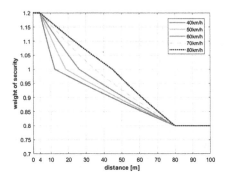

Fig. 2. The relationship between object relative distance and the weight of safety at different vehicle speeds. (In this figure, $d_m = 4\,\mathrm{m}$, $d_M = 80\,\mathrm{m}$, and $d_v < d_M$)

The weight of safety α is given to the object according to the vehicle speed v of the intelligent vehicle and its relative distance d from the object, as shown in the Eq. (5). Figure 2 shows the relationship between object relative distance and the weight of safety at different vehicle speeds.

$$
\alpha = \begin{cases}
e^{\lambda_1}, & d < d_m \\
e^{\frac{\lambda_1(d_v - d)}{d_v - d_m}}, & d_m \leq d < d_v \\
e^{\frac{\lambda_2(d_v - d)}{d_v - d_M}}, & d_v \leq d < d_M \\
e^{\lambda_2}, & d \geq d_M.
\end{cases}
\tag{5}
$$

Where d_v is the braking distance of the vehicle when the speed is v; d_m is the closest distance that can be seen by the on-vehicle camera, and the object with the distance less than d_m is invisible; d_M is the average farthest visual distance that the on-vehicle camera can clearly recognize. λ_1 and λ_2 are used to adjust the value of the weight of safety.

We believe that when intelligent vehicles are traveling at speed v, the objects outside braking distance d_v are relatively safe, so their weights of safety are relatively low, and the objects within d_v are relatively dangerous, so their weights of safety are relatively high.

The weights of safety only play a regulatory role in the weight of the object. We do not want the weight of individual objects to be very large or very small. Therefore, the weights of safety of the object which at braking distance d_v are defined as 1.0000. In this paper, the parameter $\lambda_1 = ln1.2 = 0.1823$, so the weights of safety are 1.2 when the distance between the object and the intelligent vehicle is less than or equal to d_m. The parameter $\lambda_2 = ln0.8 = -0.2231$, so the weights of safety are 0.8 when the distance between the object and intelligent vehicle is greater than or equal to d_M.

When the objects are given the weight of safety, the calculation equations of $l_{i,j}^{TP}$ and $l_{i,j}^{FN}$ becomes:

$$\begin{cases} l_{i,j}^{TP} = \sum^{k \in \mathbb{S}_{i,j}^{TP}} \alpha_k l_{i,j}, \\ l_{i,j}^{FN} = \sum^{k \in \mathbb{S}_{i,j}^{FN}} \alpha_k l_{i,j}. \end{cases} \tag{6}$$

Where α_k is the weight of safety of the kth object of the jth frame of the i-th data group calculated by Eq. (5).

It is worth noting that the true positive objects and the false negative objects correspond to the real existence objects, so they have corresponding weights of safety, but the false positive objects may not correspond to the actual objects, so we let their weight of safety equal to 1.0.

Weight of Early Detection. Our evaluation method sets the weight of early detection to evaluate whether the algorithm can find the object in time.

When the object suddenly breaks into the road region, only when the object is detected at the first time can intelligent vehicle have enough time to take countermeasures. In addition, when the speed is high, the required braking distance is also relatively long. If the object is detected when the object is very close to the intelligent vehicle, it is often difficult to take effective braking measures. Sudden objects are often accompanied by some occlusion, and distant objects usually have low resolution, which poses some challenges to existing object detection algorithms. In recent years, some scholars have devoted themselves to the low-resolution pedestrian detection [14] and the occluded objects detection [15], which provides a train of thought for intelligent vehicles to detect objects in time. However, most object detection datasets and their evaluation method only focus on the accuracy of object detection and the real-time performance of the algorithm, but do not evaluate whether the algorithm detects the object in time. In practical application, the algorithms that achieve better results on these datasets may be difficult to find objects in time.

Therefore, we multiply the weight of early detection of $\beta = 1.2$ for the false negative before the first detection of the object. At this time, the equation for calculating the length of the false negative becomes as follows:

$$l_{i,j}^{FN} = \sum^{k \in \mathbb{S}_{i,j}^{FN}} \alpha_k \beta_k l_{i,j} \tag{7}$$

Where β_k is the weight of early detection of the kth object of the jth frame of the i-th data group.

The weight of early detection ensures that the later the object is detected for the first time, the greater the additional penalty. If the false negative of the object that is formed after the target is detected, no additional penalty will be imposed.

So, the equation for calculating f-measure becomes:

$$F\text{-}measure = \frac{\sum_i \sum_j 2l_{i,j}^{TP}}{\sum_i \sum_j (2l_{i,j}^{TP} + l_{i,j}^{FP} + l_{i,j}^{FN})} \tag{8}$$

4   Experiment

In this section, we use case study to compare our method with the traditional evaluation method and demonstrate the rationality of our method. Then, the differences between the two methods are compared by the results of the off-line test in November 2018.

The method proposed in this paper is suitable for evaluating visual object detection algorithms in the video captured by on-vehicle cameras. The objects include vehicles, pedestrians, bicycles, tricycles and motorcycles. However, limited by labeled experimental data, the experiments in this paper only discuss pedestrians, bicycles, tricycles and motorcycles.

4.1   Rationality Verification by Case Study

Rationality of Road Region. As shown in Fig. 3, there are 7 objects in the figure, objects 1–3 are in the road region, and objects 4–7 are in the outer side of the road, which has a little relationship with the safe driving of intelligent vehicles.

Fig. 3. Typical data about road region.

Suppose there are two object detection algorithms applied to intelligent vehicles. Algorithm A only detects objects 1, 2, and 3. Algorithm B detects all objects.

As shown in Table 1, using the traditional evaluation method to evaluate all the objects in the figure, the f-measure of algorithm B is higher than algorithm A. However, for the practical application of the intelligent vehicle, there is no difference in the performance between algorithm A and algorithm B. The evaluation method proposed in this paper (only considering the road region) considers that the f-measure of the two algorithms are the same, which is obviously more reasonable.

Table 1. Comparison of our method and traditional method on road region

	Traditional evaluation method				Our evaluation method			
	TP	FP	FN	F1	TP	FP	FN	F1
Algorithm A	3	0	0	0.6000	3	0	0	1.0000
Algorithm B	7	0	0	1.0000	3	0	0	1.0000

Rationality of Weight of Safety. As shown in Fig. 4, there are two objects in the figure, object 2 is closer to the intelligent vehicle. A more serious safety accident may occur if object 2 is not detected by the algorithm. Object 1 is far from the intelligent vehicle. If object 1 is not detected at this time, the probability of a safety accident is relatively low.

Fig. 4. Typical data about different weights of safety.

Suppose there are two object detection algorithms applied to the intelligent vehicle. Algorithm A only detects object 2 and algorithm B only detects object 1.

Table 2. Comparison of our method and traditional method on weight of safety

	Traditional evaluation method				Our evaluation method			
	TP	FP	FN	F1	TP	FP	FN	F1
Algorithm A	1	0	1	0.6667	1.0551	0	0.8051	0.7238
Algorithm B	1	0	1	0.6667	0.8051	0	1.0551	0.6041

As shown in Table 2, using the traditional evaluation method, the f-measure of algorithm A and algorithm B are the same. The evaluation method proposed in this paper (only considering the weight of safety) needs to calculate the weight of safety of each object according to Eq. (5). The current instantaneous vehicle speed is 12.3531 m/s, the braking distance d_v is 14.9041 m, the relative distance between the object 2 and the intelligent vehicle is 11.6595 m, and the weight of safety is calculated to be 1.0551; the distance between object 1 and the intelligent vehicle is 78.1291 m, and the weight of safety is calculated to be 0.8051. Therefore, the f-measure of algorithm A is higher than algorithm B, which is consistent with the actual situation.

Rationality of Weight of Early Detection. As shown in Fig. 5, object 1 appears suddenly from the side rear of the intelligent vehicle in a 15-frame video segment. If the object is not detected in time, a serious safety accident may occur.

Fig. 5. Typical data about a sudden object.

Suppose there are two object detection algorithms applied to the intelligent vehicle. Algorithm A detects the object at frame 5 and has detected the object for the next 10 frames. Algorithm B detects the object in the first frame, but loses 4 frames in the subsequent frame. Both algorithms detect the object in 11 frames and lose it in 4 frames.

Table 3. Comparison of two methods on weight of early detection

	Traditional evaluation method				Our evaluation method			
	TP	FP	FN	F1	TP	FP	FN	F1
Algorithm A	11	0	4	0.8462	11	0	4.8000	0.8209
Algorithm B	11	0	4	0.8462	11	0	4.0000	0.8462

As shown in Table 3, since the number of missed frames of algorithm A and algorithm B are the same, the traditional evaluation method considers the f-measure of the two algorithm to be the same. The evaluation method proposed in this paper (only considering the weight of early detection) adds a weight of early detection to the false negative formed before the object is detected for the first time. Since algorithm A detects the object later, its f-measure is lower than algorithm B.

Fig. 6. Two typical data collected with large difference in vehicle speed, and their corresponding trajectory length of the data collection vehicle.

Table 4. The influence of two methods on the proportion of data

		Video A	Video B
Traditional evaluation method	Frame number	96	60
	Average number of objects per frame	1	1
	Proportion of all data	0.6154	0.3846
Our evaluation method	Average speed (m/s)	4.9845	10.2794
	Trajectory length of vehicle (m)	39.876	51.397
	Proportion of all data	0.4369	0.5631

Rationality of Evaluation by Existence Length. As shown in Fig. 6, there are two video sequences of A and B with a frame rate of $12fps$. The figure shows the 1st frame, the 48th frame, and the 96th frame of the video A, and the 1st frame, the 30th frame, and the 60th frame of the video B.

According to the traditional evaluation method, video A has a higher proportion than video B because of its large number of frames. Follow the evaluation method proposed in this paper (only considering the evaluation based on trajectory length), although the duration of the video A is longer, the speed of the data collection vehicle running is too slow, and the image repeatability of each frame of the video is too high. Therefore, after switching to the trajectory length that the vehicle has traveled, the proportion of video A is lower than that of video B (Table 4).

4.2   Rationality Verification by Off-Line Test Results

The method proposed in this paper is applied to off-line test task in November 2018. The test data has 110 groups, where each group of data has 50–96 frames, and a total of 7788 scene images with a resolution of 1280 * 1024. In the whole test dataset of the task, a total of 16,460 personnel bounding boxes were labeled, of which 13758 were ground truth, 2,702 were *ignore* regions, and there were 2 personnel objects in each picture on average. In fact, 60 groups of data were selected for the test. In the end, 9 teams submitted their results in accordance with the rules of the competition. The traditional evaluation method and the evaluation method proposed in this paper are used to evaluate them. The precision, recall and f-measure are the average of 9 teams. The results are shown in Table 5.

Table 5. Comparison of traditional method and our method on off-line test

	Precision	Recall	F1
Traditional evaluation methods	0.7232	0.4572	0.5381
Traditional methods + road region	0.7166	0.4911	0.5593
Traditional methods + the weight of safety	0.7103	0.4650	0.5405
Traditional methods + early detection	0.7232	0.4296	0.5148
Traditional methods + existence length	0.7155	0.4544	0.5301
Our evaluation method	0.6934	0.4628	0.5257

(1) In fact, 18 groups of data have objects outside the road region. According to the traditional evaluation method, if the algorithms fail to detect these objects, these objects will be recorded as false negative. Our evaluation method does not consider objects outside the road region, so the recall of the algorithms has been significantly improved.

(2) When only considering the weight of safety, we found that the precision decreased slightly, and the recall increased slightly. The reason is that most data for off-line test are collected at relatively low speeds. The lower the vehicle speed is, the smaller the braking distance is. Therefore, most objects in this dataset are outside the braking distance, and the weight of safety is less than 1.0, while the weight of safety of false positive is always 1.0, so the precision decreases. Most of the false negatives are far away from the data collection vehicle, so their weight of safety is lower. Therefore, the weight of the false negatives are often lower, leading to higher recall. It can be inferred that the precision and recall may be slightly increased on the datasets collected at relatively high speeds.

(3) Our early detection is to weight the false negatives before the first detection of the object. Therefore, the recall of the algorithms is correspondingly reduced after considering early detection.

(4) Since the existence length is only used to reasonably adjust the proportion of data, the performance of the algorithm does not change much after considering the existence length.

(5) The f-measure of the detection results evaluated by our method are lower than those evaluated by traditional evaluation method, this may mean the traditional method overestimates the visual object detection ability of intelligent vehicle.

5 Conclusion

In this paper, we propose a novel evaluation method for visual object detection of intelligent vehicles. We only evaluate the object in the road region, and convert the number of frames that the object persists into the "existence length" of the object. In addition, the corresponding weight of safety is set for each object, and the weight of early detection is set to evaluate whether the algorithm can detect the object in time. The experiments show that the traditional evaluation method overestimates the visual object detection ability of intelligent vehicle, and our evaluation method is more suitable for intelligent vehicle.

References

1. Ding, Z., Huang, X., Peng, H., Lam, H., Leblanc, D.J.: Accelerated evaluation of automated vehicles in car-following maneuvers. IEEE Trans. Intell. Transp. Syst. **99**, 1–12 (2016)
2. Zhang, C., Liu, Y., Zhang, Q., Wang, L.: A graded offline evaluation framework for intelligent vehicle's cognitive ability. In: IEEE Intelligent Vehicles Symposium (IV), pp. 320–325. IEEE (2018)
3. Li, L., Huang, W.-L., Liu, Y., Zheng, N.-N., Wang, F.-Y.: Intelligence testing for autonomous vehicles: a new approach. IEEE Trans. Intell. Veh. **1**(2), 158–166 (2016)
4. Li, L., et al.: Artificial intelligence test: a case study of intelligent vehicles. Artif. Intell. Rev. **50**(3), 441–465 (2018)
5. Zhang, Q., Hua, G., Liu, W., Liu, Z., Zhang, Z.: Auxiliary training information assisted visual recognition. IPSJ Trans. Comput. Vis. Appl. **7**, 138–150 (2015)
6. Zhang, Q., Hua, G.: Multi-view visual recognition of imperfect testing data. In: Proceedings of the 23rd ACM International Conference on Multimedia, pp. 561–570. ACM (2015)
7. Zhang, Q., Hua, G., Liu, W., Liu, Z., Zhang, Z.: Can visual recognition benefit from auxiliary information in training? In: Cremers, D., Reid, I., Saito, H., Yang, M.-H. (eds.) ACCV 2014. LNCS, vol. 9003, pp. 65–80. Springer, Cham (2015). https://doi.org/10.1007/978-3-319-16865-4_5
8. Geiger, A., Lenz, P., Stiller, C., Urtasun, R.: Vision meets robotics: the KITTI dataset. Int. J. Robot. Res. **32**(11), 1231–1237 (2013)
9. Geiger, A., Lenz, P., Urtasun, R.: Are we ready for autonomous driving? The KITTI vision benchmark suite. In: IEEE Conference on Computer Vision and Pattern Recognition, pp. 3354–3361. IEEE (2012)
10. Dollar, P., Wojek, C., Schiele, B., Perona, P.: Pedestrian detection: a benchmark. In: IEEE Conference on Computer Vision and Pattern Recognition, pp. 304–311. IEEE (2009)
11. Maddern, W., Pascoe, G., Linegar, C., Newman, P.: 1 Year, 1000km: the Oxford RobotCar dataset. Int. J. Robot. Res. (IJRR) **36**(1), 3–15 (2017)

12. Enzweiler, M., Gavrila, D.M.: Monocular pedestrian detection: survey and experiments. IEEE Trans. Pattern Anal. Mach. Intell. **31**(12), 2179–2195 (2009)
13. Everingham, M., Van Gool, L., Williams, C.K.I., Winn, J., Zisserman, A.: The PASCAL Visual Object Classes (VOC) challenge. Int. J. Comput. Vision **88**(2), 303–338 (2010)
14. Li, X., Liu, Y., Chen, Z., Zhou, J., Wu, Y.: Fused discriminative metric learning for low resolution pedestrian detection. In: 25th IEEE International Conference on Image Processing (ICIP), pp. 958–962. IEEE (2018)
15. Su, Y., Liu, Y., Cuan, B., Zheng, N.: Contour guided hierarchical model for shape matching. In: Proceedings of the IEEE International Conference on Computer Vision, pp. 1609–1617 (2015)

NLP and Recommender System

Large Scale Name Disambiguation Using Rule-Based Post Processing Combined with Aminer

Lizhi Zhang and Zhijie Ban[✉]

School of Computer Science, Inner Mongolia University, Hohhot 010021, China
zhijieban@imu.edu.cn

Abstract. Author name ambiguity has long been viewed as a challenging problem in scientific literature management, and in recent years, due to the rapid increase in academic publications, author disambiguation on large-scale academic data has become an urgent problem. In this paper, we present a rule-based post processing method combined with Aminer's framework to address the large scale author name disambiguation problem. In our method, we first introduce the Aminer's disambiguation model for author name disambiguation. Based on the Aminer's model, we propose an efficient post processing algorithm, aiming to improve the disambiguation performance by rule-based clustering. Our algorithm utilizes similarity features based on metadata information and implements two types of disambiguation rules. We carefully evaluate the proposed post processing method on real-world large data and experimental result shows that our method achieves clearly better performance (+11% in terms of F1-score) than the state-of-the-art Aminer [1] method.

Keywords: Name disambiguation · Post-processing · Clustering

1 Introduction

In the information age, people rely on search engines to quickly retrieve the information they need. Among them, the search of author entity information based on person name keywords is one of the important search contents of search engines. There are a large number of authors who shared the same name, researchers can't accurately identify and distinguish different authors, so digital literature resources face serious author name ambiguity problem.

Name disambiguation, also known as entity resolution [2, 3], web appearance disambiguation [4, 5], name identification [6], and object distinction [7] has long been viewed as a challenging problem in many domains. Two or more researchers having the same name even within the same discipline, school, or institution. Most of these methods measure the similarity between entities based on a single feature or multiple features and then use clustering algorithms for clustering. In recent years, due to the rapid increase in publication by academics, researchers attempt to solve author name disambiguation problem on a large-scale data. Aminer proposes a novel representation learning method by incorporating both global and local information and presents an

© Springer Nature Singapore Pte Ltd. 2019
K. Knight et al. (Eds.): ICAI 2019, CCIS 1001, pp. 147–158, 2019.
https://doi.org/10.1007/978-981-32-9298-7_12

end-to-end cluster size estimation method [1]. The algorithm has been deployed in Aminer's website[1] to deal with the disambiguation problem at the billion scale. Although Aminer propose an effective model to solve cluster size estimation problem on large scale data, the number of clusters predicted by the model is a little lower than the actual value for most cases. In addition, they proposed a representation learning method which projected each entity into a low-dimensional latent common space. This method takes advantage of the inverted document frequency of the word in the document feature (e.g. title, keywords, co-author, etc.) as a word weight, ignoring the importance of each feature.

In order to address the problems caused by the shortcomings of Aminer's framework, we propose a rule-based post-processing method. Our method is based on the Aminer's disambiguation results which use rule-based clustering at a different level. Our experiments show that the proposed solution achieves better performance than the Aminer's method (+11% in terms of F1-score).

The remainder of this paper is organized as follows. Section 2 surveys related work, while Sect. 3 describes the Our method. The results and evaluations are reported in Sect. 4. Section 5 concludes the paper and discusses future work.

2 Related Works

Many different solutions are proposed for the author disambiguation problem and they can be generally categorized as supervised, unsupervised and graph-based methods. The first techniques are supervised methods in which labeled training data is manually created and inputted to the classifier. Tran et al. [8] proposed a new approach which uses a deep neural network to learn features automatically for solving author name ambiguity. Additionally, they have proposed the general system architecture for author name disambiguation on any dataset. Five supervised machine learning techniques-Random Forest, Support Vector Machine, k-Nearest Neighbors, C4.5(decision tree) and naïve Bayes are proposed by Huynh et al. [9] for solving Vietnamese author name ambiguity problem.

Unsupervised methods harness some pre-defined likeness between people records in grouping them into clusters. Tang et al. [10] formalize the disambiguation problem using Markov random fields, in which the data is cohesive on both local attributes and relationships. Mann et al. [11] utilize an unsupervised clustering technique over a rich feature space of biographic facts, which are automatically extracted via a language-independent bootstrapping process. Xu et al. [12] propose a network embedding method to learn the representations of papers for author disambiguation.

Some other methods have also been proposed based on graphical theory. For example, Yin et al. [7] propose an approach called DISTINCT by combining two similarity measures: set resemble of neighbor tuples and random walk probability. They propose a method to weight different types of links in the graph. A graph-based model-ADANA is proposed by Wang et al. [13], in which they modeled pairwise factor

---

[1] https://aminer.org.

graph that can be used to integrate several types of features as well as user feedbacks into a unified model. A topological-collaborative approach is presented to solve homonyms in DLs by Amancio et al. [14] in which they used topological features of the collaborative graph along with weighted collaboration patterns among authors. Fan et al. [15] propose a novel similarity metric and adopt the affinity propagation algorithm to group intermediate results into clusters.

Our work is based on the initial disambiguation result produced by the Aminer's method, followed by rule-based clustering. We utilize co-authorship and author affiliation as the primary reference document feature. This work can significantly improve the performance of the initial result.

3   Methodology

In this section, we will introduce our method in details. Our method aims to solve the author name disambiguation problem on large scale data. We first use Aminer's method to get the initial result of disambiguation and then perform a two-step rule-based clustering to improve the disambiguation result. Figure 1 provides a visual summary of our disambiguation post-processing method.

Fig. 1. An overview of the author name disambiguation process

Formally, let a be a given name reference, and $\mathcal{D}^a = \{D_1^a, D_2^a, \ldots, D_N^a\}$ be a set of N documents associated with the author name a. We call $\mathcal{D}^a$ a candidate set of a. We use $\mathbb{I}(D_i^a)$ to denote the identity (corresponding real-world person) of D_i^a. We omit the superscript a in the following description if there is no ambiguity. The task of author disambiguation is to find Φ to partition $\mathcal{D}^a$ into a set of disjoint clusters, i.e., $\Phi(\mathcal{D}^a) \rightarrow C^a$, where $C^a = \{C_1^a, C_2^a, \ldots, C_K^a\}$.

3.1   Name Disambiguation in Aminer

In this section, we introduce the implementation of Aminer's [1] deployed solution to author name disambiguation problem in detail.

Global Metric Learning. First, embed each document D_i as a IDF-weighted Word2Vec [16] vectors x_i. Input document D_i is represented as a varied-length set of features $D_i = \{x_1, x_2, \ldots\}$, where feature are words in title and coauthor names, venue names, affiliations, etc. Each feature is a one-hot vector. Then use Word2Vec to obtain an embedding $\bar{x}_n \in \mathbb{R}^d$ for each feature x_n. The feature embedding of document D_i is $x_i = \sum_{x_n \in D_i} \alpha_n \bar{x}_n$, which is a weighted sum of the embedding of each feature in

D_i. In the equation, α_n is the inverted document frequency of feature x_n. Then they leverage labeled data to fine-tune the embedding, which enforces positive pairs to be close in the embedding space and negative pairs to be far away.

Triplet Loss. Let (D_i, D_{i+}, D_{i-}) be a triplet where D_i, D_{i+} belong to the same author and D_i, D_{i-} belong to the different author. The objective $\mathcal{L}_f$ is

$$\mathcal{L}_f = \Sigma_{(D_i, D_{i+}, D_{i-}) \in \mathcal{T}} max\{0, \delta(y_i, y_{i+}) - \delta(y_i, y_{i-}) + m\}. \tag{1}$$

Where $\mathcal{T}$ is the set of all triplets in training set, m is a margin enforced between positive pairs and negative pairs and $\delta(v_1, v_2) = \|v_1 - v_2\|$ is the Euclidean distance.

Local Linkage Learning. For a given name reference a, construct a local linkage graph $\mathcal{G}^a = (\mathcal{D}^a, \mathcal{E}^a)$, where $\mathcal{D}^a = \{D_i^a\}$ is the set of documents authored by a person named a, $\mathcal{E}^a = \{D_i^a, D_j^a\}$ is the set of edges capturing the similarity between the documents.

The linkage weight $W(D_i, D_j) = \sum_{x \in D_i \cap D_j} w_x$ between D_i and D_j as a weighted sum of all the common features. The weight $\{w_x\}$ is the inverted document frequency of feature x. They construct an edge in $\mathcal{G}$ between D_i and D_j if $W(D_i, D_j)$ is above a threshold. And then use an unsupervised auto-encoder architecture [14, 18] to learn from the local linkage graph.

Graph Auto-encoder. Let $Y = [y_1^T, \ldots, y_N^T]^T$ denotes an embedding matrix of $\mathcal{D}$ generated by the global model, $A \in \mathbb{R}^{N \times N}$ is the adjacency matrix of $\mathcal{G}$. A graph auto-encoder is comprised of a node encoder model $Z = g_1(Y, A)$ and an edge decoder model $\tilde{A} = g_2(Z)$, where $Z = [z_1^T, \ldots, z_N^T]^T$ is a node embedding matrix, $\tilde{A}$ is a predicted adjacency matrix. The objective is to minimize the reconstruction error between the predicted $\tilde{A}$ and the original adjacency matrix A.

Cluster Size Estimation. Based on the learned embeddings, the final partition of a candidate set is determined by a clustering algorithm. One common issue for the clustering problem is how to estimate the number of clusters K. They adopt an end-to-end model $h(\mathcal{D}) \to \mathbb{R}$ that takes a set of documents as input and directly estimates the number of different identities in the candidate set. First, construct a pseudo-training set as right. Then, adopt RNN as an encoder and try to map a set of embedding vectors to the true number of clusters in the set. Optimize the Mean Squared Logarithmic Error (MSLE) as follows:

$$\mathcal{L}_h = \frac{1}{N} \sum_{t=1}^{a} [\log(1 + h(\mathcal{D}_t)) - \log(1 + K_t)]^2 \tag{2}$$

After getting the number of clusters K, they use hierarchical agglomerative clustering algorithm (HAC) as the main clustering method.

3.2   Rule-Based Post Processing

The initial disambiguation result is obtained by the Aminer disambiguation framework. In each candidate set of the result, each cluster has several documents written by a real author with a higher similarity between the documents. Aminer's method uses IDF as a word weight to quantify document similarity. However, different features of the document (e.g. title, co-author names, venue names, etc.) play different roles. Paper title is a "weak" evidence for determining if two authors sharing the same name are the same person compared with "strong" evidence such as email and co-authorship [19]. The IDF is intended to suppress the negative effects of meaningless high-frequency words in a document, there is a question here that deserves to be discussed: common words are not equal to meaningless words. Similarly, the occasional appearance of low-frequency words will be treated as high-weight keywords, which over-amplify the importance of the unfamiliar words. In addition, for large scale bibliographic datasets, the cluster size estimation method which use RNN model to obtain the number of clusters is obviously better than the traditional X-means algorithm with Bayesian Information Criterion as the measurement of clustering quality [1]. Although the RNN model has a more accurate estimate for the number of clusters than other existing methods, we found that for most names, the RNN model predicted to be smaller than the actual number of clusters.

For the above defects, we proposed a rule-based post processing method, which can obviously improve the performance of disambiguation method mentioned in Sect. 3.1 In order to effectively distinguish different author, we first design an author disambiguation rule set, and then cluster each author's papers within each candidate set based on these rules. Finally, handling papers from all authors in the candidate set.

A Simple Fuzzy Matching Algorithm. In a publication, the most representative of the author's feature attributes is author name and author's affiliation, as well as the author's field and focus. We chose the co-author's name and their affiliation of a publication as the primary reference feature for disambiguation post-processing.

In the process of author name disambiguation, the disambiguation result is not good if we use an exact matching algorithm to match the feature values, because the feature values often have a certain degree of fuzziness, which is easy to cause the problem of excessive recognition. So, we adopt a simple fuzzy matching algorithm for some feature values (e.g. author affiliation), we reduce the scope of disambiguation by fuzzy matching, then use exact matching for the unique author name, thereby achieving a better disambiguation performance. We use Levenshtein ratio as the standard for matching strings, i.e.

$$Match(\alpha, \beta) = \left(1 - \frac{ldist}{lensum}\right) * \delta \tag{3}$$

where $ldist$ is the Levenshtein distance between α, β and $lensum$ is the length of the longest of two strings. δ is equal to 100 so that the match result represents the percentage of similarity.

After analyzing the actual article-level data, in general, at $match(\alpha, \beta) > \eta$ (we set η as 70), the matching effect on the author affiliation is the best. It should be noted that although the fuzzy matching algorithm in this paper cannot completely identify the same author affiliation, it has little effect on the accuracy of the final author name disambiguation result, and can also improve the recall rate of disambiguation result. In the post processing of disambiguation, the matching of the author affiliation is only an auxiliary condition, combined with the accurate matching of the author name, the final result still achieves better performance.

Author Disambiguation Rules. $\mathcal{D} = \{D_1, D_2, \ldots, D_N\}$ is the candidate set of a name, each document D_i has one or more authors $A(D_i) = \{a_i^0, a_i^1, \ldots, a_i^u\}$, in order to distinguish different authors' name from the common name in the candidate set, we describe the author that we are going to disambiguate of document D_i as a_i^x, where the name of a_i^x is the common name in the candidate set. In addition, $aff(a_i^x)$ is the affiliation of the author a_i^x, and $coauthor(D_i)$ is the coauthor set of D_i. Thus if D_i and D_j are authored by the same author, we have $\mathbb{I}(D_i^a) = \mathbb{I}(D_j^a)$. The algorithm's disambiguation rules are as follows:

Let $D_i, D_j \in \mathcal{D}$, then:

$$Match\left(aff(a_i^x), aff(a_j^x)\right) > \eta \wedge \left(coauthor(D_i) \cap coauthor(D_j)c \neq \emptyset\right) \Rightarrow \mathbb{I}(D_i) = \mathbb{I}(D_j).$$

$$(4)$$

Let $D_i, D_j \in \mathcal{D}$, then:

$$\exists\, a_i^k \in coauthor(D_i), a_j^l \in coauthor(D_j), Match\left(aff(a_i^k), aff(a_j^l)\right)\eta \wedge a_i^k = a_j^l \Rightarrow \mathbb{I}(D_i) = \mathbb{I}(D_i).$$

$$(5)$$

The disambiguation rules utilize the co-authorship and affiliated institution to be clustered as feature attributes, based on the fuzzy matching algorithm, combined with exact matching and perform secondary clustering of initial disambiguation results. Rule (4) indicates authors of the same name who are in the same institution are considered to be the same author entity if they have more than one co-authorship with the author entity of the same person name. Rule (5) indicates that for any two authors of the same name, they are considered to be the same author entity as long as they have more than one co-authorship with the authors of the same name who are in the same institution.

Rule-Based Clustering. Rule enforcement can be done in two levels: at the cluster-level and at the candidate-level.

Cluster-Level. Cluster rules are constraints that check the co-authorship and author affiliation between documents. In each candidate of the initial result, the number of clusters is lower than the actual value. Therefore, we spilt every cluster according to the author affiliation, and then merge the separate subclusters based on the co-authorship between documents.

Algorithm 1: Cluster-level Processing

Input: a cluster C_k in candidate set

Output: Processed cluster C_k'

1. Make each element in C_k be a subset, then $C_k = \{S_1, S_2, ..., S_m\}$, where $S_i = \{D_i\}$
2. $C_k' \leftarrow \emptyset$
3. **while** $C_k \neq \emptyset$ **do**
4.   Draw a S_i from C_k,
5.   **if** length $[S_i] == 1$ **then**
6. $C_k' \leftarrow C_k' \cup S_i, C_k = C_k - S_i$
7.   **end if**
8.   **foreach** element S_j in C_k **do**
9.   **if** $\exists D_i \in C_k'[S_i], D_j \in C_k[S_j]$ and $match\left(aff(a_i^x), aff(a_j^x)\right) > \eta$ **then**
10. $C_k'[S_i] \leftarrow C_k'[S_i] \cup C_k[S_j], C_k = C_k - S_j$
11.   **end if**
12.   **end for**
13. **end while**
14. **for** i $\leftarrow$ 0 to length $[C_k'] - 1$
15.   **for** j $\leftarrow$ i + 1 to length $[C_k']$
16.   **if** $\exists D_i \in C_k'[i], D_j \in C_k'[j]$ and $coauthor(D_i) \cap coauthor(D_j) c \neq \emptyset$ **then**
17. $C_k'[i] \leftarrow C_k'[i] \cup C_k'[j], C_k' \leftarrow C_k' - C_k'[j]$
18.   **end if**
19.   **end for**
20. **end for**

Candidate-Level. Candidate rules are constraints that check the co-authorship and author affiliation between clusters. After the cluster-level processing, the number of clusters in a candidate set will increase. Consider that different clusters may be associated, we first measure whether the author affiliation in the two clusters match, and then measure whether the same cluster has the same co-author, if both are true, merge the two clusters.

Algorithm 2: Candidate-level Clustering

Input: a candidate set C after cluster-level processing
Output: Clustered candidate set
  1. **for** i ← 0 to length $[C]$ - 1
  2.   **for** j ← i + 1 to length $[C]$
  3.   **if** $\exists D_i \in C[i], D_j \in C[j]$, $a_i^k \in coauthor(D_i)$

   $a_j^l \in coauthor(D_j), Match\left(aff(\,a_i^k), aff(a_j^l)\right)$

            $> \eta \wedge a_i^k = a_j^l$ **then**
               $C[i] \leftarrow C[i] \cup C[j], C \leftarrow C - C[j]$
  4.   **end if**
  5.   **end for**
  6. **end for**

---

4 Experiment

4.1 Dataset and Experiment Design

We compare our approach with Aminer's disambiguation method which is the state-of-the-art method. And we perform our experiments on a large scale dataset proposed by Aminer. The benchmark consists of 70,258 documents from 12,798 authors. Comparing to existing benchmarks, this benchmark is significantly larger (in terms of the number of documents) and more challenging (since each candidate set contains much more clusters) [1].

In our experiment, we used PairwisePrecision, PairwiseRecall, and PairwiseF1 score to evaluate. From Tang et al. [10], the measures are defined as follows:

$$PairwisePrecison = \frac{\#PairsCorrectlyPredictedToSameAuthor}{\#TotalPairsPredictedToSameAuthor}$$

$$PairwiseRecall = \frac{\#PairsCorrectlyPredictedToSameAuthor}{\#TotalPairsToSameAuthor}$$

$$PairwiseF1 = \frac{2 \times PairwisePrecision \times PairwiseRecall}{PairwisePrecision + PairwiseRecall}$$

A micro-F1 score is calculated according to all test names.

4.2 Experimental Results

The results in Table 1 show that average values of evaluation metrics on Aminer's dataset. Aminer uses a representation learning method by incorporating both global and local information and present an end-to-end cluster size estimation method. The final partition is determined by a clustering algorithm (HAC) that leverage learned embeddings and the estimated number of clusters. Our method combined Aminer's disambiguation method with rule-based clustering. Experiment proves that our method outperforms the baseline in terms of F1-score (+11.22% over Aminer). Table 2 shows the performance of our method and Aminer method on some sampled names.

Table 1. Average values of evaluation metrics

Method	Precision	Recall	Micro-F1
Aminer	72.67	46.18	56.49
Our approach	74.87	61.80	**67.71**

Table 2. Result of author name disambiguation on sampled names

Name	Aminer			Our approach		
	Prec	Rec	F1	Prec	Rec	F1
Lan Sun	79.58	41.94	54.93	93.23	51.40	66.26
Jian Du	68.32	38.51	49.25	72.71	56.18	63.38
Jianhua Lu	91.28	27.51	42.28	93.72	94.00	93.86
Jing Luo	44.60	30.58	36.28	51.81	39.59	44.88
Zhigang Zeng	92.67	57.89	71.26	92.02	98.79	95.28
Jian Feng	83.11	30.29	44.40	87.29	79.86	83.41
Hua Fu	79.87	40.60	53.83	64.07	93.02	75.87
Mei Xu	62.98	65.51	64.22	76.84	66.61	71.36
Ruijin Liao	87.28	99.11	92.82	87.23	98.23	92.40
Yuming Wang	66.46	30.09	41.42	70.32	42.02	52.60
Bo Hong	80.30	70.42	75.04	79.54	80.38	79.96
Weiming Zhu	81.47	90.32	85.67	82.90	98.71	90.12
Xu Xu	71.65	36.64	48.49	68.45	46.56	55.42
Rong Yu	86.88	20.93	33.73	92.27	32.79	48.39
Yong Tian	61.83	41.38	49.58	83.37	48.78	61.54

Table 3. Comparison of cluster size

Name	Actual	Aminer	Our method
Mei Xu	79	50	76
S Lin	144	43	117
Dandan Zhang	130	138	134
Tian Chen	71	35	73
Lei Song	159	70	159
Yanqing Wang	64	48	61
Jing Luo	174	95	181
Bo Hong	60	38	60

We also analyze the change of cluster size between Aminer and our method. We have rounded the real values predicted by the Aminer's RNN model. From Table 3 it can be concluded that after the post processing of Aminer's initial disambiguation result, the number of clusters is obviously closer to the actual value.

As shown in Fig. 2, the performance of pairwise F1 that the clustering of cluster-level (i.e. within each cluster of a name) is slightly higher than Aminer's result for most cases. And we can significantly improve the performance by first clustering within the cluster of a name and then clustering within the candidate set (i.e. All).

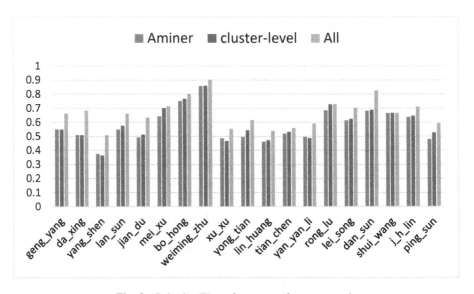

Fig. 2. Pairwise F1 performance of our approach

5   Conclusions and Future Works

In this paper, we proposed a rule-based post processing method to improve disambiguation performance. By performing a two-level clustering on the initial disambiguation results generated by the Aminer's method, the performance of disambiguation can be significantly improved on large scale data.

Although the algorithm proposed in this paper has been able to achieve better results in a specific application, there are still many places that need further improvement. When judging the change of author affiliation, we can make use of time information. Moreover, in the process of the actual application, a user feedback mechanism can be added to improve author name disambiguation.

Acknowledgements. The work is supported by the National Natural Science Foundation of China (No. 61662053).

References

1. Zhang, Y., Zhang, F., Yao, P., Tang, J.: Name disambiguation in Aminer: clustering, maintenance, and human in the loop. Presented at the KDD (2018)
2. Benjelloun, O., Garcia-Molina, H., Menestrina, D., Su, Q., Whang, S.E., Widom, J.: Swoosh: a generic approach to entity resolution. VLDB J. Int. J. Very Large Data Bases **18**, 255–276 (2009)
3. Bhattacharya, I., Getoor, L.: Collective entity resolution in relational data. ACM Trans. Knowl. Discov. Data (TKDD) **1**(1), 5 (2007)
4. Jiang, L., Wang, J., An, N., Wang, S., Zhan, J., Li, L.: Grape: a graph-based framework for disambiguating people appearances in web search. In: 2009 Ninth IEEE International Conference on Data Mining, pp. 199–208. IEEE (2009)
5. Bekkerman, R., McCallum, A.: Disambiguating web appearances of people in a social network. In: Proceedings of the 14th International Conference on World Wide Web, pp. 463–470. ACM (2005)
6. Li, X., Morie, P., Roth, D.: Identification and tracing of ambiguous names: discriminative and generative approaches. In: Proceedings of the National Conference on Artificial Intelligence, pp. 419–424. AAAI Press/MIT Press, Menlo Park and Cambridge (2004)
7. Yin, X., Han, J., Philip, S.Y.: Object distinction: distinguishing objects with identical names. In: 2007 IEEE 23rd International Conference on Data Engineering, pp. 1242–1246. IEEE (2007)
8. Tran, H.N., Huynh, T., Do, T.: Author name disambiguation by using deep neural network. In: Nguyen, N.T., Attachoo, B., Trawiński, B., Somboonviwat, K. (eds.) ACIIDS 2014, Part I. LNCS (LNAI), vol. 8397, pp. 123–132. Springer, Cham (2014). https://doi.org/10.1007/978-3-319-05476-6_13
9. Huynh, T., Hoang, K., Do, T., Huynh, D.: Vietnamese author name disambiguation for integrating publications from heterogeneous sources. In: Selamat, A., Nguyen, N.T., Haron, H. (eds.) ACIIDS 2013, Part I. LNCS (LNAI), vol. 7802, pp. 226–235. Springer, Heidelberg (2013). https://doi.org/10.1007/978-3-642-36546-1_24
10. Tang, J., Fong, A.C.M., Wang, B., Zhang, J.: A unified probabilistic framework for name disambiguation in digital library. IEEE Trans. Knowl. Data Eng. **24**, 975–987 (2012). https://doi.org/10.1109/TKDE.2011.13
11. Mann, G.S., Yarowsky, D.: Unsupervised personal name disambiguation. In: Proceedings of the Seventh Conference on Natural Language Learning at HLT-NAACL 2003, Edmonton, Canada, pp. 33–40. Association for Computational Linguistics (2003)
12. Xu, J., Shen, S., Li, D., Fu, Y.: A network-embedding based method for author disambiguation. In: Proceedings of the 27th ACM International Conference on Information and Knowledge Management - CIKM 2018, Torino, Italy, pp. 1735–1738. ACM Press (2018)
13. Wang, X., Tang, J., Cheng, H., Yu, P.S.: ADANA: active name disambiguation, December 2011
14. Amancio, D.R., Oliveira Jr., O.N., da Costa, L.F.: Topological-collaborative approach for disambiguating authors' names in collaborative networks. Scientometrics **102**, 465–485 (2015)
15. Fan, X., Wang, J., Pu, X., Zhou, L., Lv, B.: On graph-based name disambiguation. J. Data Inf. Qual. **2**, 1–23 (2011). https://doi.org/10.1145/1891879.1891883

16. Mikolov, T., Sutskever, I., Chen, K., Corrado, G., Dean, J.: Distributed Representations of Words and Phrases and their Compositionality (2013). arXiv:1310.4546 [cs, stat]
17. Kipf, T.N., Welling, M.: Variational graph auto-encoders (2016). arXiv preprint: arXiv: 1611.07308
18. Tian, F., Gao, B., Cui, Q., Chen, E., Liu, T.-Y.: Learning deep representations for graph clustering. In: Twenty-Eighth AAAI Conference on Artificial Intelligence (2014)
19. Zhu, J., Wu, X., Lin, X., Huang, C., Fung, G.P.C., Tang, Y.: A novel multiple layers name disambiguation framework for digital libraries using dynamic clustering. Scientometrics **114**, 781–794 (2018). https://doi.org/10.1007/s11192-017-2611-8

Hybrid Recommendation Algorithm Based on Weighted Bipartite Graph and Logistic Regression

Wei Song[1,2(✉)] (iD), Pengwei Shao[1], and Peng Liu[1]

[1] School of Information Science and Technology,
North China University of Technology, Beijing 100144, China
songwei@ncut.edu.cn
[2] Beijing Key Laboratory on Integration and Analysis of Large-Scale Stream Data, Beijing 100144, China

Abstract. Bipartite graph-based recommendation is a key topic in the field of recommender systems. Using logistic regression, a new recommendation algorithm based on a bipartite graph is proposed. First, the weights of the bipartite graph and the similarity of users are defined. Then, a recommendation list based on the bipartite graph is generated. Next, items in the recommendation list are re-sorted using the classification results of logistic regression. Furthermore, a balance factor is proposed to measure the accuracy and diversity of a recommender system comprehensively. Experimental results show that the recommendation results of the proposed algorithm are good.

Keywords: Personalized recommendation · Hybrid recommendation · Weighted bipartite graph · Logistic regression · Balance factor

1 Introduction

Recommender systems (RSs) [1, 11] are a key topic of multidisciplinary research. Methods in RS can be mainly categorized into content-based recommendation [4, 13], collaborative filtering [6, 12], and hybrid recommendation [3, 14] methods. These methods have a wide range of applications in education [2], e-commerce [5], social networks [10], and other fields.

A graph is a data structure with complex expression capabilities [15]. Because the structure is relatively clear, recommendation algorithms that construct bipartite graphs with the users and items as the nodes are receiving increasingly more attention. Zhou et al. [17] and Liu et al. [9] proposed user similarity calculation methods based on resource allocation in bipartite graphs, that achieved good personalized recommendation results. Huang et al. [7] used a spreading activation algorithm in the bipartite graph to address the problem of data sparsity in collaborative filtering. On the premise of distinguishing between high and low ratings, Wang et al. [16] introduced the ratio of item degree to summed item weights in a recommendation system based on a bipartite graph. Using the monotonic saturation function as the weight, Li et al. [8] proposed a bipartite graph-based recommendation algorithm. The above-mentioned works focus on either the allocation of resources or weight adjustment in bipartite graphs. Reprocessing of the recommendation list was rarely considered.

© Springer Nature Singapore Pte Ltd. 2019
K. Knight et al. (Eds.): ICAI 2019, CCIS 1001, pp. 159–170, 2019.
https://doi.org/10.1007/978-981-32-9298-7_13

In this paper, we propose a hybrid recommendation algorithm that incorporates logistic regression into a bipartite graph. The basic recommendation list is constructed using a bipartite graph and then modified by logistic regression. Furthermore, a balance factor is also defined to evaluate the recommendation results comprehensively from the perspectives of accuracy and diversity. Finally, the proposed algorithm is compared with other recent related algorithms.

2   Preliminaries

Let $U = \{u_1, u_2, \ldots, u_m\}$ be a finite set of users and $O = \{o_1, o_2, \ldots, o_n\}$ be a finite set of items. An RS is usually defined by the user-item rating matrix $R = \{R_{ij}\}_{m \times n}$, where R_{ij} is the rating of item o_j by user u_i. The RS then calculates several unrated items with high predicted ratings for the target user and recommends them.

Let $G = <V, E>$ be a graph, where V is the set of nodes and E is the set of edges, and assume V can be divided into two subsets V_1 and V_2, where $V_1 \cup V_2 = V$ and $V_1 \cap V_2 = \varnothing$. For any $e = (v_i, v_j) \in E$, if $v_i \in V_1$, $v_j \in V_2$, G is called a bipartite graph [15]. Because U and O are two disjoint sets, R can be represented as a bipartite graph using users or items as nodes and ratings as edges. Figure 1 shows an example bipartite graph consisting of three users and four items.

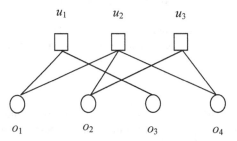

Fig. 1. Example bipartite graph

For bipartite-graph-based recommendation methods [9, 17], it is assumed that each user has a certain amount of resources (recommendation energy) at an initial point. Let e_1, e_2, and e_3 be the initial energies of u_1, u_2, and u_3 in Fig. 1, respectively. For a user u_i ($1 \leq i \leq 3$), we suppose its energy e_i is evenly allocated to those items that already have been rated by u_i. For example, the energy allocated by user u_1 to item o_1 is $e_1/2$. For the bipartite graph shown in Fig. 1, after all users have allocated their energy to all the items, the energy of the four items is shown in Fig. 2.

When all the items have been allocated energy, the recommendation energy is allocated back to the users along the edges in the bipartite graph. For example, item o_2 in Fig. 2 allocates the energy of $(e_2/3 + e_3/2)/2$ to u_2 and u_3, respectively. The resulting energy of the users in the example bipartite graph is shown in Fig. 3.

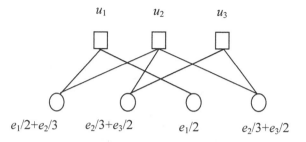

Fig. 2. Energy allocation from users to items

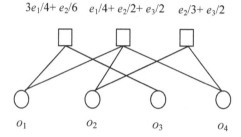

Fig. 3. Energy allocation from the items back to the users

After allocating and recovering the energy, the similarity between the users is calculated based on the energy. Then, recommendation results are generated from the users that are the most similar to the target user.

3 Proposed Algorithm

3.1 Overall Process

The overall flowchart of the proposed method, called the Recommendation based on weighted Bipartite Graph and Logistic Regression (RBGLR) method is shown in Fig. 4.

The RBGLR algorithm first deletes items with frequencies that are very low from the original data. Then, the user-item bipartite graph is constructed, and the user similarities are calculated according to the weights in the bipartite graph. User predicted ratings are approximated according to the similarity. Then, the top n items with the highest predicted ratings form the recommendation list. After that, logistic regression is used to adjust the order of the items in the recommendation list to form the final recommendation results for the target user.

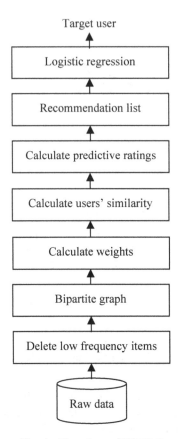

Fig. 4. Flowchart of RBGLR

3.2 Recommendation Based on a Weighted Bipartite Graph

The sigmoid function is used to add weights to the edges of the user-item bipartite graph. Here, weights indicate the relationship between users and items. For weights, three factors are considered: users, items, and user ratings of items. In the bipartite graph, the weight between u_i and o_j is calculated by

$$w_{ij} = \frac{R_{ij}}{1 + e^{-\left(D(u_i) \times D(o_j)\right)^{\alpha}}} \tag{1}$$

where R_{ij} is the rating of user u_i for o_j; $D(u_i)$ and $D(o_j)$ are the degrees of u_i and o_j in the bipartite graph, respectively; and $\alpha \in (0, 1)$ is an adjustable parameter called the *influence factor*.

Using Eq. 1, the weights are first calculated from the users to the items, and then calculated from the items back to the users. Thus, a user-item *energy matrix* $E = \{e_{ij}\}_{m \times n}$ can be constructed, where e_{ij} is the energy that u_i allocates back from o_j.

Based on the energy matrix, the similarity between user u_i and target user u_t can be calculated as follows:

$$sim(u_i, u_t) = \frac{\sum\limits_{o_j \in (R(u_i) \cap R(u_t))} \frac{e_{ij} \times e_{tj} \times \varphi_{it,j}}{D(o_j)}}{D(u_i)} \tag{2}$$

where $R(u_i)$ and $R(u_t)$ are the sets of items that are rated by u_i and u_t. In addition, $\varphi_{it,j}$ is the *similarity coefficient* of u_i and u_t with respect to o_j and is defined by

$$\varphi_{it,j} = 1 - \frac{|R_{ij} - R_{tj}|}{R_{max} - R_{min}} \tag{3}$$

where R_{max} and R_{min} are the maximum and minimum of all ratings, respectively.

In this way, the predicted rating of the target user u_t for item o_j is defined by

$$R_{tj} = \overline{R_{u_t}} + \frac{\sum\limits_{R_{ij} \neq 0} sim(u_i, u_t) \times e_{ij} \times (R_{ij} - \overline{R_{u_i}})}{\sum\limits_{R_{ij} \neq 0} sim(u_i, u_t)} \tag{4}$$

where $\overline{R_{u_t}}$ and $\overline{R_{u_i}}$ are the average ratings of u_t and u_i, respectively.

3.3   Recommendation List Modification Using Logistic Regression

Logistic regression is a statistical method for analyzing a dataset in which there are one or more independent variables that determine an outcome. The outcome is measured with a binary variable, that is, there are only "yes or no" responses. After the initial recommendation list is generated using the bipartite graph, logistic regression is exploited to reorder the recommendation list. Thus, the recommendation results are not only accurate, but also can enable users to easily select their favorite items from the first few positions in the recommendation list.

Let $DB = \{(x_i, y_i)\}$, $i = 1, 2, \ldots, n$, $x_i \in R^D$, be a data set, $x_i = \left(x_i^{(1)}, x_i^{(2)}, \ldots, x_i^{(D)}\right)$ be the input vector, and $y_i \in \{0, 1\}$ be the class label. Then, the probability model of logistic regression is as follows:

$$p(y|x, w) = (h_w(x))^y (1 - h_w(x))^{1-y} \tag{5}$$

where $w = (w_0, w_1, \ldots, w_D)$ is the weight vector of the features to be learned, where $h_w(x)$ is determined by

$$h_w(x) = g(w^T x) = \frac{1}{1 + e^{-w^T x}} \tag{6}$$

Assuming that the training samples are independent of each other, the likelihood of maximizing the training set of the logistic regression is

$$\prod_{i=1}^{N} (h_w(x_i))^{y_i} (1 - h_w(x_i))^{1-y_i} \tag{7}$$

This problem is equivalent to minimizing the negative likelihood probability or the following form of the entropy error function:

$$
\begin{aligned}
E(w) &= -ln \prod_{i=1}^{N} (h_w(x_i))^{y_i} (1 - h_w(x_i))^{1-y_i} \\
&= - \sum_{i=1}^{N} (y_i ln(h_w(x_i)) + (1 - y_i)ln(1 - h_w(x_i)))
\end{aligned} \tag{8}
$$

Thus, weights can be found using the partial derivative with respect to w in Eq. 8. Let u_t be the target user, each record in $DB_i = (x_i, y_i)$ is a binary vector representing the preference on item o_i by u_t. For $x_i = \left(x_i^{(1)}, x_i^{(2)}, \ldots, x_i^{(D)}\right)$, when o_i has feature j, $x_i^{(j)} = 1$; otherwise, $x_i^{(j)} = 0$. For the training set, Eq. 5 is used to calculate a value y'_i corresponding to item o_i, and y_i is determined by

$$y_i = \begin{cases} 1, & \text{if } y'_i \geq \bar{y} \\ 0, & \text{if } y'_i < \bar{y} \end{cases} \tag{9}$$

Let o_j and o_k be two items corresponding to two records DB_j and DB_k; if $y_j = 1$ and $y_k = 0$, o_j will be rearranged to a position above o_k in the final recommendation list; if $y_j = y_k$, the positions of the two items will be unchanged in the final recommendation list.

3.4 Algorithm Description

Based on the above discussion, the proposed RBGLR method is described in Algorithm 1.

Algorithm 1	RBGLR
Input	Raw data set, target user u_t, low frequency item threshold T, influence factor α, neighbor similarity threshold S
Output	Set of recommendation items
1	Delete items that have a frequency lower than T;
2	Construct a user-item bipartite graph;
3	Calculate the weights of the edges in the bipartite graph using Eq. 1;
4	**for** each user $u_i (i \neq t)$ **do**
5	Calculate the similarity between u_i and u_t according to Eq. 2;
6	**if** $sim(u_i, u_t) \geq S$ **then**
7	**for** each o_j that are not rated by u_t **do**
8	Calculate the predicted rating of u_t to o_j according to Eq. 4;
9	**end for**
10	**end if**
11	**end for**
12	Recommendation list L is formed from the top-n items with the highest predicted ratings of u_t, and the items in L are arranged in descending order of predicted rating;
13	Classify items in L using logistic regression;
14	Reorder the items in L according to the classification result to obtain a new recommendation list L';
15	Recommend L' to u_t.

4   Experimental Results

4.1   Datasets

The hetrec2011-movielens-2 k dataset (http://ir.ii.uam.es/hetrec2011/datasets.html) was used for the evaluation. In this dataset, there are 2,113 users and 10,197 movies. In addition, 70% of the data was used as a training set, 20% of the data was used as a test set, and 10% of the data was used to modify the logistic regression model.

4.2   Evaluation Parameters

The experimental results were evaluated from the perspectives of accuracy and diversity. Accuracy was evaluated by the mean absolute error (MAE), root mean squared error (RMSE), and average rank score (ARS); while diversity was evaluated using the Hamming distance (HD).

The MAE and RMSE are the most common parameters for evaluating RSs. They both evaluate the accuracy of the recommendation results using the difference between the users' predicted ratings and the actual ratings for the same items, and are expressed as

$$\mathrm{MAE} = \frac{\sum_{i=1}^{n} |p_i - q_i|}{n} \tag{10}$$

$$RMSE = \sqrt{\frac{\sum_{i=1}^{n} (p_i - q_i)^2}{n}} \tag{11}$$

where $\{p_1, p_2, ..., p_n\}$ is the set of predicted ratings, $\{q_1, q_2, ..., q_n\}$ is the set of actual ratings, and n represents the number of items.

The ARS evaluates the prediction accuracy based on the position of the items that are actually selected by users in the recommendation list. Assuming that user u_i actually selects item o_j and the position of o_j in u_i's recommendation list L_i is p_j, the relative position of o_j in L_i is defined as follows:

$$RP_{ij} = p_j/LEN \tag{12}$$

where LEN is the length of L_i. When more items near the head of recommendation list are actually selected by the user, namely those at lower relative positions, the recommendation results are considered to be more accurate. The average rank score for all users is defined as

$$ARS = \frac{\sum_{i=1}^{m} \sum_{j=1}^{n} RP_{ij}}{m \times n} \tag{13}$$

Smaller values of the ARS indicate that the overall recommendation results are more accurate.

The HD evaluates the diversity of prediction results based on the number of common items in different users' recommendation lists. The HD between users u_i and u_j is

$$H_{ij} = 1 - |O_{ij}|/LEN \tag{14}$$

where O_{ij} represents the set of items common to the recommendation list of u_i and the recommendation list of u_j, $|O_{ij}|$ represents the number of elements in O_{ij}, and LEN represents the length of the recommendation list. If the two recommendation lists are identical, then $H_{ij} = 0$; if the two recommendation lists do not have any of the same items, then $H_{ij} = 1$. The average HD for all users is the HD of the entire system, calculated as follows:

$$HD = \frac{\sum_{i \neq j} H_{ij}}{m \times (m-1)} \tag{15}$$

where m is the number of users. Larger HD values indicate that the recommendation results are more diverse.

4.3   Parameters Settings

As shown in Algorithm 1, the low-frequency item threshold T, influence factor α, and neighbor similarity threshold S are the parameters of the RBGLR algorithm. In this regard, we use experiments to first determine the approximately optimal range of these three parameters, and then determine the final values through further refinement. The three sets of experiments given in this section are all experimental results within the approximate optimal range (e.g., T between 0.1 and 0.2).

Low Frequency Item Threshold

The algorithm first deletes some of the items that have very low frequencies to reduce the effects of these items on the recommendation results. Table 1 shows the changes in MAE and RMSE when threshold T is varied.

Table 1. Changes in MAE and RMSE for different values of T

	0.10	0.12	0.14	0.16	0.18	0.20
MAE	0.427	0.384	0.323	0.335	0.340	0.361
RMSE	0.799	0.925	1.018	1.070	1.110	1.220
Mean	0.613	0.655	0.671	0.703	0.725	0.791

Table 1 shows that when the values of T change from 0.10 to 0.20, the value of MAE gradually decreases at first, reaching a minimum when $T = 0.14$, and then increases gradually. In contrast, the values of RMSE continue to increase. Here, the mean value is used to determine the optimal values of these two parameters. In this study, T was set to 0.10 because the mean value of MAE and RMSE in Table 1 is the minimum at this threshold.

Influence Factor

Table 2 shows the changes in MAE and RMSE for different values of α. It can be seen from Table 2 that when the influence factor is 0.01, the mean value of MAE and RMSE is minimized, so α is set to 0.01 in the experiments.

Table 2. Changes of MAE and RMSE for different values of α

	0.002	0.004	0.006	0.008	0.01
MAE	0.372	0.372	0.371	0.371	0.371
RMSE	1.227	1.227	1.227	1.227	1.226
Mean	0.7995	0.7995	0.799	0.799	0.7985

Neighbor Similarity Threshold

Neighbor similarity threshold S is used to appropriately ignore the neighbors with lower similarities in the experiments. The setting of S can improve both the accuracy and efficiency of the algorithm. Table 3 shows the changes in MAE and RMSE for different values of S. Similarly, $S = 0.15$ in the subsequent experiments.

Table 3. Changes in MAE and RMSE for different values of S

	0.11	0.12	0.13	0.14	0.15	0.16	0.17
MAE	0.376	0.367	0.358	0.367	0.307	0.342	0.364
RMSE	1.200	1.105	1.063	0.992	0.856	0.867	0.911
Mean	0.788	0.736	0.711	0.680	0.582	0.605	0.638

4.4 Experimental Comparison

We evaluated the performance of our RBGLR algorithm and compared it with collaborative filtering (CF) [12], a recommendation algorithm based on a bipartite graph (BG) [16], and a recommendation algorithm based on a weighted bipartite graph (WBG) [8]. The results are compared in Table 4.

Table 4. Comparison of the recommendation results of different algorithms

	RBGLR	CF	BG	WBG
MAE	0.301	1.440	1.131	0.675
RMSE	0.812	1.801	1.768	0.892
ARS	0.421	0.602	0.590	0.581
HD	0.890	0.703	0.857	0.912

Table 4 shows that the proposed RBGLR algorithm obtains the best results with respect to the three accuracy evaluation parameters MAE, RMSE and ARS. According to Eq. 13 in Sect. 4.2, when more items near the head of the recommendation list are selected by users, the ARS is smaller. Therefore, the evaluation based on ARS reveals that by adjusting the order of the items in the recommendation list by logistic regression, RBGLR makes it easier for users to find their favorite items from among the first few positions in the recommendation list.

In addition, the value of HD obtained by RBGLR is higher than those of CF and BG. This shows that the RBGLR algorithm can recommend more diverse results while taking into account accuracy.

According to Table 4, although the RBGLR algorithm is superior to WBG with respect to accuracy, its diversity parameter HD is slightly inferior to that of WBG. To this end, we define the balance factor (BF) to further compare these two methods.

$$BF = Accuracy/HD \qquad (16)$$

Here, $Accuracy \in$ {MAE, RMSE, ARS} and can only take the value of one of these three parameters at the same time.

It can be seen from Sect. 4.2 that, in a good recommendation method, the three accuracy parameters (MAE, RMSE and ARS) tend to be low, while the diversity parameter HD tends to be high. Thus, a smaller BF, indicates that the recommendation results are good. Figure 5 compares the BFs of RBGLR and WBG.

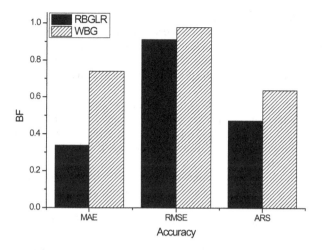

Fig. 5. Comparison on balance factors of RBGLR and WBG

As Fig. 5 shows, whether *Accuracy* uses MAE, RMSE, or ARS, the proposed RBGLR algorithm has BFs that are better than those of WBG. Especially for the most commonly used MAE, the BF of RBGLR is only 45.70% that of WBG. This indicates that the RBGLR algorithm properly balances the accuracy and diversity of the recommendation results thanks to its combination of a bipartite graph and logistic regression.

5   Conclusion

A new hybrid recommendation algorithm called RBGLR that is based on a bipartite graph was proposed. In contrast to the previous related methods, the RBGLR algorithm not only defines the weights and similarity of users, but also adjusts the order of the items in the recommendation list through logistic regression. The BF proposed in this study can be used to determine the trade-off between the accuracy and diversity of the recommendation results. Experimental results show the effectiveness and superiority of the proposed algorithm.

Acknowledgments. This work was partially supported by the Great Wall Scholar Program (CIT&TCD20190305), High Innovation Program of Beijing (2015000026833ZK04), and Beijing Urban Governance Research Center.

References

1. Aggarwal, C.C.: Recommender Systems. Springer, Cham (2016). https://doi.org/10.1007/978-3-319-29659-3
2. Dahdouh, K., Dakkak, A., Oughdir, L., Ibriz, A.: Large-scale e-learning recommender system based on Spark and Hadoop. J. Big Data **6**(1), 2 (2019)

3. Dong, X., Yu, L., Wu, Z., Sun, Y., Yuan, L., Zhang F.: A hybrid collaborative filtering model with deep structure for recommender systems. In: Proceedings of the Thirty-First AAAI Conference on Artificial Intelligence, pp. 1309–1315. AAAI, Palo Alto (2017)
4. Gu, Y., Zhao, B., Hardtke, D., Sun, Y.: Learning global term weights for content-based recommender systems. In: Proceedings of the 25th International Conference on World Wide Web, pp. 391–400. ACM, New York (2016)
5. Guo, G., Zhang, J., Thalmann, D., Yorke-Smith, N.: Leveraging prior ratings for recommender systems in e-commerce. Electron. Commer. R. A. **13**(6), 440–455 (2014)
6. Huang, L., Wang, C.-D., Chao, H.-Y., Lai, J.-H., Yu, P.S.: A score prediction approach for optional course recommendation via cross-user-domain collaborative filtering. IEEE Access **7**, 19550–19563 (2019)
7. Huang, Z., Chen, H., Zeng, D.: Applying associative retrieval techniques to alleviate the sparsity problem in collaborative filtering. ACM T. Inform. Syst. **22**(1), 116–142 (2004). https://doi.org/10.1145/963770.963775
8. Li, Z.-D., Luo, Q., Shi, L.L.: Weighted bipartite network recommendation based on increasing similarity coefficient. Comput. Sci. **43**(7), 259–264 (2016). (in Chinese)
9. Liu, J.-G., Wang, B.-H., Guo, Q.: Improved collaborative filtering algorithm via information transformation. Int. J. Mod. Phys. C **20**(2), 285–293 (2009)
10. Mahboob, V.A., Jalali, M., Jahan, M.V., Barekati, P.: Swallow: resource and tag recommender system based on heat diffusion algorithm in social annotation systems. Comput. Intell. **33**(1), 99–118 (2017)
11. Najafabadi, M.K., Mohamed, A.H., Mahrin, M.N.: A survey on data mining techniques in recommender systems. Soft. Comput. **23**(2), 627–654 (2019)
12. Schafer, J.B., Frankowski, D., Herlocker, J., Sen, S.: Collaborative filtering recommender systems. In: Brusilovsky, P., Kobsa, A., Nejdl, W. (eds.) The Adaptive Web. LNCS, vol. 4321, pp. 291–324. Springer, Heidelberg (2007). https://doi.org/10.1007/978-3-540-72079-9_9
13. Soares, M., Viana, P.: Tuning metadata for better movie content-based recommendation systems. Multimedia Tools Appl. **74**(17), 7015–7036 (2015). https://doi.org/10.1007/s11042-014-1950-1
14. Tarus, J.K., Niu, Z., Kalui, D.: A hybrid recommender system for e-learning based on context awareness and sequential pattern mining. Soft. Comput. **22**(8), 2449–2461 (2018)
15. Trudeau, R.J.: Introduction to Graph Theory. Dover Publications, New York (2013)
16. Wang, Q., Duan, S.-Y.: Improved recommendation algorithm based on bipartite networks. Appl. Res. Comput. **30**(3), 771–774 (2013). (in Chinese)
17. Zhou, T., Ren, J., Medo, M., Zhang, Y.-C.: Bipartite network projection and personal recommendation. Phys. Rev. E **76**(4), 046115 (2007)

Academic Paper Recommendation Based on Clustering and Pattern Matching

Jinpeng Chen and Zhijie Ban[(✉)]

School of Computer Science, Inner Mongolia University, Hohhot 010021, China
banzhijie@imu.edu.cn

Abstract. With the rapid growth of the scholarly literature, finding relevant and influential articles is becoming increasingly important. Research shows that a scholar's past works represent his latent interests. However, its effectiveness in recommending scholarly papers has not been well explored in the existing studies. In this paper, we propose an academic paper recommendation model, called CPM, which mainly mines researcher's published works for improving scholarly navigation. Firstly, a scholar's papers are divided into different interest points by clustering technologies. Then, scholar's information needs are represented in terms of pattern equivalence classes. Finally, matching degree and preference degree are integrated to rank the candidate papers. Experimental results on real datasets demonstrate that CPM outperforms (5.6% in terms of NDCG@5 and 8.1% in terms of MRR) the baseline method.

Keywords: Paper recommendation · K-means clustering ·
Pattern equivalence class

1 Introduction

Recently, the fast growing repository of scientific literature has posed many challenging issues in literature management [1]. Preliminary statistics show that there are more than 1 million researchers, 3 million publications, and 32 million citation relationships in Computer Science [2]. Several systems have been developed for academic search, for instance DBLP, Google Scholar and Xueshu Baidu. However, it is still difficult for scholars to search relevant literature because the publications of matching their queries are largely irrelevant to their latent information needs. Academic paper recommendation is an effective technology to tackle this problem. It is mainly applied in the digital library, the paper-sharing platform and the academic social website.

Traditional academic recommendation models are based on user historical behaviors, such as publishing a paper, marking a paper as favorite, rating a paper, making a comment, and tagging a paper [3, 4]. But these approaches pay a high price for collecting user behavior information. Meanwhile, the construction of personalized recommendation model is a slow process. To overcome the limitation of behavior-based approaches, the methods based on citation network or co-author network have been used to calculate the correlation between documents [5–9]. However, a scholar's

© Springer Nature Singapore Pte Ltd. 2019
K. Knight et al. (Eds.): ICAI 2019, CCIS 1001, pp. 171–182, 2019.
https://doi.org/10.1007/978-981-32-9298-7_14

published papers offer the most specific and accurate description about his research interests. At the same time, it is easy to acquire a researcher's all publications without his participation. Sugiyama et al. [10–12] recognized the important user need and proposed the scholarly paper recommendation via user's recent research interests. Scholar's needs were commonly constructed by vector space model in these papers. But their methods ignore structural relationship between words. On the other hand, all papers of a scholar are used to build one interest model. However, in fact, a scholar's research theme has variety.

A scholar's interests can be diverse and he often has multiple research directions. For instance, Jiawei Han, a well-known academic expert, is not only interested in data mining but also database. Therefore, we propose to divide one scholar's published papers into different groups by clustering technologies. Each group represents the scholar's one interest point. Each interest point reflects one research direction of the researcher. On the other hand, it is impossible for a scholar to attention equally his research directions. Thus, we also define the preference degree of scholar's research interest to quantify his attention on one research direction.

We consider that pattern-based topic model is more accurate to represent topics than word-based representations (i.e. LDA) because the former includes structural relationship between words. In other words, the pattern-based topic consists of many word sequence patterns while word-based topic only consists words. For example, one topic of LDA is {machine, learning, artificial, intelligence, academic, network} while pattern-based topic can be expressed as {{machine learning}, {artificial intelligence}, {academic network}}. Therefore, it's obvious that pattern-based method is more distinctive to represent topics than single terms for describing papers. Moreover, the writing of academic papers is relatively formal and rigorous. Its idea or method is careful thought and conceived. As a result, its topics are more specificity so that frequent patterns can more exact express the technical terms than single word. Therefore, this paper models one scholar's need based on pattern equivalence class [13]. Furthermore, Gao et al. propose to select the most representative and discriminative patterns, which are called maximum matched patterns, to represent topics [13]. However, this method only emphasizes on the longest matched patterns while its sub patterns are ignored. These sub patterns also represent user's interest. So we propose to choose all matched patterns in pattern equivalence class to compute similarity between a candidate paper and scholar information needs.

In Sect. 2, we discuss the related work about some state-of-the-art academic recommendation models and related techniques. Section 3 presents the details of our proposed models. Then, extensive experiments on the proposed models and the baseline model have been conducted in Sect. 4. Finally, Sect. 5 concludes the whole work.

2 Related Works

Academic paper recommendation is one important branch of recommendation system, which helps scholars to find relevant papers in the tens of millions papers. Scholarly recommendation systems usually obtain user information needs from user historical behaviors because user's online behaviors and context information often indicate the user's interests. Wang et al. [3] proposed a method that models user historical behavior, which built preference model for each user through collecting the operations on scientific papers of online users and carrying on the detailed analysis. Then, the personalized recommendation model was constructed based on content filtering and statistical language model. Cui et al. [4] examined the use of tagging for research paper recommendation. This mechanism recommends a research paper according to a set of tags which researchers have labeled. He et al. [8] given context-aware citation recommendation, and built a context- aware citation recommendation prototype. This system is capable of recommending the bibliography to a manuscript and providing a ranked set of citations to a specific citation placeholder. Wang et al. [14] proposed the merits of traditional collaborative filtering and probabilistic topic modeling. It provided an interpretable latent structure for users and items, and could form recommendations about both existing and newly published articles. Some studies are based on the citation and co-author networks. Tang et al. [1, 15] presented a unified topic model to simultaneously model topical aspects of different objects in the academic network. Bolelli et al. [16] proposed an information theoretic approach towards measuring the significance of individual words based on the underlying link structure of the document collection. They generated a non-uniform weight distribution of the feature space which was used to augment the original corpus-based document similarities. Liang et al. [17] improved recommendation performance by defining local relation strength and global relation strength. Collaborative filtering has proven to be valuable for recommending items in many different domains. Mcnee et al. [5] explored the use of collaborative filtering to recommend research papers and the citation web between papers to create the ratings matrix. Specifically, they tested the ability of collaborative filtering to recommend citations that would be suitable additional references for a target research paper. Sugiyama et al. [10–12] considered the academic achievement of a scholar could embody the scholar interests. They proposed a generic scholarly recommendation model by analyzing scholar's past publications. Pan et al. [18] proposed a recommendation method with a heterogeneous graph in which both citation and content knowledge were included. Shah et al. [19] proposed a recommendation system using association mining over clustering. Wang et al. [20] proposed two paper recommendation algorithms by studying the community structure of the citation-collaboration network, which comprehensively considered several metrics such as textural similarity, author similarity, closeness, and influence. Son and Kim [21] proposed a method which was based on multilevel citation networks that compared all the indirectly linked papers to the paper of interest to inspect the structural and semantic relationships among them.

In summary, a scholar's published papers are very important for modeling the scholar's information needs. But the property isn't systematically explored in most existing studies. Therefore, this paper mainly studies the effect of a scholar's past works for recommending papers to the scholar.

3   Academic Paper Recommendation Based on Clustering and Pattern Matching

3.1   Overview

Figure 1 shows the architecture of our recommendation system, which includes three components. The first component partitions a researcher's published papers into different interest groups by clustering technology. It outputs a scholar's k interest sets. Each interest set includes the scholar some papers which have the same research theme. The second component constructs scholar need model based on pattern equivalence class. Papers in the same interest set are used to construct one LDA model. Enhanced patterns are generated by deleting repeating words from the LDA topics and mining frequent patterns. Closed itemsets and their generators form pattern equivalence class. Scholar need model is generated from many topic pattern equivalence classes. The third component computes each candidate paper's score by pattern matching degree and preference degree. The matching degree measures similarity between a candidate paper and scholar need model. The preference degree is a scholar's attention on one research direction.

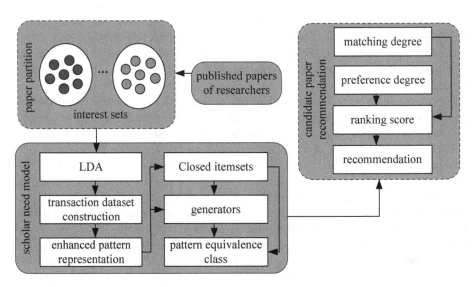

Fig. 1. The architecture of CPM

3.2   Partition of Researcher's Published Works

A scholar's papers are classified according to the documents' similarity between them. In other words, different paper groups are created by partitioning one scholar's past works. Suppose the scholar r has k studying interests, one paper is related to one research interest. We divide his papers into k nonempty subsets which are the pairwise disjoint subsets. Each subset is called one interest point, whose elements indicate the scholar's research interest. In order to partition the paper set PS of the scholar r, we construct the corresponding vector space model for every paper in the PS. Suppose the scholar r has published n papers, let $PS = \{X_1, X_2, ..., X_n\}$ where X_i represents the vector space model of the paper i and $X_i = (x_{i1}, x_{i2}, ..., x_{iM})$. In this paper, we measure the distance between elements using the following formula [22]:

$$d_{ij} = \sqrt{\sum_{m=1}^{M} (x_{im} - x_{jm})^2} \tag{1}$$

where d_{ij} is the distance between the paper i and the paper j, M is the vector dimensionality. We use k-means clustering method because it is easily realized and runs fast. Then, PS is divided into different groups and can be defined as:

$$PS = \bigcup_{i=1}^{k} IP_i, IP_i \neq \emptyset, IP_i \bigcap IP_j = \emptyset, 1 \leq i \neq j \leq k \tag{2}$$

where IP_i is an interest point and k is the interest number. Then, the scholar r has k interest points, each of which consists of some papers.

3.3   Scholar Need Model Based on Pattern Equivalence Class

This part mainly studies how to model one scholar's interests based on his published papers. Our basic principles are to mine frequent patterns from the papers and utilize these frequent patterns to build scholar need model. The frequent patterns refer to combined sequences of multiple different words. They may be phrases, terms, or even an inexplicable combination of words. Frequent pattern-based topics are better than word-based because frequent patterns are more able to distinguish two different papers than words. The followings are the modeling process of scholar needs:

Construct Transactional Dataset and Enhanced Pattern Representation. The LDA model can be used to produce the word-topic assignments of all papers in one scholar's interest point IP_i. The Fig. 2 (a) illustrates an example about four papers and three topics, where t_j ($j = 1, 2, ..., 11$) represents a word. It shows that every paper includes three topics and every topic includes some words.

Topic transactional datasets are constructed by removing repeated words in one topic of one paper and ignoring the sequence of words. For example, topic transactional dataset TD_3 equals to $\{\{t_2, t_7\}, \{t_1, t_{11}, t_{10}\}, \{t_1, t_7, t_{10}, t_{11}\}, \{t_1, t_{11}, t_{10}, t_2\}$ in the Fig. 2 (b) where TD_3 is generated from topic Z_3 in the Fig. 2(a).

Enhanced representation datasets denote one topic can be represented by a set of all frequent patterns whose support degree $SUPP(X)$ is greater than or equal to a given threshold. Figure 2(c) shows an example of enhanced representation datasets for TD3 when the threshold is 2.

Pattern Equivalence Class. For an enhanced transactional dataset, the construction of pattern equivalence class includes the following three steps [13]:

Mining Closed Itemset. An itemset y is a closed itemset if there no exists itemset g such that y is proper subset of g and their support degrees are equal. The Fig. 2(d) gives an example about closed itemsets.

Mining Generators of Closed Itemset. Suppose y is a closed itemset, an itemset g is a generator of y if g is proper subset of y and their support degrees are equal. The Fig. 2 (e) shows the generators of closed itemsets. For example, the generators of closed itemset $\{t_1, t_{11}, t_{10}\}$ include the patterns $\{t_{11}, t_{10}\}$, $\{t_1, t_{10}\}$ and $\{t_{10}\}$.

Mining Pattern Equivalence Class. One pattern equivalence class is the union of all generators of a closed itemset y and y. The Fig. 2(f) illustrates some examples about pattern equivalence classes. For example, $\{\{t_1\}, \{t_{11}\}, \{t_1, t_{11}\}\}$ is one pattern equivalence class. The frequency of patterns in an equivalence class is the same and indicates the statistical significance of pattern equivalence class. For example, the frequency of equivalence class $\{\{t_1\}, \{t_{11}\}, \{t_1, t_{11}\}\}$ in the Fig. 2(f) is 3 because their support degrees are 3 as is seen in the Fig. 2(c).

paper	Topic Z_1	Topic Z_2	Topic Z_3
p_1	t_1,t_3,t_5,t_1	t_1,t_8,t_{10}	t_2,t_7,t_2
p_2	t_2,t_4,t_6,t_2	t_7,t_9,t_1,t_9	t_1,t_{11},t_{10}
p_3	t_2,t_1,t_7,t_1	t_7,t_4,t_4,t_2	t_1,t_7,t_{10},t_{11}
p_4	t_2,t_7,t_8	t_9,t_8,t_1	t_1,t_{11},t_7,t_2

(a) an example of word-topic assignments

paper	TD_1	TD_2	TD_3
p_1	$\{t_1,t_3,t_5\}$	$\{t_1,t_8,t_{10}\}$	$\{t_2,t_7\}$
p_2	$\{t_4,t_6,t_2\}$	$\{t_7,t_9,t_1\}$	$\{t_1,t_{11},t_{10}\}$
p_3	$\{t_2,t_1,t_7\}$	$\{t_7,t_4,t_2\}$	$\{t_1,t_7,t_{10},t_{11}\}$
p_4	$\{t_2,t_7,t_8\}$	$\{t_9,t_8,t_1\}$	$\{t_1,t_{11},t_7,t_2\}$

(b) transaction datasets (TD) generated from (a)

support degree	enhanced representation $(supp(x) >= 2)$
supp=2	$\{t_2\}, \{t_{10}\}, \{t_2,t_7\}, \{t_1,t_7\}, \{t_1,t_{10}\},$ $\{t_{10},t_{11}\}, \{t_7,t_{11}\}, \{t_1,t_{11},t_{10}\}$
supp=3	$\{t_1\}, \{t_7\}, \{t_{11}\}, \{t_1,t_{11}\}$

(c) pattern enhanced representation of TD_3

closed itemsets
$\{t_1,t_{11}\}$, $\{t_7\}$, $\{t_1,t_{11},t_{10}\}$,
$\{t_2,t_7\}$, $\{t_1,t_7\}, \{t_7,t_{11}\}$

(d) closed itemsets generated from (c)

closed itemset	generators
$\{t_1,t_{11}\}$	$\{t_1\}, \{t_{11}\}$
$\{t_7\}$	
$\{t_1,t_{11},t_{10}\}$	$\{t_{11},t_{10}\}, \{t_1,t_{10}\}, \{t_{10}\}$
$\{t_2,t_7\}$	$\{t_2\}$
$\{t_1,t_7\}$	
$\{t_{11},t_7\}$	

(e) generators of closed itemset from (d)

equivalence classes
1 $\{\{t_1\}, \{t_{11}\}, \{t_1,t_{11}\}\}$
2 $\{\{t_7\}\}$
3 $\{\{t_1,t_{11},t_{10}\}, \{t_{11},t_{10}\}, \{t_1,t_{10}\}, \{t_{10}\}\}$
4 $\{\{t_2,t_7\}, \{t_2\}\}$
5 $\{\{t_1,t_7\}\}$
6 $\{\{t_{11},t_7\}\}$

(f) equivalence classes generated from (e)

Fig. 2. An example of constructing equivalence classes

Pattern Equivalence Class-Based Scholar Need Model. Let U_i is one equivalence class set which is generated from topic transactional dataset TD_i, U_i is defined by the following formula:

$$U_i = \{EC_{i1}, EC_{i2}, \ldots, EC_{iq}\} \tag{3}$$

where q is the number of pattern equivalence class and EC_{ij} ($j = 1,2,\ldots,q$) is one pattern equivalence class. Then, pattern equivalence class model PEC_i of interest point IP_i for one scholar r is defined by the following formula:

$$PEC_i = \{U_1, U_2, \ldots, U_m\} \tag{4}$$

where m is the topic number of IP_i. The scholar's need model ECM is defined by the following formula:

$$ECM = \{PEC_1, PEC_2, \ldots, PEC_k\} \tag{5}$$

where k is the number of the scholar's interest points.

3.4   Pattern Matching-Based Paper Ranking

Matching Degree. Matching degree measures the similarity between a candidate paper CP and scholar interest point. For a candidate paper CP, its model is $\{t_1, t_2, \ldots, t_n\}$ where t_i is one word. we compute the matching degree between CP and interest point model PEC_i using the following formula:

$$S(CP, PEC_i) = \sum_{i=1}^{m} \alpha_i f(CP, U_i) \tag{6}$$

where PEC_i has m topics. α_i is generated by the LDA model and is the topic distribution probability. U_i is equivalence class set that is generated from topic transactional database TD_i. $f(CP, U_i)$ is the similarity degree between CP and U_i. We consider all matched patterns which are included in PCE_i and CP. It is computed as follows:

$$f(CP, U_i) = \sum_{j=1}^{qi} \sum_{EC_{ij} \in U_i, u \in EC_{ij}, u \subseteq CP} |u|^{0.5} g(u) \tag{7}$$

where q_i is the number of equivalence class for the ith topic. EC_{ij} belongs to equivalence class set U_i. u is one element of equivalence class EC_{ij} and is also subset of CP. $g(u)$ is the statistical significance of EC_{ij}. $|u|$ is the length of the pattern u and $|u|^{0.5}$ means that its length is important for computing similarity degree.

Preference Degree. Suppose the interest number of scholar r is k, the preference degree about scholar r for interest point IP_i is defined by the following formula:

$$pd(IP_i) = \begin{cases} 1, & IP_i = IP_{main} \\ \frac{1}{1+e^{-\theta|IP_i|}}, & IP_i \neq IP_{main} \end{cases} \qquad (8)$$

where the interest point IP_{main} is the main interest point of the scholar r, which includes the scholar's newest published paper. θ is attention coefficient whose effect is to regulate the growth rate of preference degree and $\theta \in [0, 1]$. $|IP_i|$ is the paper number of IP_i. Formula (8) shows that one scholar generally pays more attention to his newest paper. If IP_i isn't main interest point, the value of preference degree increases with its size.

Paper Ranking. The basic steps are firstly to recommend n papers for every interest set. Then, kn papers are sorted according to their scores and the top n papers as the final recommendation results. The score of one candidate paper CP is computed by Eq. (9):

$$\text{score(CP)} = \text{pd}(IP_i) \cdot S(CP, PEC_i) \qquad (9)$$

The effect of preference degree $pd(IP_i)$ is to revise matching degree between interest point model and the candidate paper.

4   Experimental Results

This section discusses the experimental setup and results. The results show that our method significantly outperforms the baseline model for the three measures.

4.1   Experimental Setup

Data. The data of the experiments is based on the published papers of National University of Singapore SPRD (Scholar Paper Recommendation Dataset) [10]. This data has collected all of papers in the proceedings of ACL-ARC (Annual Meeting of the Association for Computational Linguistics-Anthology Reference Corpus) from 2000 to 2006 and our experiments is based on the senior dataset.

Evaluation Measures and Baseline Methods. We conduct various experiments to evaluate the effectiveness of the proposed approaches. Top-5 and Top-10 papers are respectively recommended for each senior scholar and we focus on NDCG [10] as the main evaluation index, MRR [10] as an auxiliary index. In all experiments, we conduct evaluations for all methods in terms of NDCG@5, NDCG@10 and MRR. For comparison purpose, we implement the baseline method BM [10]. We implement our two models CPM and CPM-PD. CPM integrates matching degree with preference degree for paper ranking. CPM-PD only uses matching degree during recommendation.

4.2   Results

Interest Point Number. In order to determine the number of scholar's interest, we compare NDCG@5, NDCG@10, and MRR values under the condition of different interests in the experiments. We set the interest number as integer and it is in the range of [1, 5]. The experimental results of CPM-PD are shown in the Fig. 3. CPM-PD has the same trend under three recommended measures as can be seen from the Fig. 3. NDCG@5, NDCG@10 and MRR first descend and then ascend with the increasing of interest point number. When the interest point number is 2, all measures reach the maximum value. We can conclude that the average clustering number of all scholars is 2 from this experiment.

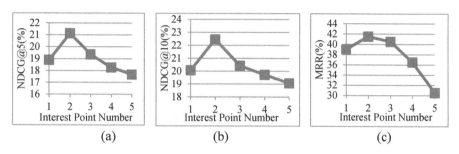

Fig. 3. Performance vs. interest point number

Preference Coefficient. According to the results of the above experiments, the interest point number is set as 2 in the experiment. NDCG@5, NDCG@10, and MRR is respectively compared under the condition of different preference coefficient. The experimental results of CPM are shown in the Fig. 4.

From the Fig. 4, we can see that the recommendation performance changes with the changing of preference coefficient. This indicates that preference coefficient has an important influence on the performance. Preference coefficient is determined by observing the changing trend of the performance curve. In other words, the three measures reach the maximum value when the coefficient is 0.9 as can be seen from the Fig. 4.

Performance Comparison. According to the results of the above experiments, interest point number is fixed to 2 and preference coefficient is set as 0.9. We compare the performance of different methods including BM, CPM-PD and CPM. The experimental results are shown in Table 1.

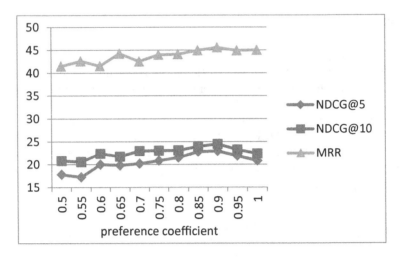

Fig. 4. Performance vs. preference coefficient

Table 1. Performance of different methods

Methods	NDCG@5(%)	NDCG@10(%)	MRR(%)
BM	17.4	21.4	37.5
CPM-PD	21.1	22.5	41.5
CPM	23.0	24.5	45.6

From the Table 1, we can see that our method CPM-PD outperforms the baseline model for all the three measures. The comparisons shows taking multiple interest points into consideration can generate more accurate scholar need model. This result also demonstrates that using the proposed pattern equivalence class to represent scholar's interest needs achieves better results than only using words (BM). At the same time, we also can see that our method CPM achieves better performance among three models, while our two methods CPM and CPM-PD consistently outperform BM. The improvement achieved by the CPM against the BM in terms of NDCG @5 and MRR is 5.6 and 8.1%, respectively. This indicates that preference degree is very important for accurate recommending academic papers. Therefore, we conclude that the CPM is an exciting achievement in representing the scholar's need model because it uses pattern-based topic model and clustering technology. Additional, matching strategy and preference degree also contributes to the accurate candidate paper ranking.

5 Conclusions

In this paper, a novel model based on clustering and pattern matching technologies has been proposed to recommend articles for researchers. In order to determine interest point, we cluster a scholar's papers. Then, the papers in the same set are incorporated to build a pattern equivalence class-based scholar need model. Finally, we integrate matching degree with preference degree to rank the candidate papers. Experimental results on one real dataset demonstrate the effectiveness of the proposed method.

As our future work, we plan to incorporate additional information such as the citation network, co-author network and user behavior to further improve recommendation performance.

Acknowledgements. The work is supported by the National Natural Science Foundation of China (No. 61662053).

References

1. Tang, J., Zhang, J., Jin, R.: Topic level expertise search over heterogeneous networks. Mach. Learn. J. **82**, 99–110 (2011)
2. Tang, J., Yao, L., Zhang, D., Zhang, J.: A combination approach to web user profiling. ACM Trans. Knowl. Discovery Data **5**(1), 1–38 (2010)
3. Wang, Y., Liu, J., Dong, X., Liu, T., Huang, Y.: Personalized paper recommendation based on user historical behavior. In: Zhou, M., Zhou, G., Zhao, D., Liu, Q., Zou, L. (eds.) NLPCC 2012. CCIS, vol. 333, pp. 1–12. Springer, Heidelberg (2012). https://doi.org/10.1007/978-3-642-34456-5_1
4. Cui, T., Tang, X., Zeng, Q.: Usage of tagging for research paper recommendation. In: The 3rd International Conference on Advanced Computer Theory and Engineering, 2:439-2:442, IEEE (2010)
5. Mcnee, S.M., Albert, I., Cosley, D., et al.: On the recommending of citations for research paper. In: The 2002 ACM Conference on Computer Supported Cooperative Work, pp. 116–125. ACM (2002)
6. Gori, M., Pucci, A.: Research paper recommender systems: a random-walk based approach. In: The IEEE/WIC/ACM International Conference on Web Intelligence, pp. 778–781. IEEE (2006)
7. Chen, C.-H., Mayanglambam, S.-D., Hsu, F.Y., et al.: Novelty paper recommendation using citation authority diffusion. In: The 2011 International Conference on Technologies and Applications of Artificial Intelligence, pp. 126–131. IEEE (2011)
8. He, Q., Pei, J., Kifer, D., et al.: Context-aware citation recommendation. In: The 19th International Conference on World Wide Web, pp. 421–430 (2010)
9. Xue, H., Guo, J., Lan, Y., et al.: Personalized paper recommendation in online social scholar system. In: IEEE/ACM International Conference on Advances in Social Networks Analysis and Mining, pp. 612–619. IEEE (2014)
10. Sugiyama, K., Kan, M.Y.: Scholarly paper recommendation via user's recent research interests. In: The 10th Annual Joint Conference on Digital Libraries, pp. 29–38. ACM (2010)
11. Sugiyama, K., Kan, M.Y.: Serendipitous recommendation for scholarly papers considering relations among researchers. In: The 11th Joint International Conference on Digital Libraries, pp. 307–310. Canada (2011)
12. Sugiyama, K., Kan, M.Y.: Exploiting potential citation papers in scholarly paper recommendation. In: The 13th ACM/IEEE Joint Conference on Digital Libraries, pp. 153–162 (2013)
13. Gao, Y., Xu, Y., Li, Y.: Pattern-Based topics for document modelling in information filtering. IEEE Trans. Knowl. Data Eng. **27**(6), 1629–1642 (2015)

14. Wang, C., Blei, D.M.: Collaborative topic modeling for recommending scientific articles. In: The 17th ACM SIGKDD International Conference on Knowledge Discovery and Data Mining, pp. 448–456 (2011)
15. Tang, J., Zhang, J., Yao, L., et al.: ArnetMiner: extraction and mining of academic social networks. In: The International Conference on World Wide Web, China, pp. 990–998 (2008)
16. Bolelli, L., Ertekin, S., Giles, C.L.: Clustering scientific literature using sparse citation graph analysis. In: Fürnkranz, J., Scheffer, T., Spiliopoulou, M. (eds.) PKDD 2006. LNCS (LNAI), vol. 4213, pp. 30–41. Springer, Heidelberg (2006). https://doi.org/10.1007/11871637_8
17. Liang, Y., Li, Q., Qian, T.: Finding relevant papers based on citation relations. In: Wang, H., Li, S., Oyama, S., Hu, X., Qian, T. (eds.) WAIM 2011. LNCS, vol. 6897, pp. 403–414. Springer, Heidelberg (2011). https://doi.org/10.1007/978-3-642-23535-1_35
18. Pan, L., Dai, X., Huang, S., Chen, J.: Academic paper recommendation based on heterogeneous graph. In: Sun, M., Liu, Z., Zhang, M., Liu, Y. (eds.) CCL 2015. LNCS (LNAI), vol. 9427, pp. 381–392. Springer, Cham (2015). https://doi.org/10.1007/978-3-319-25816-4_31
19. Shah, J.M., Sahu, L.: Recommendation based on clustering and association rules. In: IJARIIE-ISSN(O)-2395–4396, vol.1, no. (2) 2015
20. Wang, Q., Li, W., Zhang, X., Lu, S.: Academic paper recommendation based on community detection in citation-collaboration networks. In: Li, F., Shim, K., Zheng, K., Liu, G. (eds.) APWeb 2016. LNCS, vol. 9932, pp. 124–136. Springer, Cham (2016). https://doi.org/10.1007/978-3-319-45817-5_10
21. Son, J., Kim, S.B.: Academic paper recommender system using multilevel simultaneous citation networks. Decis. Support Syst. **105**, 24–33 (2018)
22. Fan, J., Xie, W.: Distance measure and induced fuzzy entropy. Fuzzy Sets Syst. **1104**(2), 305–314 (1999)

The Methods for Reducing the Number of OOVs in Chinese-Uyghur NMT System

Mewlude Nijat, Askar Hamdulla$^{(\boxtimes)}$, and Palidan Tuerxun

School of Information Science and Engineering,
Xinjiang University, Urumqi 830046, China
askar@xju.edu.cn

Abstract. Recently, neural machine translation (NMT) has made significant achievements in multiple language pairs, and surpassed traditional statistical machine translation. However, NMT has strict restrictions on vocabulary, which leads to out-of-vocabulary (OOV) problems. Agglutinative languages, such as Uyghur, which are rich in morphological changes, theoretically have unlimited vocabulary, and we confront with serious OOV problems with these languages in NMT. In a quest of how to reduce OOVs in Chinese - Uyghur pair NMT, we present two different solutions on this study: In the first solution, with regard of the key feature of agglutinative languages, the declension, we segment Uyghur words into stems and affixes. The NMT test we ran on the stem-affix segmented data showed that the number of OOVs reduced from an original 1526 to only 121. In the second solution, we ran a Similar Word Replacement test on low-frequency words from Chinese corpus after training and achieved an even more reduced OOV result of 98. The mass reduction of OOVs from 1.5 thousand to only a hundred signifies the effectiveness of the solutions in this study.

Keywords: Machine translation ·
Chinese-Uyghur machine translation ·
Neural machine translation (NMT)

1 Introduction

Machine Translation is a sub-field of computational linguistics that investigates the use of software to translate text or speech from one language to another. The strict restrictions of neural machine translation on vocabulary leads to out-of-vocabulary (OOV) problems. Agglutinative languages like Uyghur, which are rich in morphological changes, theoretically have unlimited vocabulary, and we confront with serious OOV problems with these languages in NMT. In Chinese – Uyghur pair NMT, OOVs exist for several reasons:

(1) There is basically no morphological change in Chinese, but Uyghur is an agglutinative language, in which new words made by attaching affixes to the stem.

© Springer Nature Singapore Pte Ltd. 2019
K. Knight et al. (Eds.): ICAI 2019, CCIS 1001, pp. 183–195, 2019.
https://doi.org/10.1007/978-981-32-9298-7_15

(2) In theory, Uyghur have infinite vocabulary, but unfortunately, digital language resources are short, and it's challenging because both of the statistical machine translation and NMT rely on large-scale data sets.
(3) Chinese is SVO language, while Uyghur is non-strict SOV, the different typological systems the two languages belong increase the difficulty of word alignment.
(4) Because of high computational complexity of the of NMT decoding, the vocabulary size is limited, generally around 30,000–50,000 is considered ideal. The vocabulary limitation which doesn't cope with the rich morphological change, always results in unknown words (UNK) in translation.

In order to solve to this issue of reducing the number of OOVs in Chinese-Uyghur NMT, we present two different solutions on this study: (1) With regard of the key feature of agglutinative languages, we segmented Uyghur words into stems and affixes, and compare the original data to the segmentation one; (2) we ran a Similar Word Replacement test on low-frequency words from Chinese corpus using the open source word vector tool word2vec. The experimental results showed effectiveness and practicality of the solutions.

2   Machine Translation of Aggulinative Language

Uyghur is an agglutinative language, and it is challenging to reach a high accuracy in machine translation from a non-agglutinative language (such as Chinese) into an agglutinative language. Thanks to the previous research that have been done on machine translation of other agglutinative languages, which we reference in our NMT study on Chinese – Uyghur pair. Since Chinese doesn't have morphological change, we face a problem of morpheme correspondence between source and target language. To solve this issue, paper [1] proposes a new combination of post-processing morphological prediction and morpheme-based translation. Paper [2] discusses analysis of the difficulties encountered in dealing with the machine translation of Malayalam due to it's agglutinative-ness, especially the verb forms; This analysis helps us to reduce the data sparsity of machine translation systems. The machine translation between Chinese and Mongolian also faces the same problems of morphological change and lack of data just like Chinese - Uyghur NMT did. Paper [3] proposes a method of combining the source-side syntax information with the target-end morphological information to simultaneously reduce the differences in word order and morphology. Paper [4] builds a Chinese - Mongolian Morphemes And Mongolian Morphemes - Mongolian Machine Translation System by dividing Mongolian into morphemes, and the experimental results show that this method can mitigate the impact of data sparseness and morphological differences on Chinese - Mongolian machine translation, it can also be used in translations of morphologically asymmetrical and low-resource language pairs. The researchers solved this problem that in the translation from agglutinative to non-agglutinative language due to differences in morphology and syntactic structure, as well as poor word alignment and data sparseness caused by shortage of corpus resources, in a variety of ways, including pre-processing, post-processing, sentence reordering, and part-of-speech tagging.

3   Memory-Augmented NMT

Memory-augmented translation (M-NMT) for enhanced memory is a model proposed by Zhang et al. [5]. The model consists of two parts: One is a typical
attention-based NMT module, the other is the memory module. The end result
is a combination of the outputs of the two parts. The simple working principle is as follows: First, the global memory in the model (the lower right corner
of Fig. 1) is used to record multiple possible translations corresponding to the
source words, the global memory here is obtained from the SMT vocabulary;
Secondly, based on the source words in each input sentence, dynamically select
the appropriate elements from the global memory to form local memory (local
memory, as shown in the right middle of Fig. 1); then, select the most likely
translation like NMT; finally, the target word in local memory may correspond
to multiple source words. In order to save memory, a merge operation (merged
memory, as shown in the upper right corner of Fig. 1) is proposed, combine
elements with the same target word into one element. The frame structure of
M-NMT is shown in Fig. 1. It consists of two parts: a typical attention-based
NMT module on the left and a memory module on the right. The end result is
a combination of the outputs of the two modules.

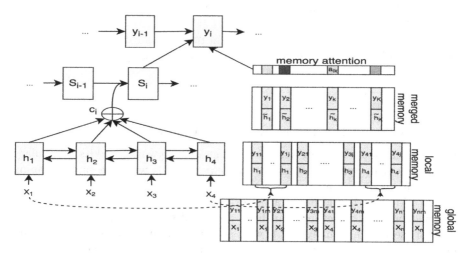

Fig. 1. M-NMT framework

The M-NMT model solves several problems. First, the word list from SMT
may contain unreasonable mappings, so they first remove unreliable mappings;
second, they propose a merge operation to save memory and load more target
words as much as possible; Third, the memory module and the neural model
module can be jointly trained. They train the neural model module and keep
it unchanged, and then train the memory module, so that step by step avoids
memory shortage and over-fitting. We used the M-NMT model that Feng et al.
implemented in Tensorflow framework, and reduced the parameters to improve
the model efficiency without affecting performance [5].

4   Uyghur Morphology

Uyghur is formed by 8 vowels and 24 consonants, for a total of 32 letters. The Uyghur alphabet is different from the writing rules of common languages such as Chinese and English. Uyghur letters are not only in one form, but each letter may have several different forms. For example, the stand-alone type usually appears at the end of a word. After the word can't be connected, the front-link appears before the letter can be connected, the middle-link appears in the middle of the two letters that can be connected, and the post-join appears after the letter can be connected [6].

Uyghur language can form new words by attaching affixes after stems. There are rich morphological changes and certain affixing rules. And the affixes of different parts of speech (such as verbs and names) are obviously different, and the number of verbs is more, so the combination is flexible, and the number of overlays is also high [7]. Uyghur as an agglutinative language has a high degree of flexibility in the combination of morphemes. Although the number of stems and affixes is limited, in theory, it can be combined to generate an infinite number of words. Most Uyghur words appear only once in the corpus. Uyghur language achieves rich syntactic and semantic functions by adding affixes to the stems [8]. As shown in Fig. 2, the same stem "school" has different meanings due to the addition of different affixes:

Therefore, the Uyghur language is rich in morphological changes, and there are theoretically unlimited number of words, which leads to more serious OOV problems in neural machine translation.

5   Word Vector

5.1   Synonymous in Chinese and Word Vector Representation

Chinese is one of the most developed languages in the world, and its vocabulary is extremely rich. Synonym refers to a group of words with the same meaning or similarity, which is a horizontal combination system of words reflecting a semantic. Modern Chinese has a large number of synonyms. One concept or same phenomenon sometimes can have dozens of synonyms. For example, the action of "seeing" has many words with similar meanings. Synonyms are group words whose rational meanings are basically the same, with a single auxiliary meaning or usage.

Synonyms are generally artificial dictionaries. And Word2vec is a deep learning tool that trains a dictionary-like model by training on a large amount of text. The semantic distance between two words (appearing in the training corpus) is obtained from the model. The shorter the similar distance, the shorter the distance between two synonyms or synonyms in the model. Word2vec is Google's open source tool used for calculating word vectors in 2013. The core of Word2vec is the word vector, that is, each word has a corresponding vector. When calculating the similarity of two words, it is actually calculating the cosine of two vectors. After the introduction of the word vector, two words with different shapes but similar meanings or related words can be identified.

Chinese	Uyghur	Stem + affix	English
学校	مەكتەپ	مەكتەپ	school
学校（复数）	مەكتەپلەر	مەكتەپ+لەر	schools
在学校	مەكتەپتە	مەكتەپ+تە	in school
学校的	مەكتەپىنىڭ	مەكتەپ+نىڭ	school's
往学校	مەكتەپكە	مەكتەپ+كە	to school
直到学校	مەكتەپىكىچە	مەكتەپ+كىچە	until school
从学校	مەكتەپىدىن	مەكتەپ+دىن	from school
...	...	...	

Fig. 2. Changes brought by different affixes

5.2   Google Word2vec

Word2vec is Google's open source tool used for calculation of word vectors in 2013, and can be effectively trained on hundreds of billions of words [9,10]. Through training, word2vec can simplify the text into a vector in K-dimensional space, and the space vector corresponds to the text semantics. Therefore, the output of word2vec - word vector can be used for related work in natural language processing, such as clustering, part of speech analysis, finding similarities, etc. Behind the word2vec algorithm is a shallow neural network, and when the word2vec algorithm or model is mentioned, it refers to the CBOW (Continuous Bag-Of-Words) model and the Skip-gram model used to calculate the word vector.

CBOW and Skip-Gram Models. The CBOW model predicts the probability of the current word based on the n words before and after, and the hidden layer sums the word vectors of the preceding and succeeding words, so the number of nodes is equal to the dimension of the word vector. Skip-gram, on the other hand, predicts the probability of n words before and after the current word, so the vector represents the context distribution. Both models use ANN as the classification algorithm. At the beginning, each word is a random N-dimensional vector. The best vector representation of each word can be obtained through training. The model is shown in Fig. 3 below:

The window size in the figure is 2. The CBOW model predicts the probability based on the first two words and the last two words of the current word, which is actually addition of context vector from the input layer to the hidden layer. Skip-Gram predicts the probability based on the current word. Each model introduces two optimization algorithms: Hierarchical Softmax and Negative Sampling.

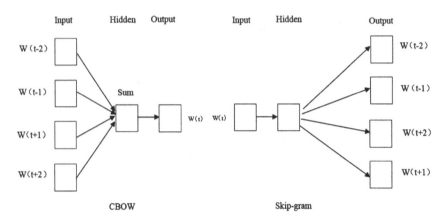

Fig. 3. CBOW & Skip-gram model

Hierarchical Softmax and Negative Sampling

Hierarchical SoftMax was first introduced into the language model by Bengio in 2005 [10]. It decomposes complex normalized probabilities into a series of conditional probability products. This will normalize the probability of V words into the probability fitting of logV words, reduces the computational complexity of the target probability, and increases the coupling between words. But Hierarchical SoftMax combined with Huffman Coding can meet the needs of the application. Negative sampling was originally derived from the algorithm of Noise-Contrastive Estimation [11], which aims to deal with the parameter estimation problem of probabilistic models that cannot be normalized. The algorithm transforms the likelihood function of the model. Experiments show that this down sampling technique can significantly improve the accuracy of word vectors in low frequency words.

6   Experiment

6.1   Experimental Data

Original Data Set

Since Chinese - Uyghur Machine Translation does not have a public data set, we built a parallel corpus, on which we conducted a preprocess due to the differences in the source, method of acquisition, the variety and the format of the corpus. The preprocessing of corpus mainly includes clauses, content filtering, handling of illegal characters, digital substitution, spell-check and proofreading. The specific preprocessing work is roughly as shown in Fig. 4:

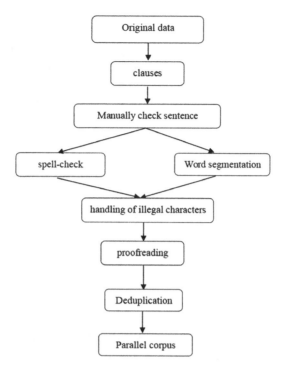

Fig. 4. Corpus preprocessing flow chart

After completing all the steps above, the total amount of data in the library finalized to be 180110 pairs. The entire corpus is divided into three data sets, the 178,110 sentence pairs in Training Set are used to train the translation model, the 1,000 sentence pairs in Development Set are used to adjust the model parameters, and the 1,000 pairs in Test Set is used for test the model. Each source sentence (Chinese) in the corpus corresponds to only one reference translation (Uyghur). In order to ensure the contrast of the experiment, the internal components of Moses are used to further standardize the text data, including the three steps of truecaser, truecasing and cleaning. And the sentence length is limited to 50 when cleaning. The statistics of the original data obtained after cleaning are shown in Table 1. The remaining sentence pairs in Training set is 177,944, and still 1,000 pairs in each of Development Set and the Test Set.

The length of the sentence pair is counted as shown in Fig. 5. Whether it is from the statistics in the table or from the pictures, most of the sentences in the corpus have little difference in the length of the sentences (that is, most Chinese and Uyghur sentences have similar numbers), which is beneficial to the system learning alignment.

Table 1. Original data set partitioning

Data set		Sent.	Words	Unique words
Training set	Chinese	177944	2535042	116675
	Uyghur		2661240	143189
Development set	Chinese	1000	14200	5384
	Uyghur		14920	6429
Test set	Chinese	1000	14235	5407
	Uyghur		15001	6431

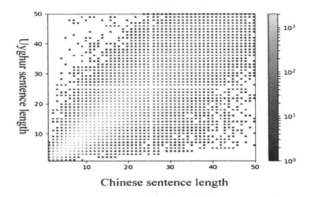

Fig. 5. Original data sentence length statistics

Experimental Data After Uyghur Stem Affixation

The Uyghur wording of "stem + affix" leads to the fact that most of the vocabulary itself is a rare word, which is easy to face data sparsity in natural language processing tasks. In NMT, OOVs are more likely to occur due to the vocabulary restrictions. In order to alleviate this phenomenon, we tried to segment the "stem + affix" form of the original data target vocabulary, which is the Uyghur vocabulary. Part of it is to keep the 10,000 words of the highest frequency unchanged, and to segment other words.

Segmentation will increase the total number of words and correspondingly increases the sentence length limit to 80 when cleaning. Statistics performed on the cleaned data resulted in 176,478 pairs for Training Set, 992 pairs for Development Set, and 988 pairs for Test Set. The statistical results are shown in Table 2 below:

Table 2. Statistics after segmentation

Data set		Sent.	Words	Unique words
Training set	Chinese	176478	2482210	115531
	Uyghur		4679916	36009
Development set	Chinese	992	13900	5296
	Uyghur		26160	2995
Test set	Chinese	988	13804	5246
	Uyghur		25928	2966

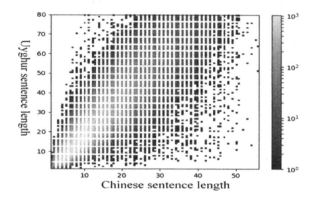

Fig. 6. Statistics of data sentence length after segmentation

The statistics of the length of the sentence pairs are shown in Fig. 6. It is found that the length of the target sentence is generally larger than the source.

Observing the data, it was found that although the total number of words on the target side increased a lot, the size of the Training Set vocabulary was reduced to one-third of the original data, and the size of the Development Set and the Test Set vocabulary was reduced by half. It is worth noting that when the vocabulary size is set to 30,000, it covers 93.6% of the source training corpus and 92.0% of the target training corpus in the original training data; After the segmentation, due to the increase in the length of the sentence, the computational complexity is increased, resulting in insufficient memory, so the vocabulary size is half of the original, which is 15,000. At this time, the source vocabulary covered 88.7% of the training corpus and the target end covered 99.5%.

6.2 Experiment and Analysis

In 2017, Feng et al. moved the Theano based model in the paper [12] to Tensorflow, and reduced the relevant parameters without affecting the performance, which improved the efficiency of the translation model and greatly reduced the training time [5]. The encoder and decoder used in the model in this paper are single-layer GRU units, and the encoder is a bidirectional RNN. The vocabulary size, the number of hidden layer units, and the number of output units are set to 30,000, 1000, and 500, respectively. The training batch size is set to 60, the optimization algorithm is Adam [13], and the learning rate is set to 0.0005. The decoding is based on the Beam Search Algorithm and the beam size is set to 12. Since the data set in the text is relatively small, the maximum number of updates is set to 300,000 times in this study (less than one-third of the number of updates in the paper [12]). The M-NMT model is a combination of NMT and memory structure. When the memory structure is trained, that is, memory attention, the NMT component remains unchanged. Therefore, the training and decoding settings are the same as NMT, and the interpolation factor is set to 1/3. The neural machine translation vocabulary size is set to 15,000. Then, the Chinese - Uyghur M-NMT comparison experiment was carried out with the original data and the segmented data under the same parameter settings. The experimental results are shown in Table 3 below:

Table 3. Experimental results for the original data and segmented data

Data	1-gram	2-gram	3-gram	4-gram	BLEU	OOV
M-NMT(unsegmented)	52.3	33.4	24.7	19.4	29.43	1526
M-NMT(segmented)	63.5	46.1	37.9	32.7	33.26	121

In order to further reduce the OOVs, the low-frequency Uyghur vocabulary in the corpus is segmented in the form of "stem + affix" beforehand. Then, the Chinese - Uyghur M-NMT comparison experiment of the original data and the segmentation data was performed under the same parameter settings. Looking at the original (unsegmented) experimental results, M-NMT combines the advantages of SMT and NMT, effectively reducing the problem of unregistered words and avoiding over-fitting to some extent. Looking at the experimental results after the word is segmented into "stem + affix", the OOVs are reduced from original 1526 to 121. Partial segmentation methods increase computational complexity and require more system memory, but the experiment proves that the method: (1) reduces the vocabulary size; (2) reduces the OOV problems; and (3) improves the translation results.

Obviously, the reduction in vocabulary size and training batches has a large effect on the experimental results. However, if the parameter settings are the same, the experimental results after segmentation are way better than the original data in both regards of faithfulness and fluency. If the model is not limited by system memory, the parameter settings can be better, and the translation results will obviously be better than the current values. All in all, partial segmentation of Uyghur vocabulary, while increasing computational complexity, does improve the quality of translation.

6.3   Word2vec Experiment and Analysis

The choice of model parameters is based on the current corpus. By analyzing the parameters, and after many attempts, this paper uses the 180,110 Chinese sentences from the corpus to train the CBOW model. Main parameter settings of the model as follows: The vector dimension size is set to 200, the learning rate alpha is 0.1, the frequency threshold min-count is set to 1, and the window size window is set to 5. In fact, there are many other adjustable parameters which can be configured according to corpus in actual operation. Then we partially replace the words in the source language (Chinese) Test Set that are not in the vocabulary, which are the words that are not in the NMT vocabulary. The specific steps are as shown in Fig. 7:

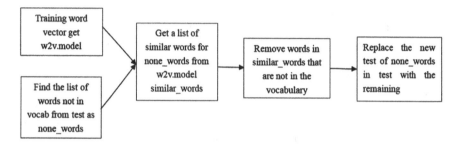

Fig. 7. Replacement process

The probability of the similarity of most words was found to be in the range of 0.3 0.4. The possible reasons are as follows: (1) the corpus in the text is too small for the training word vector; (2) What is needed in the text is the similar words to the low-frequency ones, and the tool itself performs poorly on low-frequency words. The translation results after the final test replacement have not been improved. The OOVs in the results were separately counted in order to analyze the effect of similar word substitution on the translation results of the M-NMT model. The results are shown in Table 4:

Table 4. Change in OOVs after similar words replacement

OOV on original data	OOV after the replacement of similar words	OOV reduction	OOV on segmented data	OOV after the replacement of similar words	OOV reduction
1443	1431	12	121	98	23

Although the quality of the word vector is not very high, the results of the comparison before and after the similar word substitution proved that the replacement operation can reduce the OOVs for both the original data and the segmented data. This presents the applicability of replacement method discussed in this paper. We performed similar word replacement experiment based on neural machine translation and used the open source word2vec tool to train the word vectors based on Chinese corpus in the text, then obtained a Chinese - Uyghur M-NMT result in which OOVs are much reduced. The experimental results shows the effectiveness and applicability of the replacement method.

7   Conclusion

A highly accurate Chinese - Uyghur machine translation is not an easy task. This is mostly due to the differences in morphological and syntactic structure between the two languages, and the morphological complexity of Uyghur itself. The statistical-based approach is currently the most commonly used method in Uyghur machine translation. Neural machine translation for Uyghur is also making improvements along with the development of corpus construction and deep learning.

M-NMT combines the advantages of SMT and NMT, by effectively reducing the OOV problems and avoiding over-fitting, to some extent. This experiment run a Chinese – Uyghur NMT test by segmenting the low-frequency vocabulary in the Uyghur corpus, then performing a comparison analysis on the original data and the partial segmentation data under the same parameter settings. The test results prove the applicability of the segmentation method, which significantly improved the translation quality. In the experiment, we used the open source word2vec tool to train the word vectors, and used similar word replacement method to replace the words that are not in the vocabulary with the similar ones from the Test Set, and obtained a result with massively reduced OOVs. Due to the agglutinating nature of Uyghur language, the affixation is the major barrier to overcome in NMT. Performing stem-affix segmentation can yet be regarded as a solution to reduce OOV issues that has always been more prominent. In addition to that, similar word replacement might be an ideal solution for OOV problems, but it still needs to train high quality word vector.

Acknowledgement. This work supported by Natural Science Foundation of China (61562081).

References

1. Clifton, A., Sarkar, A.: Combining morpheme-based machine translation with post-processing morpheme prediction. In: The 49th Annual Meeting of the Association for Computational Linguistics: Human Language Technologies, Proceedings of the Conference, Portland, Oregon, USA, 19–24 June, 2011, pp. 32–42 (2011). DBLP
2. Jayan, V., Bhadran, V.K.: Difficulties in processing malayalam verbs for statistical machine translation. Int. J. Artif. Intell. Appl. **6**(3), 13–24 (2015)

3. Chen, L., Li, M., Zhang, J., Zhu, Z., Yang, Z.: A statistical method for translating chinese into under-resourced minority languages. In: Shi, X., Chen, Y. (eds.) CWMT 2014. CCIS, vol. 493, pp. 49–60. Springer, Heidelberg (2014). https://doi.org/10.1007/978-3-662-45701-6_5
4. Yang, Z., Li, W., Chen, L., et al.: A Hanwen statistical machine translation method based on morpheme media. Chin. J. Inf. Sci. **31**(4), 57–62 (2017)
5. Feng, Y., Zhang, S., Zhang, A., et al.: Memory-Augmented Neural Machine Translation (2017)
6. Buick, G.A., Limit, M.A.: A preliminary study on the method of segmentation of uyghur words. Chin. J. Inf. Sci. **18**(6), 62–66 (2004)
7. Mieradilijiang, M.: Research on Uyghur stem extraction and part-of-speech tagging based on Morfessor. In: Xinjiang University
8. Abdukrem, H., Yang, L., Maosong, S.: Performance comparison of neural machine translation system in Uyghur-Chinese translation. J. Tsinghua Univ. (Sci. Technol.) **8**, 878–883 (2017)
9. Mikolov, T., Sutskever, I., Chen, K., Corrado, G., Dean, J.: Distributed Representations of Words and Phrases and their Compositionality, 17 October 2013. arXiv.org
10. Morin, F., Bengio, Y.: Hierarchical Probabilistic Neural Network Language Model. Aistats (2005)
11. Mnih, A., Kavukcuoglu, K.: Learning word embeddings efficiently with noise-contrastive estimation, pp. 2265–2273 (2013)
12. Bahdanau, D., Cho, K., Bengio, Y.: 2015 Neural machine translation by jointly learning to align and translate. In: ICLR (2015)
13. Kingma, D., Adam, B.J.: A Method for stochastic optimization. Comput. Sci. **410**(19), 1879–1902 (2014)

Machine Learning Algorithms

Joint Kernel Low-Rank Graph Construction and Subspace Learning

Xin Zhu[1], Qingxi Li[1], Hongyu Bian[2], and Yong Peng[1,3](✉)

[1] School of Computer Science and Technology, Hangzhou Dianzi University,
Hangzhou 310018, China
yongpeng@hdu.edu.cn
[2] Acoustic Science and Technology Laboratory, Harbin Engineering University,
Harbin 150001, China
bianhongyu@hrbeu.edu.cn
[3] Guangxi Key Laboratory of Multi-source Information Mining & Security,
Guangxi Normal University, Guilin 541004, China

Abstract. Subspace learning aims to retain the desirable data properties and reduce the data dimensionality by projecting high dimensional data into low dimensional subspace, which is a hot spot in machine learning and pattern recognition communities. In graph based subspace learning, the quality of constructed graph greatly affects the subsequent projection matrix learning. A common case is when data is noisy or even grossly corrupted, the constructed graph usually cannot well respect to the inner structure of data. Additionally, the widely used two-stage strategy which learns the projection matrix on a fixed data graph isolates the two closely related stages. To this end, we propose a joint kernel low-rank graph construction and subspace learning (KLGSL) model to alleviate the aforementioned disadvantages based on the theory of low-rank matrix recovery and spectral regression. In KLGSL, the kernel low-rank representation is used to characterize the possible nonlinear structure in data and the objectives of kernel low-rank representation and spectral regression co-evolves to optimum; therefore, the subspace learning can be efficiently performed on the recovered data contributed by the low-rank learning and it merges the two separated processes of graph construction and subspace learning into one whole. The KLGSL model objective can be efficiently optimized under the augmented Lagrange multiplier method framework. We evaluate the high performance of KLGSL by conducting extensive experiments on representative benchmark data sets and the results show that the low-rank learning can greatly facilitate the process of subspace learning.

Keywords: Kernel low-rank representation · Subspace learning ·
Joint learning framework · Robustness · Kernel spectral regression

1 Introduction

High dimensional data such as image, video frame and gene expressions usually contains a lot of redundancy information which might not be helpful to pattern

© Springer Nature Singapore Pte Ltd. 2019
K. Knight et al. (Eds.): ICAI 2019, CCIS 1001, pp. 199–211, 2019.
https://doi.org/10.1007/978-981-32-9298-7_16

recognition. Therefore, a more succinct representation of data is necessary to obtain by subspace learning methods which project data from a high dimensional feature space to a low dimensional subspace. Depending on the availability of labeled samples, the subspace learning methods can be roughly categorized into supervised, semi-supervised and unsupervised ones.

If having no or only a few labeled data samples, we usually explore the structure of data to facilitate the projection matrix learning. Among existing data structure exploration methods, graph is an effective way to characterize the data structure by denoting the samples with vertices and the connections between data pairs with weights. Besides the subspace learning, graph-based methods were widely used in diverse fields such as spectral clustering, feature selection, and semi-supervised learning. Though many graph-based learning methods obtained promising results, there still be some limitations within existing methods such as (1) most of graph methods construct the graph first and then perform learning tasks which leads to a two-stage process; (2) some of the graph construction methods are sensitive to noise which may not well adapt to diverse data sets; and (3) they missed considering the possible nonlinear structure involved in data.

To alleviate or partially solve the above mentioned limitations in graph-based subspace learning, in this paper, we combine the advantages from kernel low-rank representation (KLRR) [14,15,17,18] and spectral regression and propose a nonlinear joint graph construction and projection matrix learning model, term JKGSL, in which (1) the graph can be constructed via the representation coefficient matrix in KLRR to exploit the possible nonlinear structure of data; (2) the two sub-tasks corresponding to graph construction and projection matrix learning can co-evolve to optimum; and (3) the model is flexible due to the least squares formula to implement subspace learning in spectral regression.

The rest of this paper is organized as follows. In Sect. 2, we review and briefly describes the basic works including the spectral regression and the low-rank learning. In Sect. 3, we proposed our model formulation and optimization procedure. Experimental studies are presented in Sect. 4. And finally conclude this paper in Sect. 5.

2   Related Works

In this section, we give an introduction to two related works, spectral regression and low-rank learning.

2.1   Spectral Regression

From the perspective of graph-based subspace learning, supposing that $\mathbf{y} = [y_1, y_2, \cdots, y_n]^T$ is the map from the graph to the real line and $\mathbf{W} \in \mathbb{R}^{n \times n}$ is a symmetric matrix with w_{ij} as the weight of edge between vertices i and j and $\mathbf{D}$ is a diagonal matrix whose entries are column (or row, since $\mathbf{W}$ is symmetric) sums of $\mathbf{W}$, we usually solve the following objective [2]

$$\mathbf{y}^* = \arg \max_{\mathbf{y}^T \mathbf{D} \mathbf{y} = 1} \mathbf{y}^T \mathbf{W} \mathbf{y} = \arg \max \frac{\mathbf{y}^T \mathbf{W} \mathbf{y}}{\mathbf{y}^T \mathbf{D} \mathbf{y}} \tag{1}$$

and the optimal $\mathbf{y}$'s are the maximum eigenvectors of eigen-problem

$$\mathbf{Wy} = \lambda \mathbf{Dy}. \tag{2}$$

If we choose a function in reproducible kernel Hilbert space (RKHS), *i.e.*, $y_i = f(\mathbf{x}_i) = \sum_{j=1}^{n} \alpha_j k(\mathbf{x}_j, \mathbf{x}_i)$, $k(\mathbf{x}_j, \mathbf{x}_i)$ is the Mercer kernel of RKHS $\mathcal{H}_K$. (1) can be rewritten as [2]

$$\alpha^* = \arg\max \frac{\mathbf{y}^T \mathbf{Wy}}{\mathbf{y}^T \mathbf{Dy}} = \arg\max \frac{\alpha^T \mathbf{KWK}\alpha}{\alpha^T \mathbf{KDK}\alpha}, \tag{3}$$

where $\mathbf{K}$ is $m \times m$ gram matrix with $k_{ij} = k(\mathbf{x}_i, \mathbf{y}_j)$ and $\alpha = [\alpha_1, \cdots, \alpha_n]^T$. The optimal α's are the eigenvectors corresponding to the maximum eigenvalue of eigen-problem

$$\mathbf{KWK}\alpha = \lambda \mathbf{KDK}\alpha. \tag{4}$$

Theorem 1. *Let $\mathbf{y}$ be the eigenvector of eigen-problem (2) with eigenvalue λ. If $\mathbf{K}\alpha = \mathbf{y}$, then α is the eigenvector of eigen-problem in (4) with the same eigenvalue λ.*

Theorem 1 show that instead of solving the eigen-problem (4), the linear embedding functions can be acquired through two steps:

(1) Solve the eigen-problem in (2) to get $\mathbf{y}$.
(2) Find $\mathbf{K}\alpha = \mathbf{y}$. In practice, an exact solution may not exist. We need to do regularized kernel least square regression. That is, we need to find $\alpha \in \mathbb{R}^{n \times 1}$ which is the solution to the linear equation $(\mathbf{K} + \delta \mathbf{I})\alpha = \mathbf{y}$.
(3) Let $\Theta = [\alpha_1, \cdots, \alpha_{c-1}] \in \mathbb{R}^{n \times (c-1)}$ is transformation matrix. The samples can be embedded into $c - 1$ dimensional subspace by $\mathbf{x} \to \mathbf{z} = \Theta^T \mathbf{k}(:, \mathbf{x})$, where $\mathbf{k}(:, \mathbf{x}) \triangleq [k(\mathbf{x}_1, \mathbf{x}), \cdots, k(\mathbf{x}_n, \mathbf{x})]^T$.

It can be easily verified that function $f(\mathbf{x}) = \sum_{i=1}^{n} \alpha_i^k k(\mathbf{x}, \mathbf{x}_i)$ is the solution to the following regularized least square problem

$$\min_{f \in \mathcal{H}_K} \sum_{i=1}^{n} (f(\mathbf{x}_i) - y_i^k)^2 + \delta \|f\|_K^2, \tag{5}$$

where α_i^k is the i-th element of vector α_k.

The advantages of spectral regression are as follows

(1) The matrix $\mathbf{D}$ is guaranteed to be positive definite and therefore the eigen-problem in (1) can be stably solved. Moreover, both $\mathbf{L}$ and $\mathbf{D}$ are sparse matrices. The top c eigenvectors of eigen-problem (1) can be efficiently calculated with Lanczos algorithms.
(2) The technique to solve the least square problem is already matured and there exist many efficient iterative algorithms such as LQSR that can handle very large scale least squares problems.
(3) Since the regression was used as a building block, the regularization techniques can be easily incorporated and produce more stable and meaningful solutions, especially when there exists a large amount of features.

2.2   Low-Rank Learning

Generally, low-rank learning refers a group of strategies to constrain the variables of interest to be low-rank. The low-rank prior usually can be modeled in two different ways. One is to minimize the rank of the matrix of interest subject to some constraints. Since the exact rank minimization problem is NP-hard to handle, it is often relaxed by the corresponding tightest convex surrogate, the trace norm (or nuclear norm). The typical models belonging to this type are matrix completion , robust principal component analysis (RPCA) [1,5], low-rank representation (LRR) [10], and latent low-rank representation [11]. These models are often optimized under the augmented Lagrange multiplier method framework and the singular value thresholding operator. The alternative approach is to decompose the matrix of interest into a product of two factor matrices. Then, its rank is upper bounded by the lower rank of the two factor matrices. Therefore, the problem of recovering a low-rank matrix can be converted into estimating the two factor matrices. There are also some models in this type such as non-negative matrix factorization , maximum margin matrix factorization, fixed-rank representation [12], low-rank regression and low-rank matrix decomposition with rank cardinality upper bounded [8]. The alternating direction method is used to respectively update one variable while fixing the others till convergence is achieved. Below we give a more detailed description to the LRR model. Given a set of samples $\mathbf{X} = [\mathbf{x}_1, \mathbf{x}_2, \cdots, \mathbf{x}_n] \in \mathbb{R}^{d \times n}$, LRR aims to represent each sample as a linear combination of the bases in $\mathbf{A} = [\mathbf{a}_1, \mathbf{a}_2, \cdots, \mathbf{a}_m] \in \mathbb{R}^{d \times m}$, that is $\mathbf{X} = \mathbf{AZ}$, where $\mathbf{Z} = [\mathbf{z}_1, \mathbf{z}_2, \cdots, \mathbf{z}_n] \in \mathbb{R}^{m \times n}$ is the matrix with each $\mathbf{z}_i$ as the representation coefficient corresponding to sample $\mathbf{x}_i$. Therefore, each entry in $\mathbf{z}_i$ can be viewed as the contribution to the reconstruction of $\mathbf{x}_i$ with $\mathbf{A}$ as the dictionary. LRR seeks to find the lowest rank solution by solving the following optimization problem [10]

$$\min_{\mathbf{Z}} \text{rank}(\mathbf{Z}), \quad s.t. \ \mathbf{X} = \mathbf{AZ}. \tag{6}$$

It is NP-hard to directly optimize the *rank* norm. Therefore, we usually use the trace/nuclear norm as its closet surrogate to instead, which leads to the following objective [5]

$$\min_{\mathbf{Z}} \ \|\mathbf{Z}\|_*, \quad s.t. \ \mathbf{X} = \mathbf{AZ}, \tag{7}$$

where $\| \cdot \|_*$ means nuclear norm, *i.e.*, the sum of singular values of a certain matrix [4]. Considering the fact that available samples are usually noisy or even grossly corrupted by outliers, a more robust LRR can be formulated as

$$\min_{\mathbf{Z},\mathbf{E}} \ \|\mathbf{Z}\|_* + \lambda \|\mathbf{E}\|_{2,1}, \quad s.t. \ \mathbf{X} = \mathbf{AZ} + \mathbf{E}, \tag{8}$$

where the $\|\mathbf{E}\|_{2,1} = \sum_{j=1}^{n} \sqrt{\sum_{i=1}^{d} e_{ij}^2}$ is added to characterize the error term for modeling the sample-specific corruptions. The optimal solution to (8) is usually obtained via the inexact augmented Lagrange multiplier method [9].

3 The Proposed KLGSL Model

In this section, we introduce the proposed KLGSL model including the its model formulation and optimization.

3.1 Model Formulation

For sample $\mathbf{x}_i$, we use the implicit function $\phi(\cdot)$ to map it into the reproducing kernel Hilbert space $\phi(\mathbf{x}_i)$. We expect that the data representation $\boldsymbol{\Phi}(\mathbf{X}) = [\phi(\mathbf{x}_1), \phi(\mathbf{x}_2), \ldots, \phi(\mathbf{x}_n)]$ in the RKHS coincides with the low-rank representation, $\boldsymbol{\Phi}(\mathbf{X}) = \boldsymbol{\Phi}(\mathbf{X})\mathbf{Z} + \mathbf{E}$. By multiplying $\boldsymbol{\Phi}^T(\mathbf{X})$ on both sides of this equation, we can obtain $\boldsymbol{\Phi}^T(\mathbf{X})\boldsymbol{\Phi}(\mathbf{X}) = \boldsymbol{\Phi}^T(\mathbf{X})\boldsymbol{\Phi}(\mathbf{X})\mathbf{Z} + \boldsymbol{\Phi}^T(\mathbf{X})\mathbf{E}$. Denote $\tilde{\mathbf{E}} \triangleq \boldsymbol{\Phi}^T(\mathbf{X})\mathbf{E}$ and then we can rewrite it as

$$\mathbf{K} = \mathbf{K}\tilde{\mathbf{Z}} + \tilde{\mathbf{E}}, \tag{9}$$

where $\mathbf{K}$ is the Gram matrix in which each element $k(\mathbf{x}_i, \mathbf{x}_j) = \langle \phi(\mathbf{x}_i), \phi(\mathbf{x}_j) \rangle$ is usually determined by a certain kernel function.

Finally, we formulate the objective function of the joint Kernel Low-rank Graph construction and Subspace Learning (KLGSL) model as

$$\min_{\tilde{\mathbf{Z}}, \tilde{\mathbf{E}}, \boldsymbol{\Theta}} \|\tilde{\mathbf{Z}}\|_* + \lambda\|\tilde{\mathbf{E}}\|_{2,1} + \gamma\|\boldsymbol{\Theta}^T\mathbf{K} - \mathbf{Y}\|_F^2 + \delta\|\boldsymbol{\Theta}\|_K^2$$
$$s.t.\ \mathbf{K} = \mathbf{K}\tilde{\mathbf{Z}} + \tilde{\mathbf{E}} \tag{10}$$

where λ, γ and δ are all positive regularization parameters to respectively control the impacts of error term, spectral regression and the smoothness of the projection matrix $\boldsymbol{\Theta}$. It should be noted that the regression target matrix $\mathbf{Y} = [\mathbf{y}_1, \mathbf{y}_2, \ldots, \mathbf{y}_c] \in \mathbb{R}^{n \times c}$ in problem (10) cannot be obtained in case of the supervised learning in which it can be constructed by the label information of training samples. It could be obtained by solving the following maximum EVD problem

$$\mathbf{Wy} = \gamma\mathbf{Dy}, \tag{11}$$

where $\mathbf{W} = (|\mathbf{Z}| + |\mathbf{Z}|^T)/2$ is the learned graph affinity matrix for subspace learning, $|\cdot|$ is the absolute value elementwise operation, and $\mathbf{D}$ is the corresponding diagonal degree matrix. Columns in $\mathbf{Y}$ can be formed by stacking with eigenvectors of (11) corresponding to largest eigenvalues as columns. The number of columns of $\mathbf{Y}$ (*i.e.*, c) will be equal to the dimensionality of the target subspace.

3.2 Optimization

Optimizing (10) directly is not easy; therefore, we introduce an auxiliary variable
$\mathbf{J}$ with respect to $\tilde{\mathbf{Z}}$ to make the objective of KLGSL separable, which leads to
the following objective

$$\min_{\tilde{\mathbf{Z}},\mathbf{J},\tilde{\mathbf{E}},\boldsymbol{\Theta}} \|\mathbf{J}\|_* + \lambda\|\tilde{\mathbf{E}}\|_{2,1} + \gamma\|\boldsymbol{\Theta}^T\mathbf{K} - \mathbf{Y}\|_F^2 + \delta\|\boldsymbol{\Theta}\|_K^2 \tag{12}$$

$$s.t.\ \mathbf{K} = \mathbf{K}\tilde{\mathbf{Z}} + \tilde{\mathbf{E}},\ \tilde{\mathbf{Z}} = \mathbf{J}.$$

The corresponding augmented Lagrangian function is

$$\begin{aligned}
\mathcal{L} &= \|\mathbf{J}\|_* + \lambda\|\tilde{\mathbf{E}}\|_{2,1} + \gamma\|\boldsymbol{\Theta}^T\mathbf{K} - \mathbf{Y}\|_F^2 + \delta\|\boldsymbol{\Theta}\|_K^2 \\
&\quad + \langle\mathbf{Y}_1, \mathbf{K} - \mathbf{K}\tilde{\mathbf{Z}} - \tilde{\mathbf{E}}\rangle + \langle\mathbf{Y}_2, \tilde{\mathbf{Z}} - \mathbf{J}\rangle \\
&\quad + \frac{\mu}{2}(\|\mathbf{K} - \mathbf{K}\tilde{\mathbf{Z}} - \tilde{\mathbf{E}}\|_F^2 + \|\tilde{\mathbf{Z}} - \mathbf{J}\|_F^2). \\
&= \|\mathbf{J}\|_* + \lambda\|\tilde{\mathbf{E}}\|_{2,1} + \gamma\|\boldsymbol{\Theta}^T\mathbf{K} - \mathbf{Y}\|_F^2 + \delta\|\boldsymbol{\Theta}\|_K^2 \\
&\quad + \frac{\mu}{2}\left(\left\|\mathbf{K} - \mathbf{K}\tilde{\mathbf{Z}} - \tilde{\mathbf{E}} + \frac{\mathbf{Y}_1}{\mu}\right\|_F^2 + \left\|\tilde{\mathbf{Z}} - \mathbf{J} + \frac{\mathbf{Y}_2}{\mu}\right\|_F^2\right)
\end{aligned}$$

Below we give the updating rule for each variable with the others fixed.

– Update $\mathbf{J}$ with the others fixed. The objective associated with $\mathbf{J}$ is

$$\arg\min_{\mathbf{J}} \frac{1}{\mu}\|\mathbf{J}\|_* + \frac{1}{2}\|\mathbf{J} - (\tilde{\mathbf{Z}} + \frac{\mathbf{Y}_2}{\mu})\|_F^2, \tag{13}$$

which can be solved by the singular value thresholding operator [4] given in
the following theorem.

Theorem 2 (*Singular Value Thresholding Operator*). *Let* $\mathbf{C} \in \mathbb{R}^{m\times n}$ *and let*
$\mathbf{C} = \mathbf{U}\boldsymbol{\Sigma}\mathbf{V}^T$ *be the SVD of* $\mathbf{C}$ *where* $\mathbf{U} \in \mathbb{R}^{m\times r}$ *and* $\mathbf{V} \in \mathbb{R}^{n\times r}$ *have orthonormal*
columns, $\boldsymbol{\Sigma} \in \mathbb{R}^{r\times r}$ *is diagonal, and* $r = rank(\mathbf{C})$. *Then*

$$\mathcal{T}_\lambda(\mathbf{C}) = \arg\min_{\mathbf{J}}\left\{\frac{1}{2}\|\mathbf{J} - \mathbf{C}\|_F^2 + \lambda\|\mathbf{J}\|_*\right\}$$

is given by $\mathcal{T}_\lambda(\mathbf{C}) = \mathbf{U}\boldsymbol{\Sigma}_\lambda\mathbf{V}^T$, *where* $\boldsymbol{\Sigma}_\lambda$ *is diagonal with* $(\Sigma_\lambda)_{ii} = \max\{0, \Sigma_{ii} - \lambda\}$.

– Update $\tilde{\mathbf{Z}}$ with other variables fixed. Take the derivative of $\mathcal{L}$ with respect to
$\mathbf{Z}$ and we obtain

$$\frac{\partial\mathcal{L}}{\partial\tilde{\mathbf{Z}}} = -\mu(\mathbf{K}^T\mathbf{K} - \mathbf{K}\tilde{\mathbf{Z}} - \tilde{\mathbf{E}}) - \mathbf{K}^T\mathbf{Y}_1 + \mathbf{Y}_2 + \mu(\tilde{\mathbf{Z}} - \mathbf{J}) \tag{14}$$

While setting Eq. 14 to zero, we get the updating formula for $\tilde{\mathbf{Z}}$ as

$$\tilde{\mathbf{Z}} = (\mathbf{K}^T\mathbf{K} + \mathbf{I})^{-1}[(\mathbf{K}^T\mathbf{Y_1} - \mathbf{Y_2})/\mu - \mathbf{K}^T\tilde{\mathbf{E}} + \mathbf{J} + \mathbf{K}^T\mathbf{K}] \tag{15}$$

– Update $\boldsymbol{\Theta}$ with other variables fixed. The objective function associated with $\boldsymbol{\Theta}$ is

$$\mathcal{O}(\boldsymbol{\Theta}) : \arg\min_{\boldsymbol{\Theta}} \gamma \|\boldsymbol{\Theta}^T \mathbf{K} - \mathbf{Y}\|_F^2 + \delta \|\boldsymbol{\Theta}\|_F^2 \tag{16}$$

Take the derivative with respect to $\boldsymbol{\Theta}$ and set the equation to zero, we can obtain the updating rule for $\boldsymbol{\Theta}$ as

$$\boldsymbol{\Theta} = (\gamma \mathbf{K}^T \mathbf{K} + \delta \mathbf{I})^{-1}(\gamma \mathbf{K}\mathbf{Y}^T) \tag{17}$$

– Update $\tilde{\mathbf{E}}$ with other variables fixed. The objective associated with $\tilde{\mathbf{E}}$ is

$$\mathcal{O}(\tilde{\mathbf{E}}) : \arg\min_{\tilde{\mathbf{E}}} \lambda \|\tilde{\mathbf{E}}\|_{2,1} + \frac{\mu}{2} \|\mathbf{K} - \mathbf{K}\tilde{\mathbf{E}} - \tilde{\mathbf{E}} + \frac{\mathbf{Y}_1}{\mu}\|_F^2 \tag{18}$$

The corresponding updating rule for $\tilde{\mathbf{E}}$ can be solved by soft-shrinkage operator [10] which is defined in the theorem below.

Theorem 3 *(Soft-Shrinkage Operator). Let* $\mathbf{Q} = [\mathbf{q}_1, \mathbf{q}_2, \ldots, \mathbf{q}_i, \ldots]$ *be a given matrix and* $\|\cdot\|_F$ *be the Frobenius norm. If the optimal solution to*

$$\min_{\mathbf{E}} \lambda \|\mathbf{E}\|_{2,1} + \frac{1}{2} \|\mathbf{E} - \mathbf{Q}\|_F^2 \tag{19}$$

is $\mathbf{E}^*$, *then the i-th column of* $\mathbf{E}^*$ *is*

$$\mathbf{E}^*(:, i) = \begin{cases} \frac{\|\mathbf{q}_i\| - \lambda}{\|\mathbf{q}_i\|} \mathbf{q}_i, & \text{if } \lambda < \|\mathbf{q}_i\|; \\ 0, & \text{otherwise.} \end{cases} \tag{20}$$

The detailed step-by-step optimization procedure of KLGSL is show in Algorithm 1.

4 Experiments

In this section, we conduct experiments on three publicly available data sets to evaluate the efficacy of the proposed KLGSL model compare to the related subspace learning methods.

4.1 Data Sets and Experimental Setup

Three public available data sets, the CMU PIE, Extended Yale B and AR databases, were used in the experiments. Some sample images from each data set were shown in Fig. 1.

The basic properties of each data set were described as follows.

Algorithm 1. The optimization procedure to (10).

Input: Training samples $\{\mathbf{x}_i\}_{i=1}^n \in \mathbb{R}^d$, regularization parameters λ, γ and δ, the subspace dimensionality c;

Output: The coefficient matrix $\tilde{\mathbf{Z}}$ and projection matrix $\boldsymbol{\Theta}$.

1: **Initialization.** Variables $\tilde{\mathbf{Z}} = \mathbf{J} = \tilde{\mathbf{E}} = \mathbf{0}$, multipliers $\mathbf{Y}_1 = \mathbf{Y}_2 = \mathbf{0}$, increment step parameter $\rho = 1.1$, penalty parameter $\mu = 1.01$ and $\mu_{\max} = 10^{10}$, convergence threshold $\varepsilon = 10^{-8}$;

2: Construct the affinity matrix $\mathbf{W}$ based on the 'Heatkernel' weighting approach in which the bandwidth is set as the mean of pairwise distances of training samples;

3: Calculate the regression target $\mathbf{Y}$ by stacking the eigenvectors of (1) corresponding to the c largest eigen-values;

4: **while** not converge **do**

5:    Fix the other variables and update $\mathbf{J}$ by optimizing (13)via SVT operator defined in Theorem 2;

6:    Fix the other variables and update $\tilde{\mathbf{Z}}$ by (15);

7:    Fix the other variables and update $\boldsymbol{\Theta}$ by (17);

8:    Fix the other variables and update $\mathbf{E}$ by optimizing (18) via soft-shrinkage operator defined in Theorem 3;

9:    Update $\mathbf{Y}_1$ by $\mathbf{Y}_1 = \mathbf{Y}_1 + \mu(\mathbf{K} - \mathbf{K}\tilde{\mathbf{Z}} - \tilde{\mathbf{E}})$;

10:    Update $\mathbf{Y}_2$ by $\mathbf{Y}_2 = \mathbf{Y}_2 + \mu(\tilde{\mathbf{Z}} - \mathbf{J})$;

11:    Update μ by $\mu = \min(\rho\mu, \mu_{\max})$;

12:    Update $\mathbf{W} = (|\mathbf{Z}| + |\mathbf{Z}|^T)/2$ and calculate the spectral regression target $\mathbf{Y}$ by stacking the eigenvectors of (2) corresponding to the c largest eigenvalues;

13:    Check the convergence conditions

$$\|\mathbf{X} - \mathbf{X}\mathbf{Z} - \mathbf{E}\|_\infty < \varepsilon, \text{ and } \|\mathbf{Z} - \mathbf{J}\|_\infty < \varepsilon$$

14: **end while**

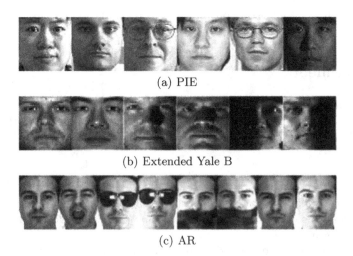

(a) PIE

(b) Extended Yale B

(c) AR

Fig. 1. Sample images respectively from the three data sets.

- The CMU PIE database [7] composes of 41368 face images collected from 68 subjects. Each subject has 13 images of different poses, 43 different illumination conditions, with four different expressions. In our experiments, we chose the subset of the near frontal pose C29 from which we randomly selected 10 images each subject to form the training set and the rest as test set. Therefore, we respectively had 680 training images and 2649 test images. Each image was resized to 32×32 pixels.
- The Extended Yale B data set [6] contains 2,414 near frontal face images from 38 subjects under different illuminations and each subject has about 64 images. In our experiments, we randomly selected 10 images each subject to train the model and the rest as test images. Each image was resized to 32×32 pixels.
- The AR data set [13] contains over 4000 color images corresponding to 126 subjects (70 men and 56 women). Images are frontal view faces with different facial expressions, illumination conditions, and occlusions (sun glasses and scarf). Each subject participated in two sessions separated by two weeks. In our experiments, we chose a subset from AR consisting of 50 men and 50 women. We selected two neutral images plus two corrupted images (with glasses or scarf) in session one as training set and three neutral images plus three corrupted images (with glasses or scarf) in session two as test set. Therefore, we had respectively 400 training samples and 600 test samples. Each image was resized to 33×24 pixels.

We compared KLGSL model with some state-of-art subspace learning methods including KLPP, kernel spectral regression (KSR) [3], sparsity preserving projection (SPP) [16] and the Low-Rank Preserving Embedding (LRPE) [19]. In the experiments, we used the training samples to learn the projection matrix which was used to project the testing samples into low-dimensional subspace for further the nearest neighbor classification.

In our proposed KLGSL model, there are three regularization parameters, λ, γ and δ. Respectively, we set $\lambda = 10$, $\gamma = 1000$, and $\delta = 0.1$ in PIE; $\lambda = 10$, $\gamma = 1000$, $\delta = 0.1$ in the extended Yale B; and $\lambda = 10$, $\gamma = 100$, $\delta = 0.01$ in AR. In SPP and LRPE, to avoid the singularity problem, we first did PCA transformation with ratio 0.98 and then conducted further operations. We set the regularization parameter λ in SPP as 10, 100, and 10 corresponding to PIE, extended Yale B and AR data sets, respectively. For LRPE, the values of regularization parameter λ are respectively 3, 8 and 5.

4.2   Experimental Results and Discussions

In the each experiments, considering the total number of each data set, we randomly selected l ($l = 5, 8, 12$ for PIE, $l = 10, 20, 30$ for Extended Yale B, $l = 5, 8, 13$ for AR) images per class from each data set to form the training set and the remaining images for testing. In terms of subspace dimensionality, we reduced all data dimensions down from 0 to 200. Each experiments were repeated five times and the average performance was calculated and reported. The experimental results are shown in Figs. 2, 3 and 4.

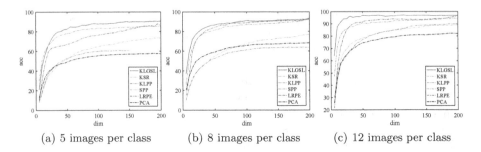

(a) 5 images per class   (b) 8 images per class   (c) 12 images per class

Fig. 2. Classification results of compared methods vs. dimensionality on PIE data set.

From the experimental results, we can roughly categorize the compared subspace learning algorithms into three different levels based on their performance.

- The performance of three subspace learning algorithms, PCA, KLPP, and KSR, are weaker in comparison with the remaining algorithms; therefore, they are at the bottom level. PCA is a linear unsupervised subspace learning algorithm, which might be not appropriate for dealing with non-Gaussian noise and characterizing the possible non-linear structure in data. Both KLPP and KSR rely on the quality of constructed graph, their performance cannot be guaranteed when the way of constructing the graph affinity matrix such as 'HeatKernel' are not adaptive and robust. They suffered from the sensitivity to noise and outliers in data.
- Thanks to the discriminative ability of ℓ_1-graph, SPP obtained much better performance improvement than PCA, KLPP and KSR. However, the sparse representation may encounter the stability problem in which SR tends to randomly select a single representative sample in dictionary and ignore other correlated ones when samples from the same class are highly correlated. Instead of computing the representation coefficient individually, low-rank representation allows to compute the coefficient matrix collectively. Therefore, the low-rank graph usually obtains much better characterization to the data structure information, leading to the improved performance of LRPE.
- Our proposed KLGSL model ranks in the top level. It shares the paradigm of joint learning the data affinity matrix and the projection matrix, which can effectively avoid the disadvantages caused by the two-stage processing strategy. By combining the advantages of KLPP and kernel spectral regression in unsupervised version, KLGSL obtained superior performance in comparison with the other subspace learning methods.

To investigate the sensitivity of the three parameters (e.g., λ, γ and δ) involved in the proposed KLGSL model, we show the accuracy of KLGSL in terms of each parameter by fixing the others. We selected 12 sample images per class from PIE data set, and set the dimension of projection matrix to 200. Figure 5 shows the experimental results on PIE data set. From this figure, we can find that γ affects the performance of KLGSL more when comparing the

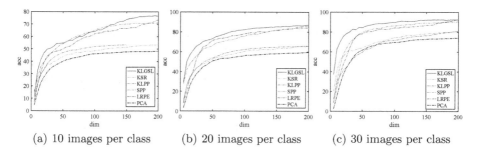

(a) 10 images per class   (b) 20 images per class   (c) 30 images per class

Fig. 3. Classification results of compared methods vs. dimensionality on Extended Yale B data set.

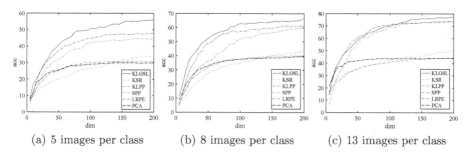

(a) 5 images per class   (b) 8 images per class   (c) 13 images per class

Fig. 4. Classification results of compared methods vs. dimensionality on AR data set.

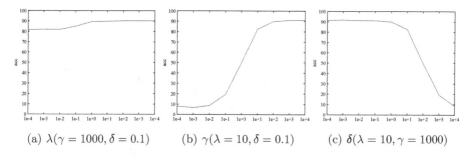

(a) $\lambda(\gamma = 1000, \delta = 0.1)$ (b) $\gamma(\lambda = 10, \delta = 0.1)$ (c) $\delta(\lambda = 10, \gamma = 1000)$

Fig. 5. Parameter sensitivity of KLGSL in terms of each parameter on PIE data set.

other two parameters since it directly controls the projection matrix learning in the KLGSL objective. Generally, KLGSL is not sensitive to the parameter λ and we can obtain a wide range for selecting the value of parameters γ and δ.

5   Conclusions

In this paper, we proposed a joint learning framework for getting a robust data affinity matrix and the projection matrix. the proposed model has several merits such as robust to outliers, considering the possible nonlinear structure of data by kernel trick, easy to optimize and flexible to being incorporated into new regularization terms. The experimental results manifest that the proposed KLGSL is an effective and robust dimensionality reduction technique. As a future work, we will investigate the extension of KLGSL by imposing the $\ell_{2,1}$-norms on the transformation matrix instead of the ℓ_2-norm to let it have the feature selection ability.

Acknowledgments. This work was partially supported by National Natural Science Foundation of China (61602140), Science and Technology Program of Zhejiang Province (2017C33049), China Postdoctoral Science Foundation (2017M620470), Opening Fund of Acoustics Science and Technology Laboratory of Harbin Engineering University (SSKF2018001,SSKF2017001), Jiangsu Key Laboratory of Big Data Security & Intelligent Processing, Nanjing University of Posts and Telecommunications (BDSIP201804), Co-Innovation Center for Information Supply & Assurance Technology, Anhui University (ADXXBZ201704), Guangxi Key Laboratory of Multi-source Information Mining & Security (MIMS18-06).

References

1. Brahma, P.P., She, Y., Li, S., Li, J., Wu, D.: Reinforced robust principal component pursuit. IEEE Trans. Neural Netw. Learn. Syst. **29**(5), 1525–1538 (2018)
2. Cai, D., He, X., Han, J.: Efficient kernel discriminant analysis via spectral regression. In: Seventh IEEE International Conference on Data Mining (2007)
3. Cai, D., He, X., Han, J.: Speed up kernel discriminant analysis. VLDB J. Int. J. Very Large Data Bases **20**(1), 21–33 (2007)
4. Cai, J.F., Candès, E.J., Shen, Z.: A singular value thresholding algorithm for matrix completion. SIAM J. Optim. **20**(4), 1956–1982 (2010)
5. Candès, E.J., Li, X., Ma, Y., Wright, J.: Robust principal component analysis? J. ACM **58**(3), 11 (2011)
6. Georghiades, A., Belhumeur, P., Kriegman, D.: From few to many: illumination cone models for face recognition under variable lighting and pose. IEEE Trans. Pattern Anal. Mach. Intell. **23**(6), 643–660 (2001)
7. Gross, R., Matthews, I., Cohn, J., Kanade, T., Baker, S.: Multi-pie. Image Vis. Comput. **28**(5), 807–813 (2010)
8. Guo, K., Liu, L., Xu, X., Xu, D., Tao, D.: GoDec+: fast and robust low-rank matrix decomposition based on maximum correntropy. IEEE Trans. Neural Netw. Learn. Syst. **29**(6), 2323–2336 (2018)
9. Lin, Z., Chen, M., Ma, Y.: The augmented Lagrange multiplier method for exact recovery of corrupted low-rank matrices. arXiv preprint arXiv:1009.5055 (2010)
10. Liu, G., Lin, Z., Yu, Y.: Robust subspace segmentation by low-rank representation. In: International Conference on Machine Learning, pp. 663–670 (2010)
11. Liu, G., Yan, S.: Latent low-rank representation for subspace segmentation and feature extraction. In: IEEE International Conference on Computer Vision, pp. 1615–1622 (2011)
12. Liu, R., Lin, Z., De la Torre, F., Su, Z.: Fixed-rank representation for unsupervised visual learning. In: IEEE Conference on Computer Vision and Pattern Recognition, pp. 598–605 (2012)
13. Martinez, A., Benavente, R.: The AR face database. Technical report, 24 CVC Technical Report (1998)

14. Nguyen, H., Yang, W., Shen, F., Sun, C.: Kernel low-rank representation for face recognition. Neurocomputing **155**, 32–42 (2015)
15. Peng, Y., Kong, W., Qin, F., Nie, F.: Manifold adaptive kernelized low-rank representation for semisupervised image classification. Complexity **2018**, 1–11 (2018)
16. Qiao, L., Chen, S., Tan, X.: Sparsity preserving projections with applications to face recognition. Pattern Recogn. **43**(1), 331–341 (2010)
17. Xiao, S., Tan, M., Xu, D., Dong, Z.Y.: Robust kernel low-rank representation. IEEE Trans. Neural Netw. Learn. Syst. **27**(11), 2268–2281 (2016)
18. Yang, S., Feng, Z., Ren, Y., Liu, H., Jiao, L.: Semi-supervised classification via kernel low-rank representation graph. Knowl.-Based Syst. **69**, 150–158 (2014)
19. Zhang, Y., Xiang, M., Yang, B.: Low-rank preserving embedding. Pattern Recogn. **70**, 112–125 (2017)

Defending Network Traffic Attack with Distributed Multi-agent Reinforcement Learning

Shiming Xia, Wei Bai, Xingyu Zhou, Zhisong Pan$^{(\boxtimes)}$, and Shize Guo

College of Command and Information System, Army Engineering University of PLA,
Nanjing 211107, Jiangsu, China
hotpzs@hotmail.com

Abstract. The DDoS attack is a serious security problem in today's Internet. To defend DDoS attacks, distributed routers need to cooperate to guarantee servers' safety. Existing rule-based methods are not ideal as these methods adjust a threshold, which lacks flexibility. The distributed DDoS defense problem can also be viewed as a multi-agent Markov decision process (MAMDP), where multiple agents intelligently coordinate to achieve a team goal. However, agents can hardly defend DDoS attacks without communication. In this paper, we propose a multi-agent reinforcement learning method for distributed routers to defend DDoS attacks via compressed communication. Our method named ComDDPG outperforms existing rule-based methods and independent reinforcement learning methods under diverse attack scenarios even with communication delay.

Keywords: DDoS · Multi-agent reinforcement learning ·
Distributed system · Compressed communication

1 Introduction

Denial of Service (DoS) attacks constitute one of the major cyber threats and among the most complicated security problems in today's Internet [1,2]. Of particular concerns are Distributed Denial of Service (DDoS) attacks, whose impact can be proportionally severe. With little or no advance warning, a DDoS attack can easily exhaust the computing and communication resources of victim servers within a short time period [3]. With the increase of Internet bandwidth and the continuous release of various DDoS hacking tools, the implementation of DDoS attacks is becoming easier, and the events of DDoS attacks are on the rise.

A DDoS attack is a highly coordinated attack, using a crowd of hosts to launch a large-scale coordinated DDoS attack against one or more targets [4]. The strategy behind it is described by the DDoS attack model [5] as shown in Fig. 1. The attackers install malicious software on vulnerable hosts to compromise them, thus capable to communicate with them. Attackers communicate

© Springer Nature Singapore Pte Ltd. 2019
K. Knight et al. (Eds.): ICAI 2019, CCIS 1001, pp. 212–225, 2019.
https://doi.org/10.1007/978-981-32-9298-7_17

with the handlers, which in turn control the hosts in order to launch a DDoS attack.

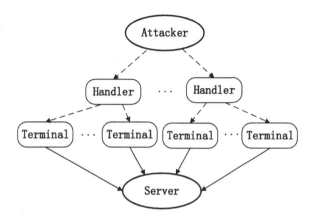

Fig. 1. DDoS attack model. Attackers communicates with the handlers, which in turn control the hosts in order to launch a DDoS attack.

The DDoS threat is challenging for many reasons [3,6]. Firstly, the DDoS traffic is distributed. Packets are generated from terminals spread all over the Internet, and then all aggregates at the victim servers. Secondly, the volume of DDoS traffic is large. It is unlikely to defend the attack by a single router near the victim servers. Thirdly, the DDoS traffic is similar to legitimate ones. The bursts due to malicious traffic are similar to the bursts by normal legitimate traffic, thus causing difficulties in detection. Therefore, it is necessary to install traffic control devices on some routing nodes farther from the server to ensure the safe of the server.

Although a centralized detection and defense mechanism with complete and global information may be qualified, as the volume of traffic increases, however, a centralized DDoS detection system becomes a communication bottleneck. The pressure of each node in decentralized DDoS defense systems is less than centralized one. Therefore, decentralized DDoS defense systems have received much attention during these years. Generally, agents need to communicate and share information among them to employ a coordinated and effective defense mechanism, as shown in [7].

A number of methods have been proposed to achieve a coordinated defense mechanism. Router Throttling [8] is a popular approach to defend against DDoS attacks, where the routers throttle traffic based on the victim servers' signals. For instance, there are two baseline methods named Server-initiated Router Throttle (SRT) algorithm and Fair Router Throttle (FRT) algorithm [8]. Both of these two policies adjust a threshold to restrict the traffic towards server within a given boundary. However, these policies just adjust a threshold to restrict the traffic towards server within a given boundary. How to learn a robust policy to decentralized control the router throttling rate under dynamic environment is the key difficulty of the DDoS defending problem.

In addition, the distributed DDoS defense mechanism can be formalized as a reinforcement learning (RL) problem, specifically multi-agent reinforcement learning (MARL) [9]. Reinforcement learning (RL) has recently been applied to solve challenging problems in single agent domains, from game playing [10,11] to robotics [12]. Unfortunately, traditional reinforcement learning approaches such as Q-Learning or policy gradient are poorly suited to multi-agent environments. However, the strategies of other agents are uncertain and changing as training progresses [13], causing the environment non-stationary from the perspective of the individual agent. Thus, it is hard for multiple agents to collaborate to learn a coordinated policy.

In this paper, we make two contributions:

1. Design a time-difference adversarial reward for multi-agent reinforcement learning, which is beneficial to accelerating training.
2. Apply multi-agent deep deterministic policies gradient (MADDPG) methods into defending DDoS attacks.

The rest of the paper is organized as follows. The DDoS problem definition and related work are presented in Sect. 2. In Sect. 3, we formally introduce our method. Experimental settings, performance evaluation, and the follow-up discussion are presented in Sect. 4. Conclusions are given in Sect. 5.

2   Denial-of-Service and Existing Defending Methods

In this section, we introduce the DDoS problem definition and some existing router throttling methods.

2.1   DDoS Problem Definition

Generally, there are three entities involved in a DDoS attack: hosts, routers, and servers. Hosts may be legitimate users or attackers. Requests from hosts over a region will be processed by a certain router, who actually has the ability to decide how many packets will be sent to servers. Finally, servers will maintain a queue and respond to requests from routers as the order of first in and first out. Each timestamp, servers will have residual resources if aggregate traffic is smaller than its capacity. But there may be unfinished packets waiting at the queue if aggregate traffic is larger than its capacity. Furthermore, if aggregate traffic is continuously excessive, congestion occurs and servers will deny the following requests.

The distributed DDoS defense mechanism can be viewed as a resource management problem. Each router will be viewed as an agent to decide how much packets will be throttled. In this paper, the aggregate traffic is defined as packets from certain region arrived at routers or servers over the last W seconds, which is called monitor window. Assume x^i is the aggregate traffic over region

i, where the router i monitors; y^i is the aggregate malicious load launched by attackers over region i at the timestamp t; finally, $a^i \in [0,1]$ is the throttling rate ($a^i = 0$ means all packets including legitimate and malicious packets will be sent to servers; $a^i = 1$ means all packets including legitimate and malicious packets will be throttled). The goal of a dense mechanism or policy is to maximize the average legitimate traffic passing rate during a time interval T as shown in (1), where N is the number of routers. We hope resources provided by servers can be utilized by users as much as possible, however, the aggregate traffic passed to servers should be less than the load limits U_s to make sure server's safety, as shown in (1).

$$\max g_t = \frac{\sum_{i=0}^{N} x_t^i (1 - a_t^i)}{\sum_{i=1}^{N} x_t^i}$$

$$s.t. \quad z_t = \sum_{i=0}^{N} (x_t^i + y_t^i)(1 - a_t^i) \leq U_s$$

(1)

Ideally, it is easy to solve the above problem by linear programming (LP) [14]. In reality, however, routers cannot easily distinguish between legitimate packets and malicious packets. What's worse, concerns like communication faults and transportation delay make exactly solving by linear programming intractable. Thus, the simple implementation based on linear programming is impossible. Nonetheless, routers can use history information and feedbacks from servers to make decisions.

2.2   Adaptive Rule-Based Policy

Here we will introduce two rule-based methods, Server-initiated Router Throttle (SRT) algorithm, and Fair Router Throttle (FRT) algorithm [8].

For Server-initiated Router Throttle (SRT) algorithm, all routers share the same throttling fraction a_t (omitting subscript i). Each router throttles a fraction a_t of traffic, to make sure the aggregate traffic towards victim servers is within the given boundary $[L_s, U_s]$. The fraction a_t is adjusted according to current server congestion. For example, the fraction a_t will increase if z_t (the total traffic transferred to victim server at moment t) is more than U_s. But the fraction a_t will decrease if z_t is less than L_s (the low-water mark of server load). In the real implementation, SRT requires a central agent to make decisions and send back the action for each router. What's more, this method is not fair as it penalizes all routers equally, regardless of whether regional traffic diversity.

For Fair Router Throttle (FRT) algorithm, all routers do not share the same throttling fraction. Instead of, each agent maintain a threshold R_t^i. Based on current aggregate traffic information z_t from servers, each router adjusts their own R_t^i. Specifically, for the router i, if $z_t > U_s$, it will decrease its threshold R_t^i; if $z_t < L_s$, it will increase its threshold R_t^i. With the threshold R_t^i, each router can make decisions by itself, thus FRT is a distributed policy.

2.3 Multi-Agent Reinforcement Learning

Reinforcement learning has been applied successfully in many fields. We firstly introduce the background of reinforcement learning and then extend it to multi-agent reinforcement learning (MARL).

Consider a standard Markov Decision Process (MDP), which is represented by a tuple: $M = \langle S, A, P, R, \gamma \rangle$, S is the state space and A is the action space. The dynamic transition is specified by probability transmit function $P(\cdot) = \Pr(s_{t+1}|s_t, a_t)$, and reward function $R(\cdot) = r(s_t, a_t)$ assigns reward for state-action pairs. By interacting with the environment, agents generate trajectories: $(s_1, a_1, r_1, s_2, \cdots, s_T)$. The goal of reinforcement learning is to learn a policy $\pi(s, a) = \Pr(a_t|s_t)$, to maximize expected cumulative rewards, as (2) shows, where discounted factor $\gamma \in [0, 1]$ is designed to make sure expected cumulative reward is bound.

$$\max E_\pi[\sum_{t=1}^T \gamma^{t-1} r_t | s_0 = s] \tag{2}$$

Reinforcement Learning assumes that at least we don't know probability transmit function P, thus we cannot get the optimal policy based on dynamic programming [15]. However, the optimal policy can be obtained by interacting with the environment to get necessary information. There are two categories to solve this problem. Value-based methods like Q-Learning [15] approximates action-value function $Q(s, a) = E_\pi[\sum_{t=1}^T \gamma^{t-1} r_t | s_0 = s, a_0 = a]$, and subsequently get the optimal action use a greedy policy, as shown in (3). In addition to value-based methods, policy gradient that involves an actor to execute policy $\pi(s, a)$ and a critic to evaluate action-value $Q(s, a)$ can also find the optimal policy [15]. We will discuss policy gradient methods later.

$$Q(s_t, a_t) = r(s_t, a_t) + \gamma \max_{a_{t+1}} Q(s_{t+1}, a_{t+1})$$
$$a_t^* = \arg\max_{a_t} Q^*(s_t, a_t) \tag{3}$$

Similarly, Multi-Agent Markov Decision Process (MAMDP) is specified by a tuple:

$$\langle N, \{O_i\}_{i=1}^N, S, \{A_i\}_{i=1}^N, P, R, \gamma \rangle$$

N is the number of agents. Different from MDP, agents can no longer get access to global and complete information $s_t \in S$. For specific agent i, $o_i^t \in O_i$ is its observation at the time t, $a_i \in A_i$ is its action at timestamp t. What's more, dynamic process and joint reward are determined by all agents, which implies that $P(\cdot) = \Pr(s_{t+1}|s_t, \{a_t^i\}_{i=1}^N)$ and $R(\cdot) = r(s_t, \{a_t^i\}_{i=1}^N)$. Although the team goal of MARL remains to maximize cumulative joint reward, but the target of one agent moves when other agents' polices change, which makes the optimization of MARL intractable.

For our problem, global state s_t includes information about each router's aggregate legitimate traffic x_t^i, aggregate malicious traffic y_t^i, servers' residual resources p_t, and unfinished packet waiting queue length q_t. Each agent can obtain its local information $o_t^i = x_t^i + y_t^i$, but can not directly distinguish x_t^i and y_t^i from o_t^i. To defend attacks, each agent chooses its action a_t^i to throttle traffic, and the team will receive a joint reward r_t based on average legitimate traffic passing rate and server congestion condition.

3   Methodology

Firstly, we introduce a centralized decision method based on Deep Deterministic Policy Gradient (DDPG) [16]. Mathematically, DDPG assumes that the agent's policy is not stochastic but deterministic. Furthermore, the policy can be differential and parameterized by π_μ. Thus we can directly optimize the policy by calculating its gradients. Specifically, each agent is also called an actor, who executes the policy and outputs actions. To optimize the policy, a virtual agent called critic, who is parameterized by $Q_\theta(o, a)$, updates the actor's policy parameters by back-propagating policy gradients. It is easy to implement this method for defending DDoS attacks. All routers send information to a central agent, who make decisions for each router.

Now we introduce our method named ComDDPG, which is based on Multi-Agent Deep Deterministic Policy Gradient (MADDPG) [17]. Different from DDPG, MADDPG involves multiple decentralized actors and one centralized critic. To simply notation, we will use bold symbol $\boldsymbol{a}$ to denote joint action tuple $(a_1, \cdots, a_n)$.

3.1   Centralized Training and Decentralized Execution

To achieve coordination, there are N individual actors with independent abilities to make decisions and one critic to evaluate their joint actions. We emphasize that the critic only exists during training, and actors do not need the critic to execute policies. Thus, agents can centralize learn coordination with global information and decentralized execute policies with local information. For optimization, the critic and actors can be trained in an online way of minimizing the following loss functions for the critic and actors, respectively shown in (4), (5). Here, K is the total number of samples.

$$L_{critic}(\theta) = \sum_{k=1}^{K}[y_t^k - Q^\theta(s_t^k, \boldsymbol{a}_t^k)]^2 \tag{4}$$
$$y_t^k = r(s_t^k, \boldsymbol{a}_t^k) + \gamma Q^\theta(s_{t+1}^k, \boldsymbol{a}_{t+1}^k)$$

$$L_{actor_i}(\mu_i) = \sum_{k=1}^{K} \nabla_{\mu_i} \pi^{\mu_i}(a_t^k|o_t^k) \nabla_{a_t^k} Q^\theta(s_t, \boldsymbol{a}_t)|_{a_t^k = \pi_{\mu_i}(o_t^k)} \tag{5}$$

We implement agent policies with a deep neural network, as shown in Fig. 2. Actor networks are built with fully connect layers and gate recurrent units (GRU) [18] to extract features from hidden state and current observation. Critic network is built with fully connect layers to approximate value function $Q(\boldsymbol{s}, \boldsymbol{a})$.

3.2   Reward Design

In a DDoS scenario, we need to strike a balance between server safety and server usage. For instance, it is important to throttle traffic to make sure the operation of the server. But if we throttle much traffic, user's demand cannot be satisfied. Thus, reward involves an equilibrium between a server and users. In experiments, we find that simple rewards, like maximizing legitimate traffic rate while receiving -1 if server overloads [19] are not effective signals to guide agents behaviors. Contrast to these simple rewards, we propose a time-difference adversarial reward, which helps agents learn quickly.

$$r'_t = r_t - r_{t-1}$$
$$r_t = g_t - \alpha q_t - \beta p_t \tag{6}$$
$$r_0 = 0, \alpha > 0, \beta > 0$$

Here r_t is the average legitimate traffic passing rate, q_t is the waiting packets queue length of the server, p_t is the residual resources of the server, and r'_t is the actual reward that agents receive. To understand this reward, our reward design encourages agents to utilize resources as many as possible by giving penalties if some resources waste. On the other hand, it is easy to see that if β is 0, agents are prone to throttle less traffic and consequently servers will overload. Thus, it is necessary to give a negative reward as to queue length. What's more, via time-difference reward, the agent will quickly learn which direction of the behavior (i.e. increase a_t^i or decrease a_t^i) is helpful to defend attacks.

Algorithm 1. ComDDPG Training

---

while Attack scenario doesn't finish **do**
    for $i = 1 : N$ **do**
      $c_t^i \Leftarrow$ Agent i prepare communication content.
      if $c_t^i > 0$ **then**
          Send c_t^i to the other agents.
      end if
    end for
    for $i = 1 : N$ **do**
      $o_t^i \Leftarrow$ Agent i observe local information.
      $a_t^i \Leftarrow$ Agent i throttle traffic based on $(o_t^i, \{c_t^j\}_{j \neq i}^N)$.
    end for
    $s_t \Leftarrow$ Critic obtain global information from the simulator.
    $r_t \Leftarrow$ Whole system updates states based on $\{a_t^i\}_{i=1}^N$.
    for $i = 1 : N$ **do**
      Agent i updates its parameters based on equation 5.
    end for
    Critic updates its parametersts based on equation 4.
    $t := t + 1$.
end while

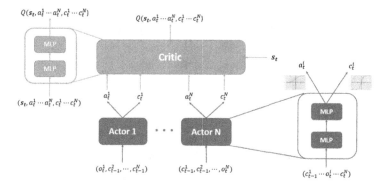

Fig. 2. Our ComDDPG neural network architecture. Actors are built with multi-layer perception (MLP), whose throttle action is activated by *sigmoid* function and communication action is activated by *tanh* function. The critic is also built with MLP, whose output value is activated by *linear* function.

4 Experiment

In this section, we evaluate the above methods under diverse DDoS attack scenarios.

There are 108 region agents to defend 30 region attackers. One attack scenario will last for 200 timestamps (the monitor window $W = 1s$). What's more, attackers' position may change during the next attack scenario. Thus, this setting is difficult for agents to learn a defend policy, which is expected to divers attacks. The training will take 100 episodes, and finally test on 10 episodes to evaluate. We emphasize that agents cannot be trained during the test stage to keep fair and examine its learned policies.

4.1 Experiment Setup

To validate the effectiveness of our proposed method, we implement our experiment based on a popular network simulator OPNET [20,21]. The simulated network topology is shown in Fig. 3. There are 27 nodes, different nodes may have different numbers of throttling routers.

During our experiment, aggregate legitimate traffic may exceed to servers capacity during a short period[1], which rejects the assumption that all congestion attributes to attacks. But with attack traffic, servers are easily to overload. Specifically, we keep the total legitimate traffic is 70%–80% of the servers capacity, the total attack traffic is 35%–45% of the servers capacity.

In addition, we suppose there is a probabilistic communication delay when communication, which is totally possible in real applications. For perception, agents will take the previous communication content to make decisions if it doesn't receive new information due to communication delay. Note that there is

---

[1] In our experiment, there is only one server.

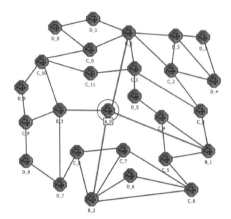

Fig. 3. The topology of our opnet internet environment. There are 27 nodes, and each node is composed of servel throttling routers. Here, a server is located in the central node, named begin with A. There are five large-scale nodes (include server node A_0) named begin with B, twelve media-scale nodes named begin with C, ten small-scale nodes named begin with D.

only distance-based delay (due to monitor window and packets transportation) and no probabilistic delay for agents to perceive its local information. Hyperparameters of centralized DDPG and ComDDPG are shown in Table 1.

Table 1. Parameters of DDPG and ComDDPG.

Hyperparameter	Reference
Memory size	7000
Critic learning rate	0.001
Action noise	0.2
Actor learning rate	0.001
Reward α	0.2
Reward β	0.2
Batch size	128
Discount factor γ	0.99
Optimizer	Adam
Regularization	l_2

4.2   Reward

Our reward function introduces p_t and q_t as shown in (6) to keep the balance of the legitimate traffic passing rate, the server usage, and the waiting queue length. To validate, packets generation distribution is Normal distribution, and attacks launch malicious traffic over 30 regions (total 108 regions) under pulse attack model (we will discuss later). ComDDPG training and testing curves under communication delay (probability is 0.2) are shown in Fig. 4. 0–20000 step is the training period, 20000–22000 step is the test period. The black line is the legitimate traffic passing rate g_t, the red line is the reward r_t, and the blue line is waiting queue length q_t with $\alpha = 0.2$, $\beta = 0.0$. We can see that, if the q_t is nearly zero, the reward is the same as the legitimate passing rate g_t. If q_t is larger, the agent will be punished and get a lower reward. Thus, agents should learn to control throttle fraction to make the waiting queue empty.

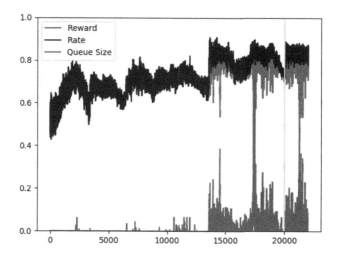

Fig. 4. Training and test period by our reward function. The increase of legitimate traffic pass rate may at the expensive of the increase of queue size, thus we need our adversarial reward function to keep the balance.

To validate the effectiveness of our reward function, we compare α and β with different values in our ComDDPG method. The legitimate traffic passing rate during training period under communication delay (probability is 0.5) is shown in Fig. 5. To exclude the influence of time-difference in reward function, these results are all under the reward function without time-difference. When $\beta = 0.0$, server usage is ignored, so the training progress increases very slowly. When $\alpha = 0.0$, the increase of legitimate traffic pass rate may at the expensive of the increase of queue size. Thus we should keep $\alpha > 0$ and $\beta > 0$, to keep the balance of the legitimate traffic passing rate, the server usage, and the waiting queue length.

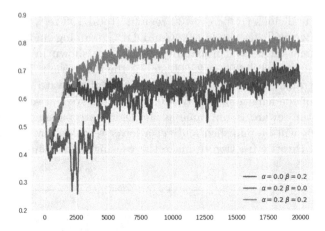

Fig. 5. Training process under different coefficients for p_t and q_t. When $\alpha > 0$ and $\beta > 0$, can keep the balance of the legitimate traffic pass rate, the server usage, and the waiting queue length.

4.3   Diverse Attacks

To validate our method outperforms under diverse attacks, we consider the following three DDoS attack types. For each sub-experiment, we only control one variable and keep the other variables the same.

1. Attackers launch an attack via different malicious packets generation distributions. Specifically, there are 3 distributions we consider in our experiment: Normal distribution, Gamma distribution, Weibull distribution.
2. Attackers launch an attack via different region numbers. We test all methods when attackers launch malicious traffic over 15 regions, 30 regions, and 50 regions, respectively.
3. Attackers launch an attack via different modes. There are three modes of attack: constant rate attack (the maximum attack rate is achieved immediately when the attack is started), increasing rate attack (the maximum attack rate is achieved gradually over 5 s), pulse attack (the attackers generate packets for 5 s with probability 0.14).

To make a fair evaluation, all algorithms cannot learn during evaluation. And we use the average results from 10 different episodes (the 10 test episodes is different from the training episodes). We observe that all algorithms will prevent servers from overloading, thus we focus on metrics based on legitimate aggregate load and threshold queue length, as shown in (7). In the following experiment, the communication delay probability is 0.2.

$$\bar{g}_{test} + [\bar{q}_{test} - 0.1, 0]_+ \tag{7}$$

We first compare our ComDDPG method with existing rule-based methods including SRT and FRT, and reinforcement learning methods including independent DQN and centralized DDPG. We can see that except Linear Programming, which is impossible to implement in reality, our ComDDPG outperforms other distributed policies, but a little worse than centralized DDPG. The independent DQN (without communication) is worst among these methods, which illustrates that communication is indispensable for learning an ideal policy.

The result comparison of different legitimate traffic generation distribution under communication delay (probability is 0.2) is shown in Table 2. We can see that centralized DDPG outperform other methods except Linear Programming method, while our method ComDDPG is a little worse than centralized DDPG. It illustrates that our ComDDPG can be robust for different legitimate traffic generation distributions.

The result comparison of the different number of attack regions is shown in Table 3. Our ComDDPG also works well, even better than centralized DDPG when the number of attack regions is 15. It demonstrates that no matter the attack traffic is relatively concentrated or distributed, our ComDDPG can do a good job.

The result comparison of different attack modes is shown in Table 4. We can see that our ComDDPG still performs well, even better than centralized DDPG under increasing rate attack mode. It illustrates that whether the attack is regular over time, our ComDDPG is robust for defending DDoS attacks.

Table 2. Comparison under different distribution.

		Distribution		
		Normal	Gamma	Weibull
Centralized policy	SRT	0.8349	0.8189	0.7676
	DDPG	**0.8531**	**0.8561**	**0.8446**
Distributed policy	DQN	0.5452	0.5469	0.5444
	FRT	0.8091	0.8424	0.7873
	ComDDPG	0.8466	0.8458	0.8308
Linear programming		0.9863	0.9891	0.9868

Table 3. Different number of router regions been attacked.

		Attack region		
		15	30	50
Centralized policy	SRT	0.8123	0.8349	0.8395
	DDPG	0.8085	**0.8531**	**0.8755**
Distributed policy	DQN	0.5444	0.5452	0.5464
	FRT	0.8134	0.8091	0.8230
	ComDDPG	**0.8312**	0.8466	0.8632
Linear programming		0.9891	0.9863	0.9885

Table 4. Different attack model.

		Attack mode		
		Constant	Increasing	Pulse
Centralized policy	SRT	0.7348	0.8541	0.8349
	DDPG	**0.8572**	0.8628	**0.8531**
Distributed policy	DQN	0.5449	0.5457	0.5452
	FRT	0.8149	0.8332	0.8091
	ComDDPG	0.8551	**0.8657**	0.8466
Linear programming		0.9853	0.9873	0.9863

5   Conclusion

In this paper, we propose a ComDDPG method for distributed routers to defend DDoS attacks. As for legitimate traffic passing rate, our method is better than the existing rule-based method and independent reinforcement learning without communication. In addition, our method can greatly reduce communication cost among distributed routers. Importantly, our method is robust for diverse attacks even under communication delay occasions. Although our method can learn a better policy than the existing methods, there is still a big gap between our method and the ideal policy, Linear Programming.

In future work, we will regard the policy of LP as the expert policy, and we will adopt imitation learning method to learn an ideal policy.

References

1. Kolias, C., Kambourakis, G., Stavrou, A., Voas, J.: DDoS in the IoT: Mirai and other botnets. Computer **50**(7), 80–84 (2017)
2. Mirkovic, J., Prier, G., Reiher, P.: Attacking DDoS at the source. In: 10th IEEE International Conference on Network Protocols, Proceedings, pp. 312–321. IEEE (2002)
3. Mirkovic, J., Reiher, P.: A taxonomy of DDoS attack and DDoS defense mechanisms. ACM SIGCOMM Comput. Commun. Rev. **34**(2), 39–53 (2004)
4. Bhavasar, D.: A survey on distributed denial of service attack and defence (2015)
5. Douligeris, C., Mitrokotsa, A.: DDoS attacks and defense mechanisms: classification and state-of-the-art. Comput. Netw. **44**, 643–666 (2004)
6. Mirkovic, J., Robinson, M., Reiher, P.L.: Alliance formation for DDoS defense. In: Proceedings of the 12th New Security Paradigms Workshop, Ascona, Switzerland, pp. 11–18 (2003)
7. Peng, T., Leckie, C., Ramamohanarao, K.: Detecting distributed denial of service attacks by sharing distributed beliefs. In: Safavi-Naini, R., Seberry, J. (eds.) ACISP 2003. LNCS, vol. 2727, pp. 214–225. Springer, Heidelberg (2003). https://doi.org/10.1007/3-540-45067-X_19
8. Yau, D.K.Y., Lui, J.C.S., Liang, F., Yam, Y.: Defending against distributed denial-of-service attacks with max-min fair server-centric router throttles. IEEE/ACM Trans. Netw. **13**, 29–42 (2005)

9. Oliehoek, F.A., Amato, C.: A Concise Introduction to Decentralized POMDPs. SpringerBriefs in Intelligent Systems, 1st edn, p. 134. Springer, Heidelberg (2016). https://doi.org/10.1007/978-3-319-28929-8

10. Mnih, V., et al.: Human-level control through deep reinforcement learning. Nature **518**, 529–533 (2015)

11. Silver, D., et al.: Mastering the game of go with deep neural networks and tree search. Nature **529**(7587), 484 (2016)

12. Levine, S., Finn, C., Darrell, T., Abbeel, P.: End-to-end training of deep visuomotor policies. J. Mach. Learn. Res. **17**(1), 1334–1373 (2016)

13. Jiang, J., Lu, Z.: Learning attentional communication for multi-agent cooperation. In: Advances in Neural Information Processing Systems, pp. 7254–7264 (2018)

14. Dantzig, G.B.: Linear programming under uncertainty. In: Infanger, G. (ed.) Stochastic Programming. ISOR, vol. 150. Springer, New York (2010). https://doi.org/10.1007/978-1-4419-1642-6_1

15. Sutton, R.S., Barto, A.G., et al.: Introduction to Reinforcement Learning, vol. 135. MIT press, Cambridge (1998)

16. Lillicrap, T.P., et al.: Continuous control with deep reinforcement learning, CoRR abs/1509.02971 (2015)

17. Lowe, R., Wu, Y., Tamar, A., Harb, J., Abbeel, P., Mordatch, I.: Multi-agent actor-critic for mixed cooperative-competitive environments. In: Proceedings of the 30th Advances in Neural Information Processing Systems, Long Beach, CA, pp. 6382–6393 (2017)

18. Sagheer, A., Kotb, M.: Time series forecasting of petroleum production using deep LSTM recurrent networks. Neurocomputing **323**, 203–213 (2019)

19. Malialis, K., Kudenko, K.: Multiagent router throttling: decentralized coordinated response against DDoS attacks. In: Proceedings of the 25th Innovative Applications of Artificial Intelligence Conference, Bellevue, Washington, pp. 1551–1556 (2013)

20. Chang, X.: Network simulations with OPNET. In: Proceedings of the 31st Conference on Winter Simulation: Simulation - A Bridge to the Future, Phoenix, AZ, pp. 307–314 (1999)

21. Miao, Y., Li, W., Tian, D., Hossain, M.S., Alhamid, M.F.: Narrowband Internet of Things: simulation and modeling. IEEE Internet Things J. **5**, 2304–2314 (2018)

Ensemble Regression Kernel Extreme Learning Machines for Multi-Instance Multi-Label Learning

Yibin Wang[1,2(✉)], Gensheng Pei[1], and Yusheng Cheng[1,2]

[1] School of Computer and Information, Anqing Normal University,
Anqing 246011, Anhui, China
wangyb07@mail.ustc.edu.cn
[2] Key Laboratory of Data Science and Intelligence Application,
Fujian Province University, Zhangzhou 363000, Fujian, China

Abstract. The multi-instance multi-label learning (MIML) framework is an extension of the multi-label learning, where each object in MIML is represented by a multi-instance bag and associated with a multi-label vector. Recently, extreme learning machine (ELM) has been widely used in multi-instance multi-label classification due to its short runtime. Simultaneously, ELM also has good classification accuracy compared to other neural network models. However, this type of ELM-based MIML classification algorithms can easily lead to overfitting problems during training and the basic ELM algorithm with random initial weights and biases is not stable. In order to solve the above problems, ensemble learning is used to overcome overfitting problems and regression kernel extreme learning machine (RKELM) as classifier instead of the basic ELM effectively can solve the problem of instability of training. In this paper, Bagging-based RKELM (BRKELM) and AdaBoost-based RKELM (ARKELM) for MIML classifications are proposed. The comparison with other state-of-the-art multi-instance multi-label learning algorithms shows that the BRKELM and ARKELM are highly efficient, feasible and stable algorithms.

Keywords: Multi-instance multi-label learning · Extreme learning machine · Bagging · AdaBoost

1 Introduction

Label learning has always been one of the hot topic researches in machine learning, of which the single-instance single-label (SISL) is currently relatively mature. At present, the study of label learning mainly focuses on single-instance multi-label (SIML or ML) [1], multi-instance single-label (MISL) [2] and multi-instance multi-label (MIML) [3]. MIML learning framework is proposed to solve the ambiguous object for multi-label classification, it also combines the advantages of multi-instance learning and multi-label learning, so its learning framework has become one of the most important research subjects.

In MIML framework, researchers have proposed various algorithms to label and predict unknown objects. For example, Zhou et al. [4] proposed a variety of multi-instance multi-label learning algorithms: MIMLBoost [3], MIMLSVM [3], MIMLkNN [5],

© Springer Nature Singapore Pte Ltd. 2019
K. Knight et al. (Eds.): ICAI 2019, CCIS 1001, pp. 226–239, 2019.
https://doi.org/10.1007/978-981-32-9298-7_18

and MIMLNN [4], etc. The above algorithms are all based on the degenerate strategy, there are mainly two commonly used degenerate strategies: (1) Convert the multi-instance multi-label problem into multi-instance single-label problem, and then convert it into single-instance single-label problem. (2) Convert the multi-instance multi-label into single-instance multi-label problem, and then convert it into single-instance single-label problem. This simplified approach has made great breakthroughs in solving MIML problem, and the execution efficiency has also been improved. It has been widely used in the classification of nature scene image [3, 6], text categorization [7], biological gene annotation [8, 9] and so on. The MIMLBoost algorithm [3] is according to the first degenerate strategy, and then uses multi-instance learning framework to solve MIML problem. The MIMLSVM algorithm [3] is according to the second degenerate strategy, then uses multi-label learning framework to solve MIML problem. However, these transformations may weaken or even lose the internal relations of processing the original MIML data, resulting in a decrease of the final classification effect. In this regard, Li et al. [10] proposed the Key Instances Sharing Among Related labels (KISAR), which considered the correlation between multi-instance and multi-label in the MIML data. To effectively process large data sets, Huang et al. [11] proposed the MIMLfast method, which first constructs a low-dimensional subspace shared by all labels, and then trains a label-specific linear model to optimize the approximate ranking loss by stochastic gradient descent.

Recently, extreme learning machine is the algorithm proposed by Huang et al. [12] to solve the problem of single hidden layer feedforward neural network. The most prominent feature of ELM is its faster speed and better generalization performance than traditional neural networks, especially single hidden layer feedforward neural networks, while ensuring learning accuracy. Yin et al. [13] used extreme learning machine to improve the multi-instance multi-label learning algorithm (MIMLELM), which solves the high computational cost caused by using SVM as a classifier, and has higher learning speed and better generalization performance. However, the MIMLELM algorithm takes the same degeneracy strategy as MIMLSVM, which uses K-Medoids clustering to convert multi-instance multi-label problems to multi-label problems. And the MIMLELM uses the basic ELM algorithm, but such an ELM needs to generate hidden node parameters at random, so the MIMLELM in this way could lead to unstable classification results. To solve this problem, Huang et al. [14] proposed kernel-based ELM algorithm, it can improve the stability of the algorithm and reduce the computational complexity instead of setting the random parameters. Simultaneously, the adaptive ensemble models of ELM algorithm proposed by Heeswijk et al. [15] gets better generalization performance. Subsequently, Liu et al. [16] proposed ELM-based ensemble learning and cross-validation to overcome the training model over-fitting problem. Samat et al. [17] combined Bagging-based and AdaBoost-based learning algorithms in ensemble learning with ELM and successfully applied the hyperspectral image classification. To some extent this method overcomes the shortcomings of ELM caused by random input weights and biases.

Summarizing the above scholars' research results, this paper proposes a regression kernel extreme learning machine algorithm based on ensemble learning to solve the MIML problems. Since the basic ELM requires random initial weights and biases, and it could cause the instability of the classification results. However, the RKELM

(Regression Kernel Extreme Learning Machine) can overcome the drawbacks of the basic ELM. And in this paper, a similarity function [10] degenerate strategy is used to convert MIML problems into ML problems. At the same time, two MIML classifiers by RKELM based on Bagging (BRKELM) and AdaBoost (ARKELM) are proposed to predict the final classification results. To some extent, these two methods not only solve the over-fitting problem in the training process but also ensure the stability and reliability of the classification model. In order to verify the performance of the proposed algorithms, we compare BRKELM and ARKELM with other state-of-the-art algorithms on multiple real-benchmark multi-instance multi-label data sets.

The rest of this paper is organized as follows. Section 2 introduces MIML framework and RKELM method. In Sect. 3, the methods of ensemble regression kernel extreme learning machines are provided. We describe the experiment detail in Sect. 4. Finally, Sect. 5 summarizes the proposed methods and indicates the future direction.

2 Methods

2.1 Multi-Instance Multi-Label (MIML)

A single instance represents an object, and an object relates only to a single label, which is called single-instance single-label learning (SISL). Let X be instance space and Y be the set of class labels. The following n_i is number of instance in X and l_i is number of labels in Y. Given the data set $\{(x_1, y_1), (x_2, y_2), \ldots, (x_n, y_n)\}$, where x_i is an instance of X, $y_i \in Y$ is the label of x_i. The purpose of SISL learning is to get the mapping relationship function $f_{\text{SISL}} : X \to Y$.

But real-world objects are multiple semantic and complex semantic. Simply using SISL to represent these objects may lose some important information, so many real-world objects are not suitable for this learning framework. To solve this problem, the researchers proposed three learning frameworks, namely, multi-instance learning (MIL), multi-label learning (MLL) and multi-instance multi-label learning (MIML). Formal definitions of these three learning frameworks are given as below.

And the MIL framework is defined as follows. Given the data set $\{(X_1, y_1), (X_2, y_2), \ldots, (X_n, y_n)\}$, where a set of instances $X_i \subseteq X$ is $\left\{x_1^{(i)}, x_2^{(i)}, \ldots, x_{n_i}^{(i)}\right\}$, $x_j^{(i)} \in X (j = 1, 2, \ldots, n_i)$, and $y \in (+1, -1)$ is label of X_i. The purpose of MIL learning is to get the mapping relationship function $f_{\text{MIL}} : 2^X \to \{+1, -1\}$.

Then, the MLL framework is defined as follows. Given the data set $\{(x_1, Y_1), (x_2, Y_2), \ldots, (x_n, Y_n)\}$, where x_i is an instance of X, and a set of labels $Y_i \in Y$ is $\left\{y_1^{(i)}, y_2^{(i)}, \ldots, y_{l_i}^{(i)}\right\}$, $y_k^{(i)} \in Y (k = 1, 2, \ldots, l_i)$, the purpose of SIML learning is to get the mapping relationship function $f_{\text{MLL}} : X \to 2^Y$.

Last, the MIML framework is defined as follows. Given the data set $\{(X_1, Y_1), (X_2, Y_2), \ldots, (X_n, Y_n)\}$, where a set of instances $X_i \subseteq X$ is $\left\{x_1^{(i)}, x_2^{(i)}, \ldots, x_{n_i}^{(i)}\right\}$, $x_j^{(i)} \in X (j = 1, 2, \ldots, n_i)$, and a set of labels $Y_i \in Y$ is $\left\{y_1^{(i)}, y_2^{(i)}, \ldots, y_{l_i}^{(i)}\right\}$,

$y_k^{(i)} \in Y (k = 1, 2, \ldots, l_i)$, the purpose of MIML learning is to get the mapping relationship function $f_{\mathrm{MIML}} : 2^X \rightarrow 2^Y$.

So, a simple way is to degrade MIML framework into MIL framework or MLL framework, then we could use multi-instance learning or multi-label learning as the bridge to solve MIML problems.

In this paper, we use a novel degenerate strategy to deal with MIML issues, and this method is based on multi-label learning as a bridge.

In the MIML framework, each example consists of a bag instances and is associated with multiple class labels. Before introducing the method, we give some important formal definitions. Let a set of data set $D = \{X_i, Y_i\}_{i=1}^n$, where a bag $X_i = \left\{ x_j^{(i)} \right\}_{j=1}^{n_i}$ has n_i instances, the label vector of X_i is $Y_i = \left\{ y_k^{(i)} \right\}_k^{l_i}$, and l_i represents the total number of examples' class labels. In order to find out the inherent relationship between each instance, we use k-means to cluster similar instances and use k-NN to find one instance closest to the cluster center as P. Referencing [10], we set the similarity function as Gaussian distance which is naturally normalized. And the similarity function is:

$$sim(X, P) = \min_{x \in X} \exp \left(-\frac{\|x - P\|_2^2}{\delta^2} \right), \qquad (1)$$

where δ is the average distance between instances in a cluster. Degenerating MIML data sets into MLL data sets is defined as:

$$\varnothing(X) = [sim(X, P_1), sim(X, P_2), \ldots, sim(X, P_K)]. \qquad (2)$$

Each instance bag is converted into a feature vector so that we can use multi-label learning to train data. Degradation algorithm is called MIML2ML (multi-instance multi-label to multi-label), the MIML2ML algorithm of details is shown in the Algorithm 1.

Algorithm 1: The MIML2ML Algorithm.

Input: D: data set, n: Number of bags, K: Number of clusters
Output: D^*: Degraded data set
Clustering:
    $D_l = \{(X_i, Y_i) | i = 1, 2, \ldots, n\}$;
    Decompose all bags into instances;
    Use k-means to cluster all instances into K clusters $\{M_1, M_2, \ldots, M_K\}$, K clusters center $\{C_1, C_2, \ldots, C_K\}$.
    for $t = 1$ to K
        Use k-NN to find an instance x^* of the cluster M_t closest to cluster center C_t;
        $P_t = x^*$.
    end for.
Mapping:
for $i = 1$ to n
    Convert data set from MIML to ML
    for $t = 1$ to K

        Compute similarity function $sim(X_i, P_t) = \min_{x \in X_i} \exp \left(-\frac{\|x - P_t\|_2^2}{\delta^2} \right)$;

    end for.
    Get a vector of instance X_i, $\varnothing(X_i) = [sim(X_i, P_1), sim(X_i, P_2), \ldots, sim(X_i, P_K)]$.
end for.
Return degraded data set: $D^* = \{(\varnothing(X_i), Y_i) | i = 1, 2, \ldots, n\}$.

2.2 Regression Kernel Extreme Learning Machine for Multi-label Learning

In recent years, neural network algorithms are widely used in regression and classification tasks. However, the traditional neural network algorithms need more network parameter settings, it is very likely to find the locally optimal solution when solving the optimal solution, so that the globally optimal solution can't be obtained. Particularly, extreme learning machine theory is an efficient algorithm for solving single-hidden layer feedforward neural network. The globally optimal solution can be solved by simply setting the number of hidden layer nodes and randomly initializing the weights and biases. At the same time, the ELM algorithm has the advantages of fast running speed and good generalization performance.

The following provisions need to be made before formalizing the ELM. Let $X = \{x_1, x_2, \ldots, x_N\}$ be feature space with N samples, where $x_i \in R^d$, $i = 1, 2, \ldots, N$, d represents feature's dimension, and let $Y = \{y_1, y_2, \ldots, y_N\}$ be label space with N samples, where $y_i \in R$, $i = 1, 2, \ldots, N$. Then, for a single-hidden layer neural network with L hidden nodes, formalization of the basic ELM is defined as

$$\sum_{j=1}^{L} \beta_j \cdot g(x_i) = \sum_{j=1}^{L} \beta_j \cdot g(\langle w_j, x_i \rangle + b_j) = o_i, \ i = 1, 2, \ldots, N, \quad (3)$$

where w_j denotes the input weight vectors, b_j denotes the bias of jth hidden neuron, $g(x)$ is an activation function, and β_j represents the output weights. The goal of a single hidden layer neural network is to minimize the output error so that it can be expressed as:

$$\sum_{i=1}^{N} \|o_i - y_i\|, \quad (4)$$

by formulas (3) and (4), the ELM is redefined as:

$$\sum_{j=1}^{L} \beta_j \cdot g(\langle w_j, x_i \rangle + b_j) = y_i, \quad (5)$$

Formula (5) can be re-expressed as:

$$H\beta = Y. \quad (6)$$

Because this paper degenerates MIML into ML, basic ELM must be adapted to ML that is a bridge for solving MIML issues. So let a multi-label data set with N samples be $\{x_i, Y_i | i = 1, 2, \ldots, N\}$, where $x_i \in R^d$ represents a d-dimensional feature vector, the function $h(x)$ maps x from the input sapce to the L-dimensional feature space, and $Y_i \in Y$ represents the set of labels associated with x_i. By formulas (3) and (5), for a multi-label task, the ELM output function is:

$$f_l(x) = H\beta = \begin{bmatrix} h(x_1) \\ \vdots \\ h(x_N) \end{bmatrix}_{N \times L} \begin{bmatrix} \beta_1^T \\ \vdots \\ \beta_L^T \end{bmatrix}. \quad (7)$$

According to the Formula (6), the hidden layer output weights is $\boldsymbol{\beta} = \boldsymbol{H}^\dagger \boldsymbol{Y}$, where $\boldsymbol{H}^\dagger$ represents Moore-Penrose generalized inverse of matrix $\boldsymbol{H}$. In order to improve the stability and generalization of the algorithm, the objective function adds the cost parameter C according to ridge regression theory. Then, the multi-label classification objective function can be defined as:

$$\min L_f = \frac{1}{2}\|\boldsymbol{\beta}\|^2 + \frac{C}{2}\sum_{i=1}^{N}\|\boldsymbol{\xi}_i\|^2,$$
$$\text{s.t. } \boldsymbol{\xi}_i^T = h(\boldsymbol{x}_i)\boldsymbol{\beta} - \boldsymbol{Y}_i^T, i = 1, 2, \ldots, N. \tag{8}$$

We can substitute the constraint Formula (8) into the objective function L_f, its equivalent unconstrained optimization problem is:

$$\min L_f = \frac{1}{2}\|\boldsymbol{\beta}\|^2 + \frac{C}{2}\sum_{i=1}^{N}\|\boldsymbol{Y} - \boldsymbol{H}\boldsymbol{\beta}\|^2. \tag{9}$$

According to Karush-Kuhn-Tucker conditions, the output weights $\boldsymbol{\beta}$ can be obtained as:

$$\boldsymbol{\beta} = \boldsymbol{H}^T\left(\frac{\boldsymbol{I}}{C} + \boldsymbol{H}\boldsymbol{H}^T\right)^{-1}\boldsymbol{Y}, \tag{10}$$

where $\boldsymbol{I}$ is a L-dimensional identity matrix. However, randomly setting weights and biases is easily influenced by the random value and results in unstable calculations. So, Huang et al. [18] proposed using kernel ELM to solve this problem, when the mapping function $h(\boldsymbol{x})$ is unknown, the kernel function $K(\boldsymbol{u}, \boldsymbol{v})$ is introduced instead of the original mapping function $h(\boldsymbol{x})$. In this paper RBF (radial basis function) kernel is adopted, the kernel ELM definition is:

$$\boldsymbol{\Omega}_{ELM} = \boldsymbol{H}\boldsymbol{H}^T : \boldsymbol{\Omega}_{ELM(i,j)} = K\left(\boldsymbol{x}_i, \boldsymbol{x}_j\right) = \exp\left(-\gamma\|\boldsymbol{x}_i - \boldsymbol{x}_j\|^2\right). \tag{11}$$

Where γ denotes the RBF kernel parameter, according to the formulas (10) and (11), the Formula (7) can be written as:

$$f_l(\boldsymbol{x}) = \boldsymbol{H}\boldsymbol{\beta} = h(\boldsymbol{x})\boldsymbol{H}^T\left(\frac{\boldsymbol{I}}{C} + \boldsymbol{H}\boldsymbol{H}^T\right)^{-1}\boldsymbol{Y} = \begin{bmatrix} K(\boldsymbol{x}, \boldsymbol{x}_1) \\ \vdots \\ K(\boldsymbol{x}, \boldsymbol{x}_N) \end{bmatrix}^T \left(\frac{\boldsymbol{I}}{C} + \boldsymbol{\Omega}_{ELM}\right)^{-1}\boldsymbol{Y}. \tag{12}$$

For multi-label learning problems, we use the ELM regression model instead of the classification model. And combining the formulas (4), (10) and (12), when both the set of labels Y and the output weight $\boldsymbol{\beta}$ of training data are known, we can predict the unseen example labels.

3   Ensemble Regression Kernel Extreme Learning Machines

3.1   Bagging-Based RKELM for MIML

Bagging, also known as bootstrap aggregating is one of the most famous ensemble learning algorithms, and it is an algorithm used to do model merging in the field of machine learning. This algorithm can improve the stability and accuracy of statistical classifiers and regressors, and it also helps the model avoid overfitting. The model deviation trained by the same learning algorithm may be larger on multiple different training data sets with the same distribution, that is, the variance of the model is relatively large. In order to solve this problem, the T training data subsets are obtained by using the bootstrap sample, and then the same weak classifier is used in the T training data subsets. Last, the final classification result is T weak classifiers results by using voting strategy. These are the core ideas of Bagging.

Algorithm 2: BRKELM

Input: *T*: iteration number, *R*: sampling ratio, X_{train}: training set, X_{test}: testing set, Ω: RBF mapping, RKELM: base classifier.

Output: Y^*: the predicted labels set of testing set

   Training: training set X_{train}
      for *t* = 1 to *T*
         Train RKELM, new training set X_t from X_{train} at sampling ratio *R*;
         Convert X_t from MIML to SIML by using Algorithm 1.
         Step 1: Compute X_t kernel matrix Ω_t according to Eq. (11);

         Step 2: Calculate the output weight $\beta_t = \left(\frac{I}{c} + \Omega_t\right)^{-1} Y_t$.

      end for.
   Testing: testing set X_{test}
      for *t* = 1 to *T*
         Convert test data from MIML to SIML by using Algorithm 1.
         Step 1: Compute X_{test} kernel matrix Ω_{test} according to Eq. (11);
         Step 2: Get the labels of testing set X_{test}: $Y^*_{test} = \Omega_{test}\beta_t$.
      end for.
   Return labels matrix: $Y^*_{test} = \{Y^*_t | t = 1,2,...,T\}$;
   The final results of predicted labels set Y^* by using voting strategy on Y^*_{test}.

3.2   AdaBoost-Based RKELM for MIML

The AdaBoost algorithm is similar to the Bagging algorithm and is also a commonly used ensemble learning algorithm. But the AdaBoost algorithm is an iterative algorithm whose core idea is to train different weak classifiers for the same training set, and then group these weak classifiers to construct a stronger classifier. In this paper, the MIML learning is degenerated into the SIML learning task by the Algorithm 1, then the RKELM is used as a weak classifier via the AdaBoost.

Different from single-label learning, the initial weights distribution D of training data in multi-label learning is $D_1^{ij} = 1/mn$, $i = 1, 2..., m$, $j = 1, 2..., n$, where m is number of samples, n is number of classes. In the training data set, the weight distribution is Dt, and the basic classifier obtained is:

$$F_t(x) : x \rightarrow \{-1, 1\}^n. \tag{13}$$

Then, in the tth iteration, the classification error ε of RKELM on the training data set is

$$\varepsilon_t = \sum_{i=1}^{m} \sum_{j=1}^{n} D_t^{ij} \left[F_t^{ij*} \neq Y_t^{ij} \right], \tag{14}$$

where * indicates the prediction result. In addition, the tth iteration of the coefficient α of the weak classifier RKELM is $\alpha_t = \frac{1}{2} ln\left(\frac{1-\varepsilon_t}{\varepsilon_t} \right)$. At this time, the weight distribution of the updated training data set is represented as:

$$D_{t+1}^{ij} = \frac{D_t^{ij} \exp\left(-\alpha_t Y_t^{ij} Y_t^{ij*} (X_t)\right)}{Z_t},$$
$$Z_t = \sum_{i=1}^{m} \sum_{j=1}^{n} D_t^{ij} \exp\left(-\alpha_t Y_t^{ij} F_t^{ij*} (X_t)\right). \tag{15}$$

In this way, the structure of stronger classifier F(x) is:

$$F(x) = sign\left(\sum_{t=1}^{T} \alpha_t F_t(x) \right). \tag{16}$$

Finally, the pseudo-code description of ARKELM for multi-instance multi-label classification is shown in Algorithm 3.

Algorithm 3: ARKELM

Input: T: Iteration number, X_{train}: training set, X_{test}: testing set, Ω: RBF mapping, RKELM: Weak learner, m: Number of samples, n: Number of classes.

Output: Y^*: the predicted labels set of testing set

  Training:
    Weight initialization: $D_1^{ij} = 1/mn$;
    for $t = 1$ to T
      Train RKELM$_t$ new training set X_t by using distribution D_t on X_{train}
      Convert X_t from MIML to SIML by using Algorithm 1;
      Step 1: Compute X_t kernel matrix Ω_t according to Eq. (11);

      Step 2: Calculate the output weight $\beta_t = \left(\frac{I}{C} + \Omega_t \right)^{-1} Y_t$;

      Step 3: Compute X_t and X_{test} kernel matrix Ω_{all} according to Eq. (11);
      Step 4: Get the labels of X_t and X_{test}: $F_t^* = \Omega_{all}\beta_t$.

      Get the error rate of RKELM$_t$ on X_t: $\varepsilon_t = \sum_{i=1}^{m} \sum_{j=1}^{n} D_t^{ij} \left[F_t^{ij*} \neq Y_t^{ij} \right]$;

      If error rate $\varepsilon_t \leq 0.5$ then choose $\alpha_t = \frac{1}{2} ln\left(\frac{1-\varepsilon_t}{\varepsilon_t} \right)$;

      Then update the weights with:

$$D_{t+1}^{ij} = \frac{D_t^{ij} \exp\left(-\alpha_t Y_t^{ij} F_t^{ij*} (X_t)\right)}{Z_t};$$

    Z_t is a normalization factor.
    end for
  Voting:
    The final predicted labels set of X_{test}, $F^* = sign(\sum_{t=1}^{T} \alpha_t F_t^* (X_{test}))$.

The computational complexity analysis of the two ensemble MIML algorithms proposed in this paper is shown on Table 1. And let N be the number of instance bags, K is the cluster center, the number of k-means iterations is t, the number of labels is m, and the number of iterations is T.

Table 1. Computational complexity analysis

Proposed algorithm	Computational complexity
BRKELM	$O(T(NtK + KmN^2))$
ARKELM	$O(T(NtK + KmN^2 + Nm^2))$

4 Experiments

4.1 Data

In this paper, the multi-instance multi-label experiments are based on 11 real-world datasets including nature scene dataset named Image [3], Reuters text dataset named Reuters [19], Microsoft Research Institute of Cambridge object recognition image database v2 named MSRCv2 [20, 21], Corel5K [11] and seven real-world biological genomes [8]. And the data set of Image consists of 2000 natural scene images with a total of 5 categories of labels: desert, mountains, sea, sunset, trees, with averaging 1.24 labels per image. The data set of Reuters considers the seven most common categories, which make up 2000 documents with an average of 1.15 labels per document. The data set of MSRCv2 contains 591 pictures and 23 categories, of which 48-dimensional features are 16-dimensional Histogram of Oriented Gradient and 32-dimensional Color Histogram. Corel5K contains 5000 images and 260 class labels, each of which is represented by an average of 9 instances. The details of the above three data sets are shown in Table 2. Then, seven real-world biological genomes include two bacteria genomes (*Geobacter Sulfurreducens*, GS), (*Azotobacter Vinelandii*, AV); two archaea genomes (*Haloarcula Marismortui*, HM), (*Pyrococcus Furiosus*, PF); and three eukaryote genomes

Table 2. The data descriptions of four MIML data sets.

Data set	Instances	Bags	Labels	Features	Domain
Image	18000	2000	5	15	images
Reuters	7119	2000	7	243	text
MSRCv2	1758	591	23	48	images
Corel5K	47065	5000	260	15	images

Table 3. The data description of genomes data sets.

Data set	Instances	Examples	Labels	Instances per bag	Labels per example
GS	1214	379	320	3.20 ± 1.21	3.14 ± 3.33
AV	1251	407	340	3.07 ± 1.16	4.00 ± 6.97
HM	950	304	234	3.13 ± 1.09	3.25 ± 3.02
PF	1317	425	321	3.10 ± 1.09	4.48 ± 6.33
SC	6533	3509	1566	1.86 ± 1.36	5.89 ± 11.52
CE	8509	2512	940	3.39 ± 4.20	6.07 ± 11.25
DM	9146	2605	1035	3.51 ± 3.49	6.02 ± 10.24

(*Saccharomyces Cerevisiae*, SC), (*Caenorhabditis Elegans*, CE), *(Drosophila Melano-gaster*, DM). Table 3 gives the details of these seven data sets, all datasets compared in the experiments are publicly at http://lamda.nju.edu.cn/CH.Data.ashx.

4.2 Experimental Results and Analyses

All experimental codes are run in MATLAB 2016a. The computer's hardware environment is Intel Core i5-7500 3.4 GHz CPU, 4G memory, and the operating system is Windows 10. In order to consider the feasibility and accuracy of algorithm verification, the experiment adopted a 10-fold cross-validation method. In this paper, we choose four common multi-instance multi-label learning evaluation indicators include Hamming Loss (HL), One-Error (OE), Ranking Loss (RL), and Average Precision (AP), and use these four kinds of evaluation indicators of MIML to verify performance of the proposed algorithm. Finally, the average value and standard deviation of the evaluation measures obtained by 10 experiments were calculated.

In this paper, two proposed algorithms BRKELM and ARKELM are compared with the state-of-the-art MIML algorithms. The parameter configuration of each comparison algorithm is proposed by the original published paper. The parameters were selected by 5-fold cross-validation on the training data. And our proposed algorithms parameters K (for k-means clustering) is 10% of the number of bags, T (for iteration number) is searched in $\{3, 5, ..., 11\}$, R (for sampling ratio) is 0.1, parameters C (for cost parameter) and γ (for RBF kernel parameter) are searched in $\{0.1, 1, 10\}$. The comparison algorithms MIMLKELM [13], MIMLkNN [5], MIMLSVM [3], KISAR [10], and MIMLfast [11] are the state-of-the-art methods in the Image, Reuters, MSRCv2, and Corel5K datasets, the experimental results are shown in Fig. 1. And in the genomes data sets, MIMLKELM, MIMLNN, and the special-purpose algorithm ENMIMLNN [8] are used for comparison, the test results are detailed in Fig. 2. Among the evaluation criteria measures, the best performance has been highlighted in boldface. In the HL, OE, and RL measures, '↓' indicates 'the lower the better', and in the AP measure, '↑' indicates 'the higher the better'.

Figure 1 shows the experimental results of four MIML data sets, where sequence number 1 to 4 represent data sets the Image, Reuters, MSRCv2 and Corel5K, respectively.

In the Image data set, BRKELM is obviously better than ARKELM, MIMLKELM, MIMLkNN, MIMLSVM, KISAR, and MIMLfast on HL, OE, and AP, and ARKELM is obviously the best on RL. In the Reuters data set, BRKELM is obviously better than MIMLKELM, MIMLkNN, MIMLSVM, and KISAR on all measures, BRKELM is ranked first on OE. Meanwhile, ARKELM is obviously better than BRKELM, MIMLKELM, MIMLkNN, MIMLSVM, and KISAR on all measures except OE. In the MSRCv2 data set, BRKELM is ranked second on HL, OE, and AP. However, ARKELM is obviously the best on all measures except RL. In the Corel5K data set, ARKELM is the best of all comparison algorithms.

Figure 2 shows the experimental results of the seven real-world biological genomes, where sequence number 1 to 7 represent data sets the GS, AV, HM, PF, SC, CE and DM. The Hamming Loss for each comparison algorithm is shown in Fig. 2(a). ARKELM and BRKELM have better performance than the MIMLKELM, MIMLNN, and ENMIMLNN. Simultaneously, in the GS, AV, PF, and DM data sets, ARKELM has the

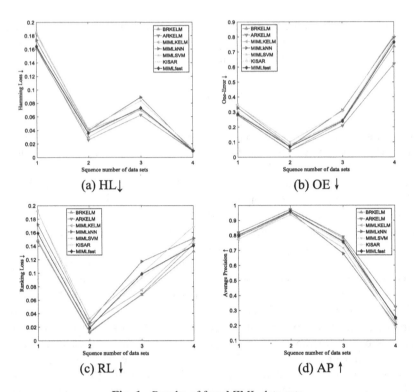

Fig. 1. Results of four MIML data sets.

best performance. And in the HM, SC, and CE data sets, BRKELM has the best performance. Figure 2(b) shows the results of the five comparison algorithms in the One-error evaluation measure. In the GS, AV, SC, CE, and DM data sets, ARKELM has better performance than the BRKELM, MIMLKELM, MIMLNN, and ENMIMLNN. The best performance algorithm of the HM and PF data sets is BRKELM.

For Ranking Loss, Figure(c) shows the results of the comparison algorithms. Among them, ARKELM and BRKELM are better than other algorithms except the GS data set. In the AV, HM, and PF data sets, ARKELM is better than other four algorithms. And BKRELMs has the best performance in the SC, CE, and DM data sets. For the last evaluation measure Average Precision, the comparison results are shown in Figure(d). In the AV, SC, CE, and DM data sets, ARKELM obtains the best performance than other four comparison algorithms. And in the GS, HM, and PF data sets, BRKELM is better than ARKELM, MIMLKELM, MIMLNN, and ENMIMLNN. Therefore, the two ensemble MIML algorithms proposed in this paper, ARKELM and BRKELM have better performance in these real-world biological genomes data sets.

In order to better test the overall performance of the proposed algorithms on the real-world genomes data sets, this paper uses the Nemenyi test [22, 23] with a significance level of 5%. If the average ranking difference of two comparison algorithms is less than or equal to the Critical Difference (CD), then the two algorithms are considered to have no significant difference, otherwise there is a significant difference.

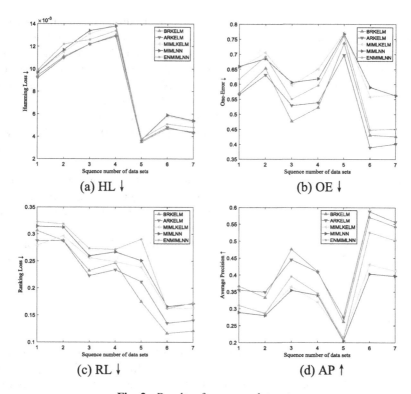

Fig. 2. Results of genomes data sets.

Figure 3 shows the comparison of each algorithm under different evaluation measures, and the CD value is 2.3056. The algorithm with no significant difference is connected with colorful solid lines. From left to right, the performance of the algorithm is reduced successively in each evaluation index subgraph.

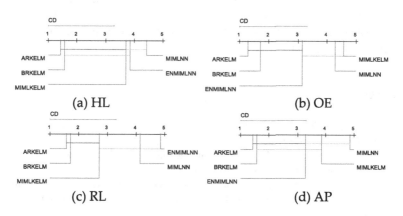

Fig. 3. The performance comparison of algorithms.

For each algorithm, there are 16 (4 comparison algorithms, 4 evaluation measures) experimental results for comparison, as shown in Fig. 3(a) to (d). The following conclusions can be drawn: (I) For ARKELM, the performance is the best on all evaluation measures. In 50% of cases, there is no significant difference with the performance of other algorithms. At the 50% level, it is statistically superior to other algorithms. (II) For BRKELM, it is ranked second on the four evaluation measures. In 37.5% of cases, the performance of BRKELM is statistically superior to other algorithms. And in 62.5% of cases, it is no significant difference with other algorithms. (III) For MIMLKELM, in 81.25% of cases, the performance is no significant difference with other algorithms. However, it is statistically inferior to other algorithms in 18.75% of cases. (IV) For ENMIMLNN, the performance is no significant difference with other algorithms in 68.75% of cases. And in 31.25% of cases, it is statistically inferior to other algorithms. (V) For MIMLNN, it is no significant difference with other algorithms at the 50% level. And it is statistically inferior to other algorithms in 50% of cases.

The analysis of the above experimental results further shows that the proposed multi-instance multi-label learning algorithms ARKELM and BRKELM in this paper have better performance in all aspects.

5   Conclusions

The real-world objects are not only multi-semantic types but are ambiguous. These are the advantages of the multi-instance multi-label learning framework. In this paper, two new algorithms of ensemble regression kernel extreme learning machines (ARKELM and BRKELM) for MIML classification are firstly proposed. And the ensemble learning methods are introduced into the training process of MIML to mitigate the over-fitting and improve the generalization performance of training model. The experimental results in MIML benchmark datasets show that the ARKELM and BRKELM are superior to the state-of-the-art MIML algorithms in multiple evaluation measures.

Currently, the ensemble regression kernel extreme learning machines for MIML are used to reduce the variance of the existing ELM based methods. However, instance imbalances and label imbalances severely affect the performance of the classifier, so the instance imbalances and label imbalances will be the next major research in multi-instance multi-label learning.

References

1. Zhang, M.L., Zhou, Z.H.: A review on multi-label learning algorithms. IEEE Trans. Knowl. Data Eng. **26**(8), 1819–1837 (2014)
2. Wu, J., Pan, S., Zhu, X., et al.: Multi-instance learning with discriminative bag mapping. IEEE Trans. Knowl. Data Eng. **30**(6), 1065–1080 (2018)
3. Zhou, Z.H., Zhang, M.L.: Multi-instance multi-label learning with application to scene classification. In: Advances in Neural Information Processing Systems, pp. 1609–1616 (2007)

4. Zhou, Z.H., Zhang, M.L., Huang, S.J., et al.: Multi-instance multi-label learning. Artif. Intell. **176**(1), 2291–2320 (2012)
5. Zhang, M.L.: A k-nearest neighbor based multi-instance multi-label learning algorithm. In: The 22nd IEEE International Conference on Tools with Artificial Intelligence (ICTAI), pp. 207–212 (2010)
6. Mercan, C., Aksoy, S., Mercan, E., et al.: Multi-instance multi-label learning for multi-class classification of whole slide breast histopathology images. IEEE Trans. Med. Imag. **37**(1), 316–325 (2018)
7. Yan, K., Li, Z., Zhang, C.: A New multi-instance multi-label learning approach for image and text classification. Multimed. Tools Appl. **75**(13), 7875–7890 (2016)
8. Wu, J.S., Huang, S.J., Zhou, Z.H.: Genome-wide protein function prediction through multi-instance multi-label learning. IEEE/ACM Trans. Comput. Biol. Bioinf. **11**(5), 891–902 (2014)
9. Wu, J., Zhu, W., Jiangm, Y., et al.: Predicting protein functions of bacteria genomes via multi-instance multi-label active learning. In: 2018 IEEE 3rd International Conference on Integrated Circuits and Microsystems (ICICM), pp. 302–307. IEEE (2018)
10. Li, Y.F., Hu, J.H., Jiang, Y., et al.: Towards discovering what patterns trigger what labels. In: The 26th AAAI Conference on Artificial Intelligence, pp. 1012–1018 (2012)
11. Huang, S.J., Gao, W., Zhou, Z H.: Fast multi-instance multi-label learning. IEEE Trans. Pattern Anal. Mach. Intell. (2018)
12. Huang, G.B., Zhu, Q.Y., Siew, C.K.: Extreme learning machine: a new learning scheme of feedforward neural networks. In: 2004 Proceedings of International Joint Conference on Neural Networks (IJCNN 2004), vol. 2, pp. 985–990 (2004)
13. Yin, Y., Zhao, Y., Li, C., et al.: Improving multi-instance multi-label learning by extreme learning machine. Appl. Sci. **6**(6), 160 (2016)
14. Huang, G.B., Zhou, H., Ding, X., et al.: Extreme learning machine for regression and multiclass classification. IEEE Trans. Syst. Man Cybern. Part B (Cybern.) **42**(2), 513–529 (2012)
15. Van Heeswijk, M., Miche, Y., Oja, E., et al.: GPU-accelerated and parallelized ELM ensembles for large-scale regression. Neurocomputing **74**(16), 2430–2437 (2011)
16. Liu, N., Wang, H.: Ensemble based extreme learning machine. IEEE Signal Process. Lett. **17**(8), 754–757 (2010)
17. Samat, A., Du, P., Liu, S., et al.: E2LMs: ensemble extreme learning machines for hyperspectral image classification. IEEE J. Sel. Top. Appl. Earth Obs. Remote Sens. **7**(4), 1060–1069 (2014)
18. Huang, G., Huang, G.B., Song, S., et al.: Trends in extreme learning machines: a review. Neural Netw. **61**, 32–48 (2015)
19. Sebastiani, F.: Machine learning in automated text categorization. ACM Comput. Surv. (CSUR) **34**(1), 1–47 (2002)
20. Briggs, F., Fern, X.Z., Raich, R.: Rank-loss support instance machines for MIML instance annotation. In: The 18th ACM SIGKDD International Conference on Knowledge Discovery and Data Mining, pp. 534–542 (2012)
21. Guo, H.F., Han, L., Su, S., et al.: Deep multi-instance multi-label learning for image annotation. Int. J. Pattern Recogn. Artif. Intell. **32**(03), 1859005 (2018)
22. Teisseyre, P.: CCnet: joint multi-label classification and feature selection using classifier chains and elastic net regularization. Neurocomputing **235**, 98–111 (2017)
23. Demšar, J.: Statistical comparisons of classifiers over multiple data sets. J. Mach. Learn. Res. **7**, 1–30 (2006)

A Variable Precision Reduction Type for Information Systems

Guilong Liu$^{(\boxtimes)}$ and Jie Liu

School of Information Science, Beijing Language and Culture University,
Beijing 100083, China
{liuguilong,liujie0829}@blcu.edu.cn

Abstract. Attribute reduction is one of the important topics in the rough set theory, and many reduction approaches based on the rough set theory have been proposed in recent decades. This paper considers two types of reductions for information systems, one is β-reduction and the other is β-variable precision reduction. We show that β-reduction coincides with β-variable precision reduction. Finally, we give an algorithm for such types of reductions by means of discernibility matrices.

Keywords: Attribute reduction · Discernibility matrix ·
Information system · Reduction algorithm ·
Variable precision rough set

1 Introduction

Rough set theory, introduced by Pawlak [16,17] in 1982, is a valid tool to handle imprecise, uncertain, and vague information. It has been widely applied in many fields such as pattern recognition, image processing, feature selection, decision support, data mining and knowledge discovery.

Attribute reduction is one of the most important research topics in rough set theory. There were several attempts [4–9,13,15,18,20] to combine attribute reduction with rough set theory in data science. In order to handle uncertain information and directly derive from the original model without any additional assumptions, Ziarko [19] introduced the concepts of variable precision rough set (VPRS, for short) and corresponding reduction. Mi, Wu and Zhang [12] introduced two types of attribute reduction to study knowledge reduction in VPRS. Liu, Hu and He [10] propose a new approach to calculate the reduction in VPRS model. Now the concept of VPRS has proven to be particularly useful in analysis of inconsistent decision tables obtained from dynamic control processes [14]. Chen and Tsang [1] proposed the concept of local reduction to describe the minimal description of a definable set by attributes of information systems. Chen and Zhao [2] considered local reduction with fuzzy rough sets for decision table, such a local reduction can identify key conditional attribute and offer a minimal description fore very single decision class. Chen and Yang [3] developed the

© Springer Nature Singapore Pte Ltd. 2019
K. Knight et al. (Eds.): ICAI 2019, CCIS 1001, pp. 240–247, 2019.
https://doi.org/10.1007/978-981-32-9298-7_19

concept of θ-local reduction in order to offer a minimal description for special θ-possible decision rules.

The notion of local reduction concerns a minimal set of attributes used in constructing rules in decision table. This paper extends the concept of θ-local reduction, we discuss two types of reductions for information systems and any given nonempty subsets, one is β-reduction and the other is β-variable precision reduction. We will show that the two types of reduction are equivalent.

This paper is organized as follows. In Sect. 2, we review some basic concepts and properties of the characteristic function of sets, binary relations and rough sets. In Sect. 3, we propose the concepts of β-reduction and β-variable precision reduction and study the relationship of them. Section 4 presents an algorithm for β-variable precision reduction. Finally, Sect. 5 concludes the paper.

2   Preliminaries

This section reviews briefly the fundamental notation and notions based on rough sets and variable precision rough sets.

Let U be a universal set and $P(U)$ be the power set of U. For any $X \subseteq U$, based on the characteristic function $\lambda_X : U \to \{0, 1\}$, that is,

$$\lambda_X(x) = \begin{cases} 1, & \text{if } x \in X, \\ 0, & \text{otherwise,} \end{cases}$$

If $U = \{x_1, x_2, \cdots, x_n\}$ is a finite set, then X corresponds to a Boolean column vector $\lambda_X = (\lambda_X(x_1), \lambda_X(x_2), \cdots, \lambda_X(x_n))^T$, where T denotes the usual transpose.

Suppose that R is an arbitrary relation on U, recall that R is called an equivalence relation if it is reflexive, symmetric and transitive. If R be an equivalence relation on U, $[x]_R$ denotes the equivalent class in R containing an element x in U. With a subset $X \subseteq U$, Pawlak defined the following sets.

$$\underline{R}(X) = \{x | x \in U, [x]_R \subseteq X\}$$

and

$$\overline{R}(X) = \{x | x \in U, [x]_R \cap X \neq \emptyset\},$$

called the lower and upper approximations of X, respectively.

If $U = \{x_1, x_2, \cdots, x_n\}$ is a finite set and R is a relation on U, then the relational matrix [11] $M_R = (m_{ij})_{n \times n}$ of R is defined by

$$m_{ij} = \begin{cases} 1, & \text{if } x_i R x_j, \\ 0, & \text{otherwise.} \end{cases}$$

There are many different types of generalization of Pawlak rough sets. For example, Ziarko [19] proposed the concept of variable precision rough set based on equivalence relations. Variable precision rough set [19] is quite interesting. We can extend Ziarko's definition of variable precision rough set as follows.

Definition 2.1 [5]. Let R be an equivalence relation on U. For $\beta \in [0,1]$, and $\emptyset \neq X \subseteq U$, we define two mappings $R^{(\beta)}, R^{(\beta+)} : P(U) \to P(U)$ as follows.

$$R^{(\beta)}(X) = \{x | x \in U, P(X | [x]_R) \geq \beta\}, \text{and}$$
$$R^{(\beta+)}(X) = \{x | x \in U, P(X | [x]_R) > \beta\},$$

where $P(X | [x]_R) = \frac{|[x]_R \cap X|}{|[x]_R|}$. Then $R^{(\beta)}(X)$ is called β-approximation of X and $R^{(\beta+)}(X)$ is called strong β-approximation of X.

Remark. (1) $(R^{(1)}(X), R^{(0+)}(X))$ is Pawlak rough approximation pair of X.
(2) If $0.5 < \beta \leq 1$, then $(R^{(\beta)}, R^{((1-\beta)+)})$ is variable precision rough set pair [19].

3 β-Reductions and Variable Precision Reductions

Recall that an information system is a pair (U, A), where $U = \{x_1, x_2, \cdots, x_n\}$ is a nonempty finite universal set, and $A = \{a_1, a_2, \cdots, a_m\}$ is a nonempty finite set of attributes. Mathematically, each attribute can be viewed as an equivalence relation on U. For each nonempty subset $B \subseteq A$, we associate an equivalence relation $R_B = \cap_{a \in B} a$. Chen, Yang and Zhang [3] proposed the concept of β-reduction for decision tables. This paper extends their definition and gives a different reduction algorithm.

Let (U, A) be an information system, nonempty subsets $X_1, X_2, \cdots, X_s \subseteq U$ and $\beta \in (0, 1]$. For $x \in U$, we denote

$$D_A^\beta(x) = \{X_i | P(X_i | [x]_A) \geq \beta\}.$$

$$\mu_A(x) = (P(X_1 | [x]_A), P(X_2 | [x]_A), \cdots, P(X_s | [x]_A))$$
$$= (\frac{|X_1 \cap [x]_A|}{|[x]_A|}, \frac{|X_2 \cap [x]_A|}{|[x]_A|}, \cdots, \frac{|X_s \cap [x]_A|}{|[x]_A|}).$$

This paper considers types of reductions that keep $D_A^\beta(x)$ and $(\mu_A(x))_\beta (\forall x \in U)$ unchanged, respectively. We also show that the two types of reductions are equivalent. Now we give their definitions.

Definition 3.1. Let (U, A) be an information system. For given nonempty subsets $X_1, X_2, \cdots, X_s \subseteq U$ and $\beta \in (0, 1]$. $\emptyset \neq B \subseteq A$, if B satisfies the following two conditions

(1) $D_A^\beta(x) = D_B^\beta(x)$ for each $x \in U$,
(2) for $B' \subset B$, then $D_A^\beta(x) \neq D_{B'}^\beta(x)$ for some $x \in U$,

then B is called β-reduction of (U, A).

Definition 3.2. Let (U, A) be an information system. For given nonempty subsets $X_1, X_2, \cdots, X_s \subseteq U$ and $\beta \in (0, 1]$. $\emptyset \neq B \subseteq A$, if B satisfies the following two conditions

(1) $(R_A^\beta(X_1), R_A^\beta(X_2), \cdots, R_A^\beta(X_s)) = (R_B^\beta(X_1), R_B^\beta(X_2), \cdots, R_B^\beta(X_s))$,

(2) for $B' \subset B$, then $R_A^\beta(X_i) \neq R_{B'}^\beta(X_i)$ for some X_i,

then B is called β-variable precision reduction of (U, A).

The relationship between β-reduction and β-variable precision reduction is characterized by the following lemma.

Lemma 3.1. Let (U, A), $X_1, X_2, \cdots, X_s$ and β be as in Definition 3.1, then the following conditions are equivalent:

(1) $X_i \in D_A^\beta(x)$.

(2) $P(X_i|[x]_A) \geq \beta$.

(3) $x \in R_A^\beta(X_i)$.

Proof. This result follows from Definitions of $D_A^\beta(x)$ and $R_A^\beta(X)$.   □

Theorem 3.1. Let (U, A), $X_1, X_2, \cdots, X_s \subseteq U$ and β be as in Definition 3.1, $\emptyset \neq B \subseteq A$, then the following conditions are equivalent:

(1) $D_A^\beta(x) = D_B^\beta(x)$ for each $x \in U$.

(2) $R_A^\beta(X_i) = R_B^\beta(X_i)$ for each $i \in \{1, 2, \cdots, s\}$.

Proof. (1) $\Rightarrow$ (2): Suppose that $x \in R_A^\beta(X_i)$, by Lemma 3.1, $X_i \in D_A^\beta(x) = D_B^\beta(x)$. This means that $P(X_i|[x]_B) = \frac{|X_i \cap [x]_B|}{|[x]_B|} \geq \beta$, hence $x \in R_B^\beta(X_i)$ and $R_A^\beta(X_i) \subseteq R_B^\beta(X_i)$. Similarly, $R_B^\beta(X_i) \subseteq R_A^\beta(X_i)$ and $R_A^\beta(X_i) = R_B^\beta(X_i)$.

(2) $\Rightarrow$ (1): For each $x \in U$, suppose that $X_i \in D_A^\beta(x)$, by Lemma 3.1, $x \in R_A^\beta(X_i) = R_B^\beta(X_i)$, thus, $P(X_i|[x]_B) = \frac{|X_i \cap [x]_B|}{|[x]_B|} \geq \beta$, so $X_i \in D_B^\beta(x)$ and $D_A^\beta(x) \subseteq D_B^\beta(x)$. Similarly, $D_B^\beta(x) \subseteq D_A^\beta(x)$ and $D_A^\beta(x) = D_B^\beta(x)$. This completes the proof.   □

Theorem 3.1 tells us that a β-reduction keeps $(R_A^\beta(X_1), R_A^\beta(X_2), \cdots, R_A^\beta(X_s))$ unchanged and vice versa. That is, β-reduction coincides with the β-variable precision reduction.

4   A Reduction Algorithms

This section will gives an algorithm for β-reduction. For given an information system (U, A), let $U = \{x_1, x_2, \cdots, x_n\}$, $A = \{a_1, a_2, \cdots, a_m\}$, nonempty subsets $X_1, X_2, \cdots, X_s \subseteq U$ and $\beta \in (0, 1]$. In order to obtain all β-reduction sets, we define the discernibility matrix $M = (m_{ij})_{n \times n}$ as follows.

$$m_{ij} = \begin{cases} \{a | a \in A, a(x_i) \neq a(x_j)\}, & (\mu_A(x_i))_\beta \neq (\mu_A(x_j))_\beta, \\ \emptyset, & otherwise, \end{cases}$$

where $(\mu_A(x_i))_\beta$ is the β-cut vector of $\mu_A(x_i)$.

Lemma 4.1. Let (U, A), $X_1, X_2, \cdots, X_s \subseteq U$ and β be as in Definition 3.1, if $(\mu_A(x_i))_\beta \neq (\mu_A(x_j))_\beta$, then $m_{ij} \neq \emptyset$.

Proof. If $m_{ij} = \emptyset$, then $[x_i]_A = [x_j]_A$ and $\mu_A(x_i) = \mu_A(x_j)$, this implies $(\mu_A(x_i))_\beta \neq (\mu_A(x_j))_\beta$. □

If $B \subseteq C$, then the restriction of R_C to $[x]_B$ is also an equivalence relation on $[x]_B$, let its quotient set be $[x]_B/R_C = \{[y_1]_C, [y_2]_C, \cdots, [y_{t_x}]_C\}$.

Lemma 4.2. Let (U, A), $X_1, X_2, \cdots, X_s \subseteq U$ and β be as in Definition 3.1, let $B \subseteq A$ and $[x]_B/R_A = \{[y_1]_A, [y_2]_A, \cdots, [y_{t_x}]_A\}$, then $\mu_B(x) = \sum_{i=1}^{t_x} \frac{|[y_i]_A|}{|[x]_B|} \mu_A(y_i)$.

Proof. Since $\frac{|X \cap [x]_B|}{|[x]_B|} = \sum_{i=1}^{t_x} \frac{|[y_i]_A|}{|[x]_B|} \frac{|X \cap [y_i]_A|}{|[y_i]_A|}$ for each subset $X \subseteq U$, we have $\mu_B(x) = \sum_{i=1}^{t_x} \frac{|[y_i]_A|}{|[x]_B|} \mu_A(y_i)$. □

Theorem 4.1. Let (U, A), $X_1, X_2, \cdots, X_s \subseteq U$ and β be as in Definition 3.1. If $\emptyset \neq B \subseteq A$, then the following conditions are equivalent:

(1) $(\mu_A(x))_\beta = (\mu_B(x))_\beta$ for each $x \in U$.
(2) If $m_{ij} \neq \emptyset$, then $m_{ij} \cap B \neq \emptyset$.

Proof. $(1) \Rightarrow (2)$: If $m_{ij} \neq \emptyset$, by the definition of m_{ij}, $(\mu_A(x_i))_\beta \neq (\mu_A(x_j))_\beta$, using hypothesis (1), $(\mu_B(x_i))_\beta \neq (\mu_B(x_j))_\beta$, thus $[x_i]_B \neq [x_j]_B$ and $m_{ij} \cap B \neq \emptyset$.

$(2) \Rightarrow (1)$: We first show that if $x R_B y$, then $(\mu_A(x))_\beta = (\mu_A(y))_\beta$. In fact, if $(\mu_A(x_i))_\beta \neq (\mu_A(x_j))_\beta$, by Lemma 4.1, $m_{ij} \neq \emptyset$, using hypothesis (2), $m_{ij} \cap B \neq \emptyset$, this means that $x_i R_B x_j$ is not true.

For each $x \in U$, let $[x]_B/R_A = \{[y_1]_A, [y_2]_A, \cdots, [y_{t_x}]_A\}$ and $y_1 = x$. Since $[x]_B = [y_i]_B = [y_j]_B$, we have $y_i R_B y_j, i, j = 1, 2, \cdots, t_x$, and $(\mu_A(y_i))_\beta = (\mu_A(y_j))_\beta$. By using Lemma 4.2, we have $(\mu_B(x))_\beta = (\mu_A(y_1))_\beta = (\mu_A(x))_\beta$ for each $x \in U$, this completes the proof. □

Corollary 4.1. Let (U, A), $X_1, X_2, \cdots, X_s \subseteq U$ and β be as in Definition 3.1, $\emptyset \neq B \subseteq A$, then B is a β-reduct of (U, A) if and only if it is a minimal subset of A satisfying $B \cap m_{ij} \neq \emptyset$ for any $m_{ij} \neq \emptyset$.

We now give the β-reduction algorithm as follows:
Algorithm. A β-reduction algorithm.
Input. An information system (U, A), $\beta \in (0, 1)$ and $X_1, X_2, \cdots, X_s \subseteq U$.
Output. All β-reduction sets.
(1) Compute $n \times s$ matrix

$$
\begin{pmatrix} \mu_A(x_1) \\ \mu_A(x_2) \\ \vdots \\ \mu_A(x_n) \end{pmatrix} = \begin{pmatrix} P(X_1|[x_1]_A) & P(X_2|[x_1]_A) & \cdots & P(X_s|[x_1]_A) \\ P(X_1|[x_2]_A) & P(X_2|[x_2]_A) & \cdots & P(X_s|[x_2]_A) \\ \vdots & \vdots & \ddots & \vdots \\ P(X_1|[x_n]_A) & P(X_2|[x_n]_A) & \cdots & P(X_s|[x_n]_A) \end{pmatrix}
$$

and its β-cut matrix.

(2) Compute the discernibility matrix $M = (m_{ij})_{n \times n}$.

(3) Transform the discernibility function f from its CNF $f = \prod(\sum_{m_{ij} \neq \emptyset} m_{ij})$ into its DNF $f = \sum_{i=1}^{t}(\prod B_i)$.

(4) $Red(C) = \{B_1, B_2, \cdots, B_t\}$ and $Core(C) = \cap_{i=1}^{t} B_i$.

End the algorithm.

Matrices are used in many applications in science and engineering, we can make full use of matrices in our proposed algorithm. In fact, let $M_A = (a_{ij})_{n \times n}$ be the relational matrix of R, then, by direct computation, we have the following formula.

$$\frac{a_{i1}}{|[x_i]_A|}\lambda_{X_j}(x_1) + \frac{a_{i2}}{|[x_i]_A|}\lambda_{X_j}(x_2) + \cdots + \frac{a_{i1}}{|[x_i]_A|}\lambda_{X_j}(x_n) = \frac{|[x_i]_A \cap X_j|}{|[x_i]_A|}.$$

Thus, we can calculate $(\mu_A(x_1), \mu_A(x_2), \cdots, \mu_A(x_n))^T$ by the following formula:

$$\begin{pmatrix} \mu_A(x_1) \\ \mu_A(x_2) \\ \vdots \\ \mu_A(x_n) \end{pmatrix} = \begin{pmatrix} \frac{1}{|[x_1]_A|} & & & \\ & \frac{1}{|[x_2]_A|} & & \\ & & \ddots & \\ & & & \frac{1}{|[x_n]_A|} \end{pmatrix} M_A(\lambda_{X_1}, \lambda_{X_2}, \cdots, \lambda_{X_s}),$$

where $|[x_1]_A|$ is the cardinality of set $[x_1]_A$.

We illustrate the above algorithm with the following example.

Example 4.1. Consider the following information system (U, A) (See Table 1), where $U = \{x_1, x_2, \cdots, x_6\}$ and $C = \{a_1, a_2, a_3, a_4, a_5\}$.

Table 1. An information system

U	a_1	a_2	a_3	a_4	a_5
x_1	1	0	0	0	0
x_2	0	1	1	1	1
x_3	0	1	0	0	1
x_4	0	1	1	0	2
x_5	0	1	0	0	0
x_6	0	1	0	0	1

Since R_A is an equivalence relation on U, by direct computation, we obtain corresponding quotient set $U/R_A = \{\{x_1\}, \{x_2\}, \{x_3, x_6\}, \{x_4\}, \{x_5\}\}$. If $X_1 = \{x_1, x_4, x_5\}$, $X_2 = \{x_2, x_3, x_4\}$, $X_3 = \{x_5, x_6\}$ and $\beta = 0.45$, then $\lambda_{X_1} = (1, 0, 0, 1, 1, 0)^T$, $\lambda_{X_2} = (0, 1, 1, 1, 0, 0)^T$, and $\lambda_{X_3} = (0, 0, 0, 0, 1, 1)^T$, moreover, we have

$$
\begin{pmatrix} \mu_A(x_1) \\ \mu_A(x_2) \\ \mu_A(x_3) \\ \mu_A(x_4) \\ \mu_A(x_5) \\ \mu_A(x_6) \end{pmatrix} = \begin{pmatrix} 1 & 0 & 0 & 0 & 0 & 0 \\ 0 & 1 & 0 & 0 & 0 & 0 \\ 0 & 0 & \frac{1}{2} & 0 & 0 & 0 \\ 0 & 0 & 0 & 1 & 0 & 0 \\ 0 & 0 & 0 & 0 & 1 & 0 \\ 0 & 0 & 0 & 0 & 0 & \frac{1}{2} \end{pmatrix} \begin{pmatrix} 1 & 0 & 0 & 0 & 0 & 0 \\ 0 & 1 & 0 & 0 & 0 & 0 \\ 0 & 0 & 1 & 0 & 0 & 1 \\ 0 & 0 & 0 & 1 & 0 & 0 \\ 0 & 0 & 0 & 0 & 1 & 0 \\ 0 & 0 & 1 & 0 & 0 & 1 \end{pmatrix} \begin{pmatrix} 1 & 0 & 0 \\ 0 & 1 & 0 \\ 0 & 1 & 0 \\ 1 & 1 & 0 \\ 1 & 0 & 1 \\ 0 & 0 & 1 \end{pmatrix} = \begin{pmatrix} 1 & 0 & 0 \\ 0 & 1 & 0 \\ 0 & \frac{1}{2} & \frac{1}{2} \\ 1 & 1 & 0 \\ 1 & 0 & 1 \\ 0 & \frac{1}{2} & \frac{1}{2} \end{pmatrix}.
$$

Thus

$$
\begin{pmatrix} \mu_A(x_1) \\ \mu_A(x_2) \\ \mu_A(x_3) \\ \mu_A(x_4) \\ \mu_A(x_5) \\ \mu_A(x_6) \end{pmatrix}_\beta = \begin{pmatrix} 1 & 0 & 0 \\ 0 & 1 & 0 \\ 0 & \frac{1}{2} & \frac{1}{2} \\ 1 & 1 & 0 \\ 1 & 0 & 1 \\ 0 & \frac{1}{2} & \frac{1}{2} \end{pmatrix}_\beta = \begin{pmatrix} 1 & 0 & 0 \\ 0 & 1 & 0 \\ 0 & 1 & 1 \\ 1 & 1 & 0 \\ 1 & 0 & 1 \\ 0 & 1 & 1 \end{pmatrix}.
$$

By symmetric property, upper triangular part of the discernibility matrix M is as follows:

$$
\begin{pmatrix} \emptyset & A & \{a_1, a_2, a_5\} & \{a_1, a_2, a_3, a_5\} & \{a_1, a_2\} & \{a_1, a_2, a_5\} \\ & \emptyset & \{a_3, a_4\} & \{a_4, a_5\} & \{a_3, a_4, a_5\} & \{a_3, a_4\} \\ & & \emptyset & \{a_3, a_5\} & \{a_5\} & \emptyset \\ & & & \emptyset & \{a_3, a_5\} & \{a_3, a_5\} \\ & & & & \emptyset & \{a_5\} \\ & & & & & \emptyset \end{pmatrix}.
$$

The discernibility function

$$
f = (a_1 + a_2)(a_3 + a_4)a_5
$$

$$
= a_1 a_3 a_5 + a_1 a_4 a_5 + a_2 a_3 a_5 + a_2 a_4 a_5.
$$

Thus $\{a_1, a_3, a_5\}, \{a_1, a_4, a_5\}, \{a_2, a_3, a_5\}$ and $\{a_2, a_4, a_5\}$ are all β-reduction results. The reduction core is $\{a_5\}$.

5   Conclusions

Many types of attribute reduction in information systems have been proposed based on rough set theory. θ-local reduction [3] is an interesting type of reduction. This paper has studied β-reduction as a generalization of θ-local reduction. In order to consider the relationship between β-reduction and variable precision reduction, we have proposed the concept of β-variable precision reduction and shown that β-reduction is equivalent to β-variable precision reduction. We also have obtained an algorithm for such types of reduction.

Acknowledgements. This work was supported by the National Natural Science Foundation of China (Grant No. 61272031) and supported by Science Foundation of Beijing Language and Culture University (The Fundamental Research Funds for the Central Universities) (Grant No. 19YJ040008).

References

1. Chen, D., Tsang, E.C.C.: On the local reduction of information system. In: Yeung, D.S., Liu, Z.-Q., Wang, X.-Z., Yan, H. (eds.) ICMLC 2005. LNCS (LNAI), vol. 3930, pp. 588–594. Springer, Heidelberg (2006). https://doi.org/10.1007/11739685_61

2. Chen, D.G., Zhao, S.Y.: Local reduction of decision system with fuzzy rough sets. Fuzzy Sets Syst. **161**, 1871–1883 (2010)

3. Chen, D., Yang, Y., Zhang, X.: Parameterized local reduction of decision systems. J. Appl. Math. **2012**, 2603–2621 (2012)

4. Ge, H., Li, L., Xu, Y., Yang, C.: Quick general reduction algorithms for inconsistent decision tables. Int. J. Approximate Reasoning **82**, 56–80 (2017)

5. Liu, G.: Matrix approaches for variable precision rough approximations. In: Ciucci, D., Wang, G., Mitra, S., Wu, W.-Z. (eds.) RSKT 2015. LNCS (LNAI), vol. 9436, pp. 214–221. Springer, Cham (2015). https://doi.org/10.1007/978-3-319-25754-9_19

6. Liu, G.: Assignment reduction of relation decision systems. In: Polkowski, L., et al. (eds.) IJCRS 2017. LNCS (LNAI), vol. 10313, pp. 384–391. Springer, Cham (2017). https://doi.org/10.1007/978-3-319-60837-2_32

7. Liu, G.L., Hua, Z., Zou, J.Y.: Local attribute reductions for decision tables. Inf. Sci. **422**, 204–217, 342 (2018)

8. Liu, G.L., Hua, Z.: Partial attribute reduction approaches to relation systems and their applications. Knowl.-Based Syst. **139**, 101–107 (2018)

9. Liu, G.L., Hua, Z.: A general reduction method for fuzzy objective relation systems. Int. J. Approximate Reasoning **105**, 241–251 (2019)

10. Liu, J.N.K., Hua, Y., He, Y.: A set covering based approach to find the reduct of variable precision rough set. Inf. Sci. **275**, 83–100 (2014)

11. Grassmann, W.K., Tremblay, J.P.: Logic and Discrete Mathematics: A computer Science Perspective. Prentice-Hall, Upper Saddle River (1996)

12. Mi, J.S., Wu, W.Z., Zhang, W.X.: Approaches to knowledge reduction based on variable precision rough set model. Inf. Sci. **159**, 255–272 (2004)

13. Min, F., Zhu, W.: Attribute reduction of data with error ranges and test costs. Inf. Sci. **211**, 48–67 (2012)

14. Mieszkowicz-Rolka, A., Rolka, L.: Variable precision rough rets in analysis of inconsistent decision tables. In: Rutkowski, L., Kacprzyk, J. (eds.) Advances in Soft Computing. Physica-Verlag, Heidelberg (2003)

15. Mieszkowicz-Rolka, A., Rolka, L.: Variable precision fuzzy rough sets. In: Peters, J.F., Skowron, A., Grzymała-Busse, J.W., Kostek, B., Świniarski, R.W., Szczuka, M.S. (eds.) Transactions on Rough Sets I. LNCS, vol. 3100, pp. 144–160. Springer, Heidelberg (2004). https://doi.org/10.1007/978-3-540-27794-1_6

16. Pawlak, Z.: Rough sets. Int. J. Comput. Inf. Sci. **11**, 341–356 (1982)

17. Pawlak, Z.: Rough Sets: Theoretical Aspects of Reasoning About Data. Kluwer Academic Publishers, Boston (1991)

18. Yao, Y., Mi, J., Li, Z.: A novel variable precision (θ, σ)-fuzzy rough set model based on fuzzy granules. Fuzzy Sets Syst. **236**, 58–72 (2014)

19. Ziarko, W.: Variable precision rough set model. J. Comput. Syst. Sci. **46**, 39–59 (1993)

20. Zhang, H.Y., Leung, Y., Zhou, L.: Variable-precision-dominance-based rough set approach to interval-valued information systems. Inf. Sci. **244**, 75–91 (2013)

AI Applications

Quantum Theories with Hypergestures: A New Approach to Musical Semantics

Guangjian Huang, Shahbaz Hassan Wasti, Lina Wei, and Yuncheng Jiang[⊠]

School of Computer Science, South China Normal University, Guangzhou, China
ycjiang@scnu.edu.cn

Abstract. Recently, theories of quantum have become a novel approach to the semantics of music. In those quantum systems, a musical note is quantized as vectors in Hilbert space, and the musical score is interpreted as a tuple. Based on Hilbert space, quantum systems parse the music in a mathematical way. However, as a new approach, even though theories of quantum can abstract semantics of music, but there remain some questions, such as how to define a state of music and how to define the elements of music composition, especially the performance of music. It turns out to be too difficult to find a solution in quantum theories. Therefore, in this paper, we introduce theories of hypergesture to give an answer. Furthermore, we define accidental space and dynamic space to abstract the information contained in musical symbols. These two spaces indicate the intensity and volume of the music and combine the mathematical form of musical notes with musical meanings. The main motivation of this paper is to bridge the gap between the mathematical musical models and the theories of music.

Keywords: Quantum · Semantics of music · Hypergesture · Musical symbol

1 Introduction

Researchers have been trying to create art such as music by Artificial Intelligent (AI) for a long time. The Google AI Magenta[1] is an exploration that helps artists and musicians to extend their processes. It devotes to providing learning algorithms for generating songs and other materials. The Google Brain team start this research project to create art such as music by machine learning, based on TensorFlow [1]. Deep learning provides many models for music [4, 6,7]. However, even though those models can create music, but they cannot understand the semantics of music [8,9]. Because those models are based on the theories of statistics, instead of theories of music. It is still difficult for a computer to understand artistic beauty and semantics of music.

Recently, theories of quantum [2,5,14] provide a new approach to the semantics of music. The field of music includes infinite notes and ways to arrange them.

---

[1] https://opensource.google.com/projects/magenta.

© Springer Nature Singapore Pte Ltd. 2019
K. Knight et al. (Eds.): ICAI 2019, CCIS 1001, pp. 251–264, 2019.
https://doi.org/10.1007/978-981-32-9298-7_20

In quantization of musical systems [3, 13, 16], the musical notes and thoughts are defined as vectors in Hilbert spaces, providing a new way to extract musical semantics.

Except for the theories of quantum, theories of hypergesture also provide another approach to the semantics of music. A famous question mentioned in [10] attracts the attention of many types of research, i.e., "If I am at s and wish to get to t, what characteristic gesture should I perform to get there?" Hypergestures in complex time can model the actions of musicians' hands when they play the piano. In piano music, what we can hear depends on the actions of musicians' hands. Therefore, hypergestures of musicians' hands can be the key to the mathematical approach to the semantics of music. In [12], researchers also devote to global functorial hypergestures over general skeleta for musical performance.

As [13] suggests, we also premise that we primarily consider artistic express-ibility of music rather than aesthetics. They believe that human neurophysiology is essential in the human creation and perception of art, particularly the beauty in art. This precondition is to make sure that the human perception of art is invariably bound by human neurophysiology. Furthermore, human perception of artistic beauty can hardly transcend natural beauty in its full exposure, when it comes to the intensity of the experience.

As a conclusion, those above traditional models [4, 6, 7] for music provided by deep learning are based on the theories of statistics, instead of the theories of music. They cannot understand the semantics of music. However, theories of quantum provide an approach to musical semantics. In this paper, we improve the quantum approach to musical semantics by theories of hypergesture. It bridges the gap between the mathematical musical model and theories of music. In Sect. 2, we briefly review the approach to the semantics of music by quantum. Section 3 is about the limitations of previous work and in the next section, we improve those limitations. In Sect. 5, we define accidental space and dynamic space for musical symbols. Finally, Sect. 6 presents our conclusion and future work.

2   Related Work

In [13], researchers quantize piano music by theories of quantum. Because the entire chromatic scale will increase the complexity of the argument with-out gaining conceptual advantages, they restrict their researches on octaves which realized by the eight white keys on the keyboard, typically written c, d, e, f, g, a, b (in C major chord). Each tone is treated as a disjoint event $\psi_i, i \in \{c, d, e, f, g, a, b\}$, as shown in Fig. 1. The sum of the probabilities of seven events equals to unity, i.e., $\sum_i P(\psi_i) = 1$. Therefore, it can be formalized in a Hilbert space H_7 with standard Euclidean scalar product and the basis of H_7 is $B = \{|\psi_c\rangle, |\psi_d\rangle \cdots |\psi_a\rangle, |\psi_b\rangle\}$. For convenience, they use the standard orthogonal basis, i.e., $(|\psi_c\rangle, |\psi_d\rangle \cdots |\psi_b\rangle) = I_7$, where I_7 is the unit matrix in H_7. Hence, any pure quantum music state can be represented as a unit vector

$|\psi\rangle \in H_7$, i.e., $|\psi\rangle = \sum_i a_i |\psi_i\rangle, i \in \{c, d, e, f, g, a, b\}$, and $\sum |a_i|^2 = 1$. Individual listeners may perceive a quantum music state in different ways, but one will hear a single tone ψ_i each time with the probability $|a_i|^2$. A musical composition can be represented by permutation of quantum music states $\{|\psi\rangle_i\}_i$.

Fig. 1. Notes in musical theory

Fig. 2. Disjoint events represent tones.

As proposed in [3], in general, any musical composition is determined by three elements: (1) a score: the syntactical component of musical compositions; (2) a set of performances: physical events that occur in space and time; (3) a set of musical thoughts: meanings for the musical phrases written in the score.

Musical thoughts are represented by superpositions of a variety of co-existent thoughts, i.e.,

$$|\mu\rangle = \sum_i c_i |\mu_i\rangle, \tag{1}$$

where $|\mu\rangle$ is an abstract object representing a musical thought and $|\mu_i\rangle$ is co-existent thought with weight $|\mu_i\rangle$. In a composite quantum system, musical thoughts have the property of holistic, i.e., the meaning of a global musical phrase determines the contextual meanings of all its parts. But for subjective feelings and situations imagined by a composer or listener are treated as extra-musical meanings. The musical thoughts can be defined as tensor products in $M space \otimes W space$, where $M space$ represents the space of musical thoughts and $W space$ represents the space of extra-musical meanings. They are represented as:

$$|M\rangle = |\mu\rangle \otimes |\omega\rangle, |\mu\rangle| \in M space, |\omega\rangle \in W space, \tag{2}$$

and a theme of music is also defined in analogous form: $|\mu\rangle = \sum_i c_i |\mu_i\rangle$, where $|\mu_0\rangle$ represents the basic theme and $|\mu_1\rangle, |\mu_2\rangle, |\mu_3\rangle \cdots$ represent ambiguous musical thoughts or its variations corresponding to the basic theme.

Semantics of music can be analyzed by quantum cognition [16]. Concepts in quantum cognition are modeled by State Context Property (SCoP) formalism. In SCoP formalism, a concept is represented by: a set of states $\mathcal{E}$, a set of relevant contexts $\mathcal{M}$, a set of properties $\mathcal{L}$, a state-transition function μ and a property evaluation function ν. Function μ describes the possibility of a state changing to another state under the influence of context. Hence, μ is defined as: $\mu : \mathcal{E} \times \mathcal{M} \times \mathcal{E} \longrightarrow [0, 1]$ and $\sum_{q \in \mathcal{E}} \sum_{p \in \mathcal{E}} \mu(q, e, p) = 1, e \in \mathcal{M}$. Function ν describes if a state $q \in \mathcal{E}$ has property $a \in \mathcal{L}$, i.e., ν is defined as: $\nu : \mathcal{E} \times \mathcal{L} \longrightarrow \{0, 1\}$, where $\nu(q,a) = 1$ means q has property a and $\nu(q,a) = 0$ means not. Therefore, a SCoP entity is a tuple $S = (\mathcal{E}, \mathcal{M}, \mathcal{L}, \mu, \nu)$.

Furthermore, [16] defines an interpretation $\mathcal{I}$ of a score S as:

$$I = (Phr, Temp, Real, Wor),$$

where $Phr = \{P_i\}_i$ is score-covering, consisted by partitions P_i of the music score. $Temp$ is a function, mapping any score-column of S to a time interval with accuracy ε. $Real$ is a map: $Phr \longrightarrow Mspace$, mapping any part $P \in Phr$ to a musical meaning $x \in Mspace$. Wor is a map: $Phr \longrightarrow Wspace$, mapping any part $P \in Phr$ to a extra-musical meaning $y \in Wspace$.

As above present, theories of quantum provide a novel efficient approach to the semantics of music. It provides a mathematical way for artificial intelligence to understand music. The main purpose of this paper is to improve the limitations of the quantum approach.

3   Limitations of Previous Work

As a novel approach to the semantics of music provided by theories of quantum, it still has the following limitations.

A limitation of [13] is the standard Euclidean scalar product of H_7. Suppose there are two different vectors $|\psi_i\rangle$ and $|\psi_j\rangle$, representing pure quantum music states, the product of them is scaler. But [3] proposes a better way. They prefer tenser product over scalar product. So the composition of vectors is still vectors, which means the composition of two music states is still a state of music. It is more reasonable.

One of the limitations of [16] is there still lack a way to define a state $q \in \mathcal{E}$ in practice.

Fig. 3. Happy birthday to you.

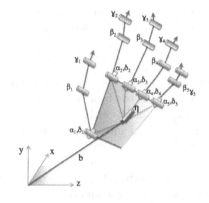

Fig. 4. Spatial coordinates of a hand.

Given a score of music, such as Fig. 3 *"Happy birthday to you"*, we have the following questions: (1) how can we define a state $q \in \mathcal{E}$ of *"Happy birthday to you"*? (2) how many sates are there in total? (3) as $\mathcal{E}$ is the set of music states, is it a finite set or infinite set?

Therefore, the details about music states for specific music are vague. However, such kind of questions as described above are basic questions to define SCoP and interpretations $\mathcal{I}$ for specific music, because $\mathcal{E}$ is an essential element of SCoP. Hence, the feasibility in practice is still uncertified.

One question that remains in [3] is how to define the semantics of performance, which is one of the major elements of a musical composition. This will lead to that we cannot know the practical form of (1) and (2) in the practical case. Such as in Fig. 3, how can we define the performance of music *"Happy birthday to you"*? The practicability still needs to be guaranteed.

The purpose of this paper is to try to answer the above questions; then we can improve the quantum approach to the semantics of music.

4   Improving Limitations of Previous Work with Theories of Hypergesture

Theories of quantum provide an efficient approach to the semantics of music. However, there remain some limitations, as analyzed in the above section. In this section, we propose some new ways to define the states and performances of music, as previous work [3,13,16] have not provided the way to define them.

4.1   The States of Music in Practice

In some cases, musical performance requires movements at infinite physical speed [10]. For instance, given a piano score, there is a consecutive notation of the third interval between middle C and E, with quaver duration. Then it is followed without rest by a third interval D and F with the same duration. First consecutive notation needs to be played with the third and fifth finger of the right hand, and second notation needs to be played with the first and third finger of the right hand. The question is that the first consecutive notation is followed without rest by the second one, which means the third finger need to move infinitely fast from C to F. In practice and physics, it is impossible. To solve the problem between symbolic notation and physical realization, they introduce complex time (C-time) in musical theory. C-time has two dimensions: real physical time dimension and imaginary time dimension. The axis iR for imaginary time is orthogonal to real-time axis R. This means the form of C-time is defined as: $t = t_r + it_i$, where t_r is real time and t_i imaginary time.

A pianist's gesture is treated as a prototype for gesturalization of performance transformations [10]. The dynamic of the pianist's hands have two aspects, i.e., spatial aspect and temporal aspect. The temporal aspect is based on the above complex time. As shown in Fig. 4, the position of a hand is defined by four angles per finger, i.e., $\alpha_i, \gamma_i, \beta_i, \delta_i$. Vector $b \in R^3$ represents the position of the carpus centre with two real numbers for the orientation of the carpus plane. Because playing the piano needs to move hands around the keyboard, so there still needs one real number for the rotational position of the carpus plane around the normal vector through the centre of the carpus plane. Hence, a gesture of a hand is denoted in a space R^{26}.

Therefore, the spatio-temporal model of a pianist's hand can be defined as a vector

$$st = (s, t) \in M_C, M_C = M \oplus C,$$

where $s \in M_C \subseteq R^{26}$ is spacial aspect and $t = t_r + it_i \in C$ is temporal aspect.

When a person listens to piano music, if we overlook the influences come from the quality of the piano itself, what listener can hear depends on the performance of the pianist's hands. This means a state of the pianist's hands directly determine the state of music. As a state of music is difficult to define, but the state of two hands can be defined as (st_1, st_2). Therefore, in this paper, we define a music state $q \in \mathcal{E}$ as follows.

Definition 1 (Music state). *A music state $q \in \mathcal{E}$ is a vector $q \in M_C^2$ with the form*

$$q = (st_1, st_2), st_i = (s_i, t) \in M_C \subseteq \mathcal{E}, i = 1, 2,$$

where st_1, st_2 are spatio-temporal models of pianist's two hands.

With this definition, we can analyze the properties of music states and give answers to the questions about music states remaining in Sect. 3.

Given a score of music, it can be played with finite movements from start to end. Recall that we analyze the semantics of music by piano, we can simulate the hypergestures of the pianist's hands. Hence, the set of states $\mathcal{E}$ is a finite set in $R^{26} \otimes R^{26}$. Then we can get the following proposition:

Proposition 1 (Finiteness). *Given a score of music, the set of states $\mathcal{E}$ is a finite set in $R^{26} \otimes R^{26}$.*

Example 1. As shown in Fig. 3, the score of "Happy birthday to you" has 25 notes. Suppose that we split the motion of a pianist's hands corresponding to notes, i.e., one action plays one note. So the state set of "Happy birthday to you" has 25 elements, i.e.,

$$\mathcal{E} = \{q_i | q_i = (st_1^i, st_2^i), i = 1, 2 \cdots 25\},$$

where q_i is the spatio-temporal model of the pianist's hands corresponding to the action to play the ith note.

In Definition 1, we use C-time as time system. Then function $Temp$ in an interpretation $\mathcal{I}$ of a score S should be redefined as a map from any score-column of S to a C-time interval. So an interpretation $I = (Phr, Temp, Real, Wor)$ is redefined as:

Definition 2 (Interpretation of a score). *An interpretation $\mathcal{I}$ of a score S is defined as:*

$$I = (Phr, Temp, Real, Wor),$$

where $Phr = \{P_i\}_i$ is score-covering, consisted by partitions P_i of the music score. $Temp$ is a function, mapping any score-column of S to a C-time interval with accuracy ε. $Real$ is a map: $Phr \longrightarrow Mspace$, mapping any part $P \in Phr$ to a musical meaning $x \in Mspace$. Wor is a map: $Phr \longrightarrow Wspace$, mapping any part $P \in Phr$ to a extra-musical meaning $y \in Wspace$.

As above, all the questions about music state presented in Sect. 3 have been solved, which means the feasibility in practice can be proofed.

4.2   A Possible Way to Define the Performance of Music

According to [3], one of the major elements of a musical composition is a set of performance: physical events that occur in space and time, but they do not provide a way to define it. In [11], researchers provide a mathematical model for gestures in musical theory. A gesture γ consists of two elements: a combinatorial 'skeleton' represented by a digraph Γ and a 'body' represented by a configuration of continuous curves $\gamma(x) : I \longrightarrow X$, where I is a unit interval and X is a topological space. Simply, hypergesture is the gesture of gestures. $I@X$ is used to represent the set that consists of continuous maps from I to X. The topological space $\overrightarrow{\Gamma@}X$ is defined as the set of gestures with skeleton Γ and body in topological space X that is canonically provided with a topology deduced from the compact-open topology on the set $I@X$.

Recall that the spatio-temporal model of a pianist's hand can be defined as a vector

$$st = (s, t) \in M_C, M_C = M \oplus C.$$

In [10], the gestural dynamic of a hand is defined as a hypergesture $h \in\uparrow \overrightarrow{@}\Delta\overrightarrow{@}M_C$, where $\uparrow$ is the skeleton digraph with two vertices and one connecting arrow. Δ is a skeleton digraph of the hand's articulated movement.

The key point is that the way to define the gestural dynamic of a hand can be generalized to define the gestural dynamic of four limbs, even to a whole person. Figure 4 shows the spatial coordinates of a hand in R^{26}. By the same way, we can define the spatial coordinates of a person. Maybe we can not describe all details about the gestural dynamic of a person, but with proper dimensions, we still can define it as a hypergesture $p \in\uparrow \overrightarrow{@}\Delta\overrightarrow{@}M_C$. Furthermore, we can simulate the performance of all moving things by hypergestures, and we can treat static things like a static gesture of moving things.

Therefore, the set of performance of music composition can be defined as a set in the form: $\{h_i| \in\uparrow \overrightarrow{@}\Delta\overrightarrow{@}M_C\}$. There are more details about theories of hypergesture in [11].

What we must state here is that the purpose of this subsection is to provide a possible approach to defining the performance of music in a mathematical way for the reason that there is still no effective mathematical approach to defining the performance of music when abstract semantics of music by quantum. As far as I know, in the field of semantics of music, there still lacks an efficient approach to defining the performance of music in a mathematical way.

5   Musical Symbols: Accidental and Dynamic Symbol

In a musical sore, there are numerous musical symbols. For instance, Figs. 5 and 6 shows two categories of familiar musical symbols: accidental and dynamic symbol. In this section, we abstract musical meanings of musical symbols by theories of quantum.

5.1   Accidental Space and Dynamic Space

In physics, frequency, volume and timbre are three characters of sound [15]. Under the premise that influences caused by the piano's quality are ignored, the timbre of the same key or same note will not have any differences. For frequency, the main approach to distinguishing different tones $|\psi_i\rangle$ defined in H_7 in Sect. 2 is frequency. But for volume, $|\psi_i\rangle$ does not have any properties of the volume. Firstly, in the same musical compositions when it goes to a climax, the volume will rise, but when it comes to the peaceful part, the volume will decrease. Secondly, different music genres have different styles of volume. Therefore, the volume contains a lot of information, which means the volume still needs to be abstracted in the semantics of music. However, in previous work mentioned in Sect. 2, they only focus on two of those three characters of sound: frequency and timbre. The semantics of volume is ignored.

In musical theory, there are two categories of familiar musical symbols that are connected with volume, i.e., accidental and dynamic symbol. For example, the sharp sign ♯ raises a note by a semitone or half-step, and a flat ♭ lowers it by the same amount, as shown in Figs. 5 and 6.

Fig. 5. (a) sharp   **Fig. 6.** (b) flat.

Similarly, another kind of familiar symbols are the dynamic symbol which indicates the relative intensity or volume of a musical line. For instance, in this paper, we consider the following dynamic symbols. (1) **ppp** (Pianississimo): extremely soft, it is specified with additional "p"s; (2) **pp** (Pianissimo): very soft, usually the softest indication in a piece of music; (3) **p** (Piano): soft, louder than pianissimo; (4) **mp** (Mezzo piano): moderately soft, louder than piano;

(5) **mf** (Mezzo forte): moderately loud, softer than forte; If no dynamic symbol appear, mezzo-forte is assumed to be the prevailing dynamic symbol level; (6) **f** (Forte): loud, used as often as piano to indicate contrast; (7) **ff** (Fortissimo): very loud, usually the loudest indication in a piece; (8) **fff** (Fortississimo): extremely loud.

In order to abstract information contained in volume, we define an accidental space **A** and a dynamic space **D**:

Definition 3 (Accidental space). *An accidental space* **A** *is defined as* $A = \{|1,n\rangle, |0,0\rangle, |0,m\rangle\}$ *where* $|1,n\rangle$ *means the number of sharp signs* ♯ *is* n; $|0,m\rangle$ *means the number of flat signs* ♭ *is* m *and* $|0,0\rangle$ *represents a note without any symbol of accidental.*

Definition 4 (Dynamic space). *A dynamic space* **D** *is defined as* $D = \{|k,n\rangle \mid 1 \leq n \leq 4, n \in Z, k = 0,1\}$ *where* $k = 0$ *means soft and* $k = 1$ *means loud.* $n = 1,2,3,4$ *is relative intensity, corresponding to dynamic symbol* **mf(mp)**, **f(p)**, **ff(pp)**, **fff(ppp)** *respectively and* $|00\rangle$ *means without any dynamic symbol. For instance,* $|01\rangle$ *and* $|11\rangle$ *represent for* **mf** *and* **mp** *respectively.*

To keep consistent with the theories of quantum, the number n and m in space **A** and **D** should be represented in the form of the binary number. In most cases, a note in music score is appended with no more than 3 accidental symbols and we only consider familiar dynamic symbols **mf(mp)**, **f(p)**, **ff(pp)**. Therefore, in this paper, for convenience, we only consider appending at most 3 signs to a note, then we let n and m be double digit. As a result, we can define **A** and **D** in unified form: $\{|ijk\rangle \mid i,j,k = 0,1\}$. If we need to take more symbols into account, it is analogous.

Definition 5 (Standard form). *The standard form of an element* φ *in accidental space* **A** *or dynamic space* **D** *is defined as* $\varphi = |ijk\rangle$ *where* $i,j,k = 0,1$.

Example 2. If the first tone in Fig. 2 is marked with two sharp signs ♯ and **mp**, it is represented as $|\varphi_c\rangle = |\psi_c\rangle \otimes |110\rangle \otimes |001\rangle, |\varphi_c\rangle \in H_7 \otimes A \otimes D$. If the second tone is only marked with **ff**, it is represented as $|\varphi_d\rangle = |\psi_d\rangle \otimes |000\rangle \otimes |111\rangle, |\varphi_d\rangle \in H_7 \otimes A \otimes D$.

We prefer to define a note in $H_7 \otimes A \otimes D$, where H_7 is Hilbert space defined in Sect. 2. Because a note defined in 13-dimension matches the three characters of voice better, i.e., frequency, volume, timbre. Another reason why we define a note in this form is that it is more convenient to analyze musical symbols which can be found in the next subsection.

5.2 Semitone and Whole Tone

In musical theory, different music has different rhythms. Interval is the key to understand the beauty of rhythm and melody in musical theory. A semitone is the smallest musical interval commonly used in Western tonal music, which is defined as the interval between two adjacent notes in a 12-tone scale. The interval between a note and the same note with a sharp or flat sign is a semitone. While a whole note is a note represented by a hollow oval note head and without note stem. For instance, the note in Fig. 5 is a whole note and the interval between the note in Fig. 5 and the penultimate note in Fig. 1 is a semitone.

Recall that in Sect. 2, we define musical tone c, d, e, f, g, a, b which correspond to sol-fa syllables *do*, *re*, *mi*, *fa*, *sol*, *la*, *ti*, *do* as $|\psi_c\rangle, |\psi_d\rangle| \cdots \psi_b\rangle$ in H_7. For convenience to count interval between notes, we rewrite the subscript as $|\psi_1\rangle, |\psi_2\rangle| \cdots \psi_7\rangle$. In musical theory, the interval between a note and the same note marked with the accidental is the number of the signs.

Example 3. The interval between c and $\sharp c$ is a semitone, and the interval between c and $\sharp\sharp c$ is two semitones. If we take $\sharp c$ as an original note and we mark a sharp with it again, i.e., $\sharp\sharp c$, the interval between $\sharp c$ and $\sharp\sharp c$ is a semitone. For flat sign $\flat$ is the same. The interval between $\flat c$ and $\sharp c$ is two semitones.

In this paper, we take a semitone as an unit of interval between different notes. In this case, when we define a note in $H_7 \otimes A \otimes D$ as:

Definition 6. *A note in a musical score is a vector* $\varphi = |\psi_n\rangle \otimes |ijk\rangle \otimes |d_1d_2d_3\rangle$ *in* $H_7 \otimes A \otimes D$.

For a note φ in $H_7 \otimes A \otimes D$, we take its standard form as $|\psi_n\rangle \otimes |ijk\rangle \otimes |d_1d_2d_3\rangle$. Then we can calculate the interval between two notes in musical score.

Definition 7. *An interval operator* T *is a function* $H_7 \otimes A \otimes D \times H_7 \otimes A \otimes D \longrightarrow [0, \infty)$ *which shows the interval between two notes.*

Proposition 2. *Given the same note with different musical symbols* $\varphi_{n_1} = |\psi_{n_1}\rangle \otimes |i_1j_1k_1\rangle \otimes |d_{11}d_{12}d_{13}\rangle$ *and* $\varphi_{n_2} = |\psi_{n_1}\rangle \otimes |i_2j_2k_2\rangle \otimes |d_{21}d_{22}d_{23}\rangle$, *so the interval between* φ_{n_1} *and* φ_{n_1} *is:*

$$T(\varphi_{n_1}, \varphi_{n_2}) = |a_1 + a_3a_2|, a_1 = (2j_1 + k_1), a_2 = (2j_2 + k_2), a_3 = 2(i_1 \oplus i_2) - 1,$$

where $\oplus$ *is addition modulo 2.*

Proof. From the definition of accidental space we can get that $a_1 = (2j_1 + k_1)$ is the number of accidental symbols appending to note φ_{n_1} and $a_2 = (2j_2 + k_2)$ is the number of accidental symbols appending to note φ_{n_2}. In musical theory, an accidental symbol represents a semitone. So if φ_{n_1} and φ_{n_2} come from the same note appended to the same accidental symbols, the interval between them is $|a_1 - a_2|$. As opposite, if φ_{n_1} and φ_{n_2} are appended to the different accidental symbols, the interval between them is $a_1 + a_2$. And by the sign of i_1 and i_2, we can know if their are appended to the same accidental symbols or not. Therefore, the interval between them are $T(\varphi_{n_1}, \varphi_{n_2}) = |a_1 + a_3a_2|$.

In Proposition 2, operator T provides a way to calculate intervals between the same note appended with different accidental symbols. The next step is to generalize it to calculate the interval between different notes. In musical theory, the interval between two adjacent whole tones is two semitones. For instance, the interval between note c and d is two semitones. However, there is an exception: the interval between tone e and tone f is one semitone. In this paper, we only analyze the intervals between the whole tone. For other kinds of tone (such as quarter note and quaver), it is similar.

Proposition 3. *Given two notes $\varphi_{n_1} = |\psi_{n_1}\rangle \otimes |i_1 j_1 k_1\rangle \otimes |d_{11} d_{12} d_{13}\rangle$ and $\varphi_{n_2} = |\psi_{n_2}\rangle \otimes |i_2 j_2 k_2\rangle \otimes |d_{21} d_{22} d_{23}\rangle$, suppose that $n_1 < n_2$, so interval between them is:*

$$T(\varphi_{n_1}, \varphi_{n_2}) = |a_1 + a_3 a_2| + b, a_1 = (2j_1 + k_1), a_2 = (2j_2 + k_2), a_3 = 2(i_1 \oplus i_2) - 1,$$

where $b = 2(n_2 - n_1)$, if $n_1 > 3$ or $n_2 < 4$; $b = 2(n_2 - n_1) - 1$, if $n_1 < 4$ and $n_2 > 3$.

Proof. Suppose that φ_{n_1} and φ_{n_2} come from the same note appended with accidental symbols. By Proposition 2, we can get the interval between them is $T(\varphi_{n_1}, \varphi_{n_2}) = |a_1 + a_3 a_2|$. But in fact, they are not the same, so we need to add the interval between the two original notes ψ_{n_1} and ψ_{n_2}, i.e., the notes after removing accidental symbols. In musical theory, the interval between two adjacent whole tones are two semitones except that the interval between tone e and tone f is one semitone. So the interval between ψ_{n_1} and ψ_{n_2} is $b = 2(n_2 - n_1)$ if it does not contain the interval between e and f, otherwise the interval is $b = 2(n_2 - n_1) - 1$. Therefore the interval between φ_{n_1} and φ_{n_2} is $T(\varphi_{n_1}, \varphi_{n_2}) = |a_1 + a_3 a_2| + b$.

Example 4. When calculate the interval between $\flat\flat c$ and $\sharp f$, because c and f are defined as ψ_1 and ψ_4 in H_7, so we can get:

$$\varphi_{n_1} = |\psi_1\rangle \otimes |010\rangle \otimes |d_{11} d_{12} d_{13}\rangle; \varphi_{n_2} = |\psi_4\rangle \otimes |101\rangle \otimes |d_{21} d_{22} d_{23}\rangle;$$
$$a_1 = 2, a_2 = 1, a_3 = 1, b = 5, T(\varphi_{n_1}, \varphi_{n_2}) = |a_1 + a_3 a_2| = |2 + 1 * 1| + 5 = 8.$$

So the interval between $\flat\flat c$ and $\sharp f$ is 8 semitones which is consistent with musical theory.

From Proposition 3 we can verify the following properties of interval operator T by simple calculation.

Proposition 4. *For any two notes φ_{n_1} and φ_{n_2} in $H_7 \otimes A \otimes D$, the interval operator T has the following properties: non-negativity: $T(\varphi_{n_1}, \varphi_{n_2}) \geq 0$; reflexivity: $T(\varphi_{n_1}, \varphi_{n_1}) = 0$; symmetry: $T(\varphi_{n_1}, \varphi_{n_2}) = T(\varphi_{n_2}, \varphi_{n_1})$.*

In space $H_7 \otimes A \otimes D$, when a note φ_n multiplied by a integral coefficient k, we take the musical meaning of $k\varphi_n$ as adjusting note φ_n with $|k|$ accidental symbols. If $k > 0$, we adjust φ_n with sharp sign $\sharp$. If $k < 0$, we adjust φ_n with flat sign $\flat$. But when $k = 0$, we take $k\varphi_n$ as remove note φ_n.

Definition 8. *For any note $\varphi_{n_1} = |\psi_{n_1}\rangle \otimes |i_1 j_1 k_1\rangle \otimes |d_{11} d_{12} d_{13}\rangle$ in $H_7 \otimes A \otimes D$, when $k = 0$, we define $k\varphi_n$ as remove note φ_n. When $k > 0$, we define $k\varphi_n$ as note φ_n appended with $|k|$ sharp signs $\sharp$. When $k < 0$, we define $k\varphi_n$ as note φ_n appended with $|k|$ flat signs $\flat$.*

Example 5. If φ_n represents the tone $\sharp c$, then $2\varphi_n$ represents the tone $\sharp\sharp\sharp c$; $-3\varphi_n$ represents the tone $\flat\flat c$ and $-\varphi_n$ represents the tone $\flat\sharp c$, i.e., the tone c.

Proposition 5. $\forall k \in Z, k > 0$, for note $\varphi_{n_1} = |\psi_{n_1}\rangle \otimes |i_1 j_1 k_1\rangle \otimes |d_{11}d_{12}d_{13}\rangle$ in $H_7 \otimes A \otimes D$, we can get

$$k\varphi_{n_1} = \begin{cases} |\psi_{n_1}\rangle \otimes |i_1 B(p)\rangle \otimes |d_{11}d_{12}d_{13}\rangle, \ if \ \ i_1 = 1; \\ |\psi_{n_1}\rangle \otimes |i_1 B(q)\rangle \otimes |d_{11}d_{12}d_{13}\rangle, \ if \ \ 2j_1 + k_1 > k \ \ and \ \ i_1 = 0; \\ |\psi_{n_1}\rangle \otimes |i_2 B(w)\rangle \otimes |d_{11}d_{12}d_{13}\rangle, if \ \ 2j_1 + k_1 < k \ \ and \ \ i_1 = 0; \end{cases}$$

where $p = 2j_1 + k_1 + k, q = 2j_1 + k_1 - k, w = k - 2j_1 - k_1, i_2 = \neg i_1$ and $\neg$ is logical function NOT. Function $B(x)$ represent the binary form of decimal number x.

proof. From Definition 8 we can get that note $k\varphi_{n_1}$ is note φ_{n_1} appended with k sharp signs $\sharp$. Then we can get $k\varphi_{n_1} = |\psi_{n_1}\rangle \otimes |ijk\rangle \otimes |d_{11}d_{12}d_{13}\rangle$. And if $i_1 = 1$, $k\varphi_{n_1}$ is appended with $p = 2j_1 + k_1 + k$ sharp signs $\sharp$. So $k\varphi_{n_1} = |\psi_{n_1}\rangle \otimes |i_1 B(p)\rangle \otimes |d_{11}d_{12}d_{13}\rangle$. But when $i_1 = 0$ and $2j_1 + k_1 > k$, $k\varphi_{n_1}$ is appended with $q = 2j_1 + k_1 - k$ sharp signs $\sharp$. Therefore $k\varphi_{n_1} = |\psi_{n_1}\rangle \otimes |i_1 B(q)\rangle \otimes |d_{11}d_{12}d_{13}\rangle$. When $i_1 = 0$ and $2j_1 + k_1 < k$, from musical theory we can know that the accidental symbols sharp sign $\sharp$ will change to flat sign $\flat$. So $k\varphi_{n_1}$ is appended with $w = k - 2j_1 - k_1$ flat signs $\flat$. Then $k\varphi_{n_1} = |\psi_{n_1}\rangle \otimes |i_2 B(w)\rangle \otimes |d_{11}d_{12}d_{13}\rangle$ where $i_2 = \neg i_1$.

As we premise before, in this paper, we only consider appending at most 6 signs to a note. So if function $B(x) > 3$, we let $B(x) = 3$.

Proposition 6. $\forall k \in Z, k < 0$, for note $\varphi_{n_1} = |\psi_{n_1}\rangle \otimes |i_1 j_1 k_1\rangle \otimes |d_{11}d_{12}d_{13}\rangle$ in $H_7 \otimes A \otimes D$, we can get

$$k\varphi_{n_1} = \begin{cases} |\psi_{n_1}\rangle \otimes |i_1 B(p)\rangle \otimes |d_{11}d_{12}d_{13}\rangle, \ if \ \ i_1 = 0; \\ |\psi_{n_1}\rangle \otimes |i_1 B(q)\rangle \otimes |d_{11}d_{12}d_{13}\rangle, \ if \ \ 2j_1 + k_1 > -k \ \ and \ \ i_1 = 1; \\ |\psi_{n_1}\rangle \otimes |i_2 B(w)\rangle \otimes |d_{11}d_{12}d_{13}\rangle, \ if \ \ 2j_1 + k_1 < -k \ \ and \ \ i_1 = 1; \end{cases}$$

where $p = 2j_1 + k_1 - k, p = 2j_1 + k_1 + k, w = -k - 2j_1 - k_1, i_2 = \neg i_1$ and $\neg$ is logical function NOT. Function $B(x)$ represent the binary form of decimal number x.

The proof of this proposition is similar to above Proposition 5.

The purpose of the above two propositions is to represent a note in standard form. It will be more convenient to calculate the interval.

Proposition 7. Given two notes $\varphi_{n_1} = |\psi_{n_1}\rangle \otimes |i_1 j_1 k_1\rangle \otimes |d_{11}d_{12}d_{13}\rangle$ and $\varphi_{n_2} = |\psi_{n_2}\rangle \otimes |i_2 j_2 k_2\rangle \otimes |d_{21}d_{22}d_{23}\rangle$, suppose that $n_1 < n_2$, the interval between $m\varphi_{n_1}$ and $n\varphi_{n_2}$ is:

$$T(m\varphi_{n_1}, n\varphi_{n_2}) = \begin{cases} t + n - m, \ if \ \ m > 0, n > 0 \ \ and \ \ m - n + t < 0; \\ t - n + m, \ if \ \ m > 0, n > 0 \ \ and \ \ m - n + t > 0; \\ t + n - m, \ if \ \ m > 0, n < 0 \ \ and \ \ m - n - t < 0; \\ -t - n + m, if \ \ m > 0, n < 0 \ \ and \ \ m - n - t > 0; \\ t + n - m, \ if \ \ m < 0, n > 0; \\ t + n - m, \ if \ \ m < 0, n < 0 \ \ and \ \ m - n + t < 0; \\ -t - n + m, if \ \ m < 0, n < 0 \ \ and \ \ m - n + t > 0. \end{cases}$$

where $t = |a_1 + a_3a_2| + b$ *and* a_1, a_2, a_3, b *is the same as Proposition 3.*

Proof. By Propositions 3, the interval between φ_{n_1} and φ_{n_2} is $T(\varphi_{n_1}, \varphi_{n_2}) = |a_1 + a_3a_2| + b = t$. Form the musical meaning of m and n in $m\varphi_{n_1}$ and $n\varphi_{n_2}$, we can get the interval between $m\varphi_{n_1}$ and $n\varphi_{n_2}$ by adjusting the interval between φ_{n_1} and φ_{n_2} with accidental symbols. And in musical theory an accidental symbol (sharp sign ♯ or flat sign ♭) represents one semitone. Therefore,

$$T(m\varphi_{n_1}, n\varphi_{n_2}) = \begin{cases} t + n - m, & if \ m > 0, n > 0 \ and \ m - n + t < 0; \\ t - n + m, & if \ m > 0, n > 0 \ and \ m - n + t > 0; \\ t + n - m, & if \ m > 0, n < 0 \ and \ m - n - t < 0; \\ -t - n + m, & if \ m > 0, n < 0 \ and \ m - n - t > 0; \\ t + n - m, & if \ m < 0, n > 0; \\ t + n - m, & if \ m < 0, n < 0 \ and \ m - n + t < 0; \\ -t - n + m, & if \ m < 0, n < 0 \ and \ m - n - t > 0. \end{cases}$$

Example 6. If φ_n represents tone ♯c, so $2\varphi_n$ represents tone ♯♯♯c. So $T(\varphi_n, 2\varphi_n) = 2$ and in musical theory, the interval between ♯c and ♯♯♯c is 2 semitones too. $-3\varphi_n$ represents tone ♭♭c and $T(\varphi_n, -3\varphi_n) = |-3| = 3$. In musical theory, the interval between ♯c and c is one semitone and the interval between c and ♭♭c is 2 semitones. So the interval between ♯c and ♭♭c is 3 semitones.

The semitone is the key to control melody in musical theory. In this subsection, we define a standard form for musical notes. With this standard form, we can translate the musical meanings of musical symbols, accidental and dynamic symbol. The properties of interval operator T connect accidental space with semitone in musical theory.

6   Conclusion and Future Work

In this paper, we bridge the gap between the mathematical musical models and the theories of music by quantum system. The theories of quantum provide a novel approach to the semantics of music. However, as a new method, it still has some limitations. In this paper, we apply the theories of hypergesture to give a solution to the limitations of quantum. We provide an approach to defining a music state and performance of music in the mathematical form by combining theories of quantum and hypergesture. Furthermore, we propose the accidental and dynamic space for the intensity and volume of the music. Those spaces combine musical states with musical meanings. We also define the interval operator to calculate the interval between whole notes. Interval is the key to understand the beauty of rhythm and melody in musical theory. It provides a way to handle the artistic beauty of music. Our future work is to improve the approach to the semantics of music provided by quantum more effectively. Another limitation of the quantized musical instrument is quantum entanglement. The entangled states cannot be determined by the entangled multipartite states. As a result, composite states cannot be represented by the product of individual quanta states. Therefore, it is hard to study the properties between composite states and the corresponding individual states.

Acknowledgments. The works described in this paper are supported by The National Natural Science Foundation of China under Grant Nos. 61772210 and 61272066; Guangdong Province Universities Pearl River Scholar Funded Scheme (2018); The Project of Science and Technology in Guangzhou in China under Grant No. 201807010043; The key project in universities in Guangdong Province of China under Grant No. 2016KZDXM024.

References

1. Abadi, M., Barham, P., Chen, J., et al.: Tensorflow: a system for large-scale machine learning. In: 12th Symposium on Operating Systems Design and Implementation, pp. 265–283 (2016)
2. Chiara, M.L.D., Giuntini, R., Leporini, R., Negri, E., Sergioli, G.: Quantum information, cognition, and music. Front. Psychol. **6**, 15–83 (2015)
3. Chiara, M.L.D., Giuntini, R., Luciani, A.R., Negri, E.: A quantum-like semantic analysis of ambiguity in music. Soft Comput. **21**(6), 1–9 (2017)
4. Dubnov, S., Assayag, G., Lartillot, O., Bejerano, G.: Using machine-learning methods for musical style modeling. Computer **36**(10), 73–80 (2003)
5. Fetter, A.L., Walecka, J.D.: Quantum theory of many-particle system. Phys. Today **25**(11), 54–55 (1972)
6. Hamel, P., Eck, D.: Learning features from music audio with deep belief networks. In: ISMIR, vol. 10, pp. 339–344 (2010)
7. Kereliuk, C., Sturm, B.L., Larsen, J.: Deep learning and music adversaries. IEEE Trans. Multimed. **17**(11), 2059–2071 (2015)
8. Koelsch, S.: Neural substrates of processing syntax and semantics in music. In: Haas, R., Brandes, V. (eds.) Music That Works, pp. 143–153. Springer, Vienna (2009). https://doi.org/10.1007/978-3-211-75121-3_9
9. Leman, M.: An embodied approach to music semantics. Music. Sci. **14**(1), 43–67 (2010)
10. Mannone, M., Mazzola, G.: Hypergestures in complex time: creative performance between symbolic and physical reality. In: Collins, T., Meredith, D., Volk, A. (eds.) MCM 2015. LNCS (LNAI), vol. 9110, pp. 137–148. Springer, Cham (2015). https://doi.org/10.1007/978-3-319-20603-5_14
11. Mazzola, G., Andreatta, M.: Diagrams, gestures and formulae in music. J. Math. Music **1**(1), 23–46 (2007)
12. Mazzola, G., Mannone, M.: Global functorial hypergestures over general skeleta for musical performance. J. Math. Music **10**(3), 227–243 (2016)
13. Putz, V., Svozil, K.: Quantum music. Soft Comput. **21**, 1467–1471 (2017)
14. Sakurai, J.J., Commins, E.D.: Modern Quantum Mechanics. Benjamin-Cummings Publishing, Addison Wesley Longman, Reading (1985)
15. Swift, R.: Sound structure in music by Robert Erickson. Perspect. New Music **14**(1), 148–158 (1975)
16. Veloz, T., Razeto, P.: The state context property formalism: from concept theory to the semantics of music. Soft Comput. **21**, 1505–1513 (2017)

Development of Automatic Grounding Wire Working Robot for Substation

Yuewei Tian[1], Weijie Xu[2(✉)], Wenqiang Zou[1], Yubin Wang[3], and Shouyin Lu[2]

[1] Guizhou Power Grid Co., Ltd., Guiyang Power Supply Bureau, Guiyang, China
603771491@qq.com, zwqgoodboy@163.com
[2] Shandong Jianzhu University,
School of Information and Electrical Engineering, Jinan, China
xuweijie0121@qq.com, sdznjqr@163.com
[3] State Grid Technical College, Jinan, China

Abstract. Installing and removing ground wires is a common substation switch operation. However, due to the high installation position of the substation equipment, the position of the grounding point is mostly high, and the grounding line has a large cross-sectional area, long length and heavy weight. It is often difficult to complete the connecting work because of the insufficient rigidity of the operating rod when it is directly connected to the ground by using the operating rod. Because of the heavy grounding wire itself and the long insulating rod, the operators on the spot stand on the herringbone ladder to carry out the hanging and dismantling operation, which is time-consuming and labor-consuming. To solve the above problem, this paper will develop a mobile operating platform technology suitable for substation complex road conditions, and applicable to substation folding and telescopic mixing robot hanging grounding wire lifting work platform technology, and based on intelligent control system substation robot multi-sensor information fusion technology. Developed an automatic hanging grounding wire robots for the voltage level below $500\,kV$, ensure the personal safety of operators, improve the intelligent level of substation operation.

Keywords: Substation hanging ground wire · Robot ·
Hydraulic arm · Insulation protection system

1 Introduction

Grounding is to prevent sudden energization of all aspects of the power outage at the work site, or to drain the residual charge and induced charge of the disconnected portion of the device. Grounding is a reliable safety measure and effective means to protect workers from electric shock. It is one of the technical measures to ensure safe operation on electrical equipment. Temporary grounding

The paper is supported by Shandong province key R&D program (2017CXGC0905, 2017CXGC0918 and 2017CXGC1505).

wire is an important tool for workers in power system in order to ensure the unexpected voltage on the equipment or line that has been cut off. Installation and removal of temporary grounding wire is an important maintenance work in power system.

At present most substation equipment maintenance, the operation and maintenance personnel are sent to the top of equipment to hang ground wire by the aerial work vehicle, which is inefficient and dangerous. Due to the high installation position of the substation equipment and the grounding wire has a large cross-sectional area, a long length, and a heavy weight. It is difficult to connect the ground wire by the operating rod because of the insufficient rigidity of the operating rod. And the grounding wire is heavy and the operating rod is long. During the process of hanging and dismantling the grounding wire on the herringbone ladder, the grounding wire shakes seriously, which consumes a lot of time and energy. Therefore, there is a certain risk in the operation of artificially hanging grounding wires. Once the operation is wrong, the grounding wires will fall, which will cause casualties. If the ground wire falls into contact with other equipment, the consequences will be more serious, which may cause short-circuit burning equipment, and injuries [1–8].

With the development of robotics technology, it is of great practical significance to combine robotics technology with substation grounding wire operation, and to use robotics technology instead of manual completion of substation grounding wire operation, so as to overcome the drawbacks of personnel grounding operation at high altitude [9].

2   Substation Automatic Mounting and Dismantling Operation Robot Structure

The robot system mainly includes: the substation automatic grounding wire working robot body, the multi degree of freedom hydraulic mechanical arm, the special working tool and the robot control system. The substation automatic grounding wire working robot body provides power and carrier for operation [10, 11]. The high power density hydraulic manipulator is to clamp the special tools to complete the automatic hanging grounding wires. The function of the control system is to realize the functions of walking, working and protection, and there are two ways of remote control and close-range visual control.

2.1   Crawler Movable Chassis Design

Tracked mobile chassis structure is adopted in live maintenance robot of substation. The crawler mobile chassis consists of chassis, crawler frame, driving wheel, load-bearing wheel, crawler, tension buffer device and hydraulic leg [12–15]. It can realize barrier-free movement in a variety of complicated road conditions and providing a movable platform for maintenance operations.

The prototype of automatic grounding wire working robot for Substation is shown in Fig. 1. Table 1 gives a main performance parameters:

Fig. 1. Prototype of automatic grounding wire working robot.

Table 1. Main performance parameters.

Demand	Parameter
Machine quality (excluding robotic arm)	2000 kg
Robot walking speed	0–4 km/h
Life time	≥4 h
Maximum vertical wall height	200 mm
Communication method	EtherCAT, Optical fiber, WiFi
Battery capacity	30 kWh

2.2 Insulation Lifting System Research

The insulation lifting part is mainly composed of main arm, upper arm, telescopic arm and small flying arm [16–18]. The working robot arm is installed at the end of the small flying arm to facilitate the clamping working tool for live working.

The lower and upper sections of the arm are made of metal materials, and the middle section is made of imported insulating hollow rods. The three-section arm is connected by bolts. The maximum pitch angle of the variable-amplitude cylinder, which acts as the actuator to drive the arm. Metal material is used at the front and end of the telescopic arm, and insulating material is used at the middle part. The front end of the inner arm of the telescopic arm is made of metal and the end of the telescopic arm is made of insulating material. A roller device is installed at the end of the outer arm to support the inner arm and reduce the sliding friction coefficient. The telescoping of the inner arm is driven by a linear cylinder. The small flying arm plays the role of fine-tuning the high power density hydraulic arm, which is driven by a linear cylinder.

The insulating material is epoxy resin glass fiber reinforced material, which is a mature and widely used composite material [19,22,23]. This material is a mature material with wide application. It has the characteristics of light in mass, high strength, corrosion resistance, excellent electrical performance, good man- ufacturability and excellent insulation performance. It has the characteristics of material design and shielding electromagnetic wave. Now it has become an irreplaceable important material in the development of national economy, national defense and science and technology.

3   Electrical Control System

The electric control system of live-working robot in substation consists of four parts: hydraulic arm, moving and lifting mechanism, high altitude camera, live working tools. The hydraulic arm is controlled by a hydraulic servo controller [20, 21, 24, 25]. Moving and lifting mechanism, high altitude camera, live working tools are controlled by bus control device. The robot control system transmits signals through wireless, which ensures that the operator is completely isolated from the high voltage electric field.

The electrical control system of the substation live working robot consists of an industrial computer and a PLC system. The industrial computer adopts industrial-grade computer to form a powerful human-computer interaction system, and completes the substation live working task by manipulating the main hand remote wireless telecontrol mechanical arm clamping special equipment or tools [26, 27]. PLC selects Omron CP1H, and there are 3 selection switches on the touch screen, which are manual control, automatic control and remote control. When remote control is selected, the servo valve is controlled by the master. In the automatic control, the servo valve is controlled by the PLC logic program. In the manual control, the servo valve is controlled by the operator directly touching the screen, as shown in Fig. 2.

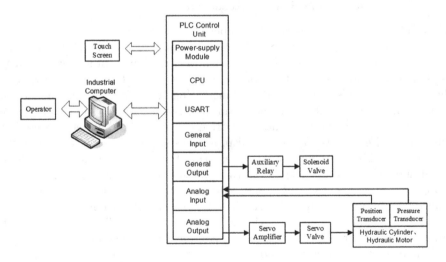

Fig. 2. Schematic diagram of hardware connection of live working robot in substation.

4   Hydraulic Control System

The hydraulic control principle of the robot is shown in Fig. 3. The DC motor is used as the power source, and the coupling is connected with the hydraulic pump through the coupling. Considering the special environment requirements for the live working of the substation, the hydraulic pump adopts the gear double pump, and all adopt the insulating hydraulic oil. The hydraulic oil enters the hydraulic control oil path of the mechanical arm from the fuel tank through the oil suction filter, and then flows back to the oil tank through the oil return filter. The one-way valve controls the oil return, and the safety valve realizes the overload protection function. The pressure gauge is connected to the oil passage through the pressure gauge switch, and displays the oil pressure in real time.

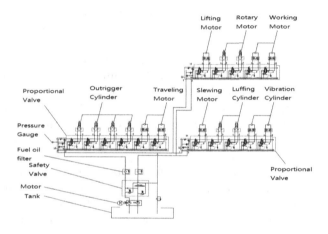

Fig. 3. Insulated lifting arm schematic.

The lower electromagnetic reversing valve group A mainly controls the telescopic movement of the front and rear legs of the robot and the walking movement of the moving platform of the working machine, and is connected with the hydraulic pump through the central rotary joint and the distribution valve [28–30].

Group B tops solenoid valve controlling the main boom tops, the connecting arm, forearm pitch and insulating work platform leveling mechanism and working platform moving working machine, tops controlled by switching button, a dispensing valve connected to the hydraulic pump.

Central swivel joint coupling element is a hydraulic rotary platform and the chassis of the passage, the rotating platform to ensure that any angle of rotation, but also the normal running motor with oil.

5   Insulation Protection System

The working robot is designed to be an open substation equipment. The working environment is complex and the design work functions are numerous, which puts higher requirements on the robot system. In the mechanical design, function and operation of the robot to meet safety requirements by optimizing the mechanical structure, improved insulation materials, increasing the length of the insulation or the like. In the design of the control system, the modular operation design, multi-control unit processing and multi-sensor fusion system are used to meet the robot operation function and safety protection requirements [31–33].

6   Communication and Monitoring System

All control systems between the working machine and the remote control command vehicle transmit signals wirelessly, ensuring that the operator is completely isolated from the high voltage electric field. The operator uses the video information returned by the camera to complete the type identification, posture analysis, and operation decision of the target through monitoring. The stereo camera can obtain 360° horizontal environment information in the image. With the feature of omnidirectional vision, the stereo camera is responsible for all-round real-time monitoring of the working platform, robot arm, tool system and high-voltage equipment. When the robotic work platform reaches the working position, more precise positioning is performed by the tracking camera, as shown in Fig. 4. Each camera is converted into a digital signal by a data acquisition system, transmitted to the control room through an optical fiber, and processed by the industrial computer to obtain visual information of the scene. In the safe area, the operator observes the working posture of the robot arm in real time through the image display screen, and uses the master-slave control type to control the main hand remote wireless remote control arm clamping special tool to contact the high voltage equipment or the line live working [34–36]. As shown in Fig. 5, the staff are testing the prototype and testing the automatic grounding wire working robot to hanging the grounding wires in the substation. In this test, the control mode is remote control mode.

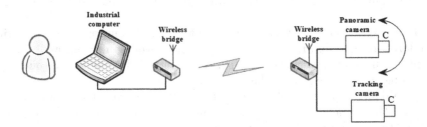

Fig. 4. Schematic diagram of video surveillance system for live working robot.

Fig. 5. The prototype test.

7   Conclusion

According to the 220 kV live working process and working environment of the substation, a new type of substation live working robot is designed. The robot device is considered from the perspective of operation safety of operators, so that robots can be used to complete live substation operation tasks instead of manual labor. The operators are located in a safe area and avoid direct contact with high-pressure equipment, which greatly improves the safety performance of live substation operation of operators and plays a positive role in reforming the operation mode of substations in China.

References

1. Lu, S., Ma, P., et al.: Development of high voltage live working robot. Autom. Electr. Power Syst. **9**, 56–58 (2003)
2. Zhang, H., Su, B., Song, H., Wei, X.: Development and implement of an inspection robot for power substation. In: 2015 IEEE Intelligent Vehicles Symposium (IV), Seoul, pp. 121–125 (2015). https://doi.org/10.1109/IVS.2015.7225673
3. Huang, X.L., Wu, X.-B., Chen, F.-D., Han, X., Feng, X.-M., Lian, L.: 500 kV substation robot patrol system. In: 2017 IEEE 3rd Information Technology and Mechatronics Engineering Conference (ITOEC), Chongqing, pp. 105–109 (2015). https://doi.org/10.1109/ITOEC.2017.8122390
4. Pinto, J.K.C., Masuda, M., Magrini, L.C., Jardini, J.A., Garbelloti, M.V.: Mobile robot for hot spot monitoring in electric power substation. In: 2008 IEEE/PES Transmission and Distribution Conference and Exposition, Chicago, IL, pp. 1–5 (2008). https://doi.org/10.1109/TDC.2008.4517245
5. Shengfang, L., Xingzhe, H.: Research on the AGV based robot system used in substation inspection. In: 2006 International Conference on Power System Technology, Chongqing, pp. 1–4 (2004). https://doi.org/10.1109/ICPST.2006.321495
6. Roncolatto, R.A., et al.: Robotics applied for safety improvement in the power distribution lines maintenance. In: 2008 IEEE/PES Transmission and Distribution Conference and Exposition, Chicago, IL, pp. 1–7 (2008). https://doi.org/10.1109/TDC.2008.4517242

7. Siebert, L.C., Toledo, L.F.R.B., Block, P., Bahlke, D.B., Roncolatto, R.A., Cerqueira, D.P.: A survey of applied robotics for tree pruning near overhead power lines. In: Proceedings of the 2014 3rd International Conference on Applied Robotics for the Power Industry, Foz do Iguassu, pp. 1–5 (2014). https://doi.org/10.1109/CARPI.2014.7030070

8. Hu, Y., Chen, J.: The research of the automatic washing-brushing robot of 500 kV DC insulator string. In: Proceedings of the 6th IEEE International Conference on Transmission and Distribution Construction and Live-Line Maintenance, ESMO 1993, Las Vegas, NV, USA, pp. 411–424 (1993). https://doi.org/10.1109/TDCLLM.1993.316230

9. Fonseca, A., Abdo, R., Alberto, J.: Robot for inspection of transmission lines. In: 2012 2nd International Conference on Applied Robotics for the Power Industry (CARPI), Zurich, pp. 83–87 (2012). https://doi.org/10.1109/CARPI.2012.6473356

10. Wu, Z.: Mechanical Design Practical Manual, 2nd edn. Chemical Industry Press, Beijing (2008)

11. Shanghai Electric Power Company: 10 kV Overhead Distribution Line Live Working Instructions. China Electric Power Press, Beijing (2007)

12. Fadaeenejad, M., Saberian, A.M., Fadaee, M., et al.: The present and future of smart power grid in developing countries. Renew. Sustain. Energy Rev. **29**(7), 828–834 (2014)

13. Siano, P.: Demand response and smart grids—a survey. Renew. Sustain. Energy Rev. **30**(2), 461–478 (2014)

14. Lu, S., Zhang, Y., Su, J.: Mobile robot for power substation inspection: a survey. IEEE/CAA J. Autom. Sin. **99**, 1–18 (2017). https://doi.org/10.1109/JAS.2017.7510364

15. Katrasnik, J., Pernus, F., Likar, B.: A survey of mobile robots for distribution power line inspection. IEEE Trans. Power Deliv. **25**(1), 485–493 (2010)

16. Podnar, G.W.: The Uranus mobile robot. Autonomous Mobile Robots, Ann. Rep., pp. 127–129 (1985)

17. Ghariblu, H., Moharrami, A., Ghalamchi, B.: Design and Prototyping of Autonomous Ball Wheel Mobile Robots. INTECH Open Access Publisher (2011)

18. Nagatani, K., Tachibana, S., Sofne, M., et al.: Improvement of odometry for omnidirectional vehicle using optical flow information. In: Proceedings of the 2000 IEEE/RSJ International Conference on Intelligent Robots and Systems (IROS 2000), vol. 1, pp. 468–473. IEEE (2000)

19. Massachussets Institute of Technology. The MIT intelligent wheelchair project Developing a voice-commandable robotic wheelchair[EB/OL], 22 July 2014. http://rvsn.csail.mit.edu/wheelchair/

20. Ye, C., Ma, S.: Development of an omnidirectional mobile platform. In: International Conference on Mechatronics and Automation, ICMA 2009, pp. 1111–1115. IEEE (2009)

21. Kang, J.W., Kim, B.S., Chung, M.J.: Development of omni-directional mobile robots with Mecanum wheels assisting the disabled in a factory environment. In: International Conference on Control, Automation and Systems, ICCAS 2008, pp. 2070–2075. IEEE (2008)

22. Maruyama, Y.: Robotic applications for hot-line maintenance. Ind. Robot Int. J. **27**(5), 357–365 (2000)

23. Gonçalves, R.S., Carvalho, J.C.M.: Review and latest trends in mobile robots used on power transmission lines. Int. J. Adv. Robot. Syst. **10**(1), 1–14 (2013)

24. Aracil, R., Pinto, E., Ferre, M.: Robots for live-power lines: maintenance and inspection tasks. IFAC Proc. Vol. **35**(1), 13–18 (2002)

25. Maruyama, Y., Maki, K., Mori, H.: A hot-line manipulator remotely operated by the operator on the ground. In: Proceedings of the Sixth International Conference on Transmission and Distribution Construction and Live Line Maintenance, ESMO 1993, pp. 437–444. IEEE (1993)

26. Yano, K., Maruyama, Y., Morita, K., et al.: Development of the semi-automatic hot-line work robot system Phase II. In: Proceedings of the Seventh International Conference on Transmission and Distribution Construction and Live Line Maintenance, ESMO 1995, pp. 212–218. IEEE (1995)

27. Takaoka, K., Yokoyama, K., Wakisako, H., et al.: Development of the fully-automatic live-line maintenance robot-Phase III. In: Proceedings of the IEEE International Symposium on Assembly and Task Planning, pp. 423–428. IEEE (2001)

28. Simas, H., Barasuol, V., Kinceler, R., et al.: Kinematic conception of a hydraulic robot applied to power line insulators maintenance. In: 20th COBEM International Congress of Mechanical Engineering, pp. 739–748 (2009)

29. Allan, J.F., Beaudry, J.: Robotic systems applied to power substations - a state-of-the-art survey. In: 2014 3rd International Conference on Applied Robotics for the Power Industry (CARPI), pp. 1–6. IEEE (2014)

30. Li, L., Li, D., Li, Y., et al.: A state-of-the-art survey of the robotics applied for the power industry in China. In: 2016 4th International Conference on Applied Robotics for the Power Industry (CARPI), pp. 1–5. IEEE (2016)

31. Zhang, Y., Lu, S.Y.: Application of fuzzy control algorithm in power substation inspection robot. Int. J. Sci. Prog. Res. **27**(01), 1–7 (2016)

32. Taheri, H., Qiao, B., Ghaeminezhad, N.: Kinematic model of a four Mecanum wheeled mobile robot. Int. J. Comput. Appl. **113**(3), 6–9 (2015)

33. De Witte, D.: Precieze controle van een omni-directionele robot. Master's thesis, Universiteit Gent (2011)

34. Amoozgar, M.H., Sadati, S.H., Alipour, K.: Trajectory tracking of wheeled mobile robots using a kinematical fuzzy controller. Int. J. Robot. Autom. **27**(1), 49–59 (2012)

35. Guo, S., Jin, Y., Bao, S., et al.: Accuracy analysis of omnidirectional mobile manipulator with mecanum wheels. Adv. Manuf. **4**(4), 363–370 (2016)

36. Rone, W.S., Ben-Tzvi, P.: Continuum robot dynamics utilizing the principle of virtual power. IEEE Trans. Robot. **30**(1), 275–287 (2014)

Intelligent Control System for Substation Automatic Grounding Wire Operation Robot

Yuan Fu[1], Rongbin Tan[2(✉)], Jiaxing Lu[3], Yuewei Tian[1], and Shouyin Lu[2]

[1] Guizhou Power Company Guiyang Power Supply Bureau, Guiyang 550000, China
21647119@qq.com, 603771491@qq.com
[2] Shandong Jianzhu University, Jinan 250101, China
trb1995@qq.com, sdznjqr@163.com
[3] Department of Computer Science, Stevens Institute of Technology,
Hoboken, NJ 07030, USA
281944888@qq.com

Abstract. This paper proposes an intelligent control system for the development of substation automatic grounding wire working robot. The system consists of two parts: robot motion control subsystem and robot visual servo subsystem. The system realizes autonomous control algorithm of robot operation process through robot vision servo technology, which combines stereo vision algorithm with equipment tracking algorithm. The object tracking algorithm and Kalman filter algorithm are used to extract and track the equipment area, and the binocular stereo vision algorithm is used to obtain the three-dimensional information of the equipment accurately, and the position information is fed back to the robot control system, which forms a servo control system. The system can realize unmanned operation of live working in the unattended situation, and improve the automation level and work efficiency of the operation.

Keywords: Substation ·
Grounding wire connection and removal robot · Control system

1 Introduction

Grounding is one of the technical [1–6] measures to ensure safe operation on electrical equipment. Installation and removal of the temporary grounding wire is an important maintenance work [7–10] in the power system. At present, when most of the substation equipment needs to be repaired and grounded, it is necessary to use the aerial work vehicle to send the operation and maintenance personnel to the top of the equipment to complete the connection of the grounding wire. There is a certain risk in manually hanging the grounding wire. Once the operation is wrong, the grounding wire will fall, which will cause loss or personnel casualties.

The paper is supported by Shandong province key R&D program (2017CXGC0905, 2017CXGC0918 and 2017CXGC1505).

In this paper, we propose a kind of substation automatic hanging and disassembling grounding wire operation robot system, which combines the technology [11,12] of motion control system and visual servo system and completes this work by robot instead of manual using robot technology [13,14] can greatly reduce the risk of workers and has a great practical significance for avoiding the disadvantages of manual operation at high altitude.

2   Substation Automatic Mounting and Dismantling Grounding Wire Working Robot

The substation automatic grounding wire operation robot system [15–17] mainly includes: mobile platform, multi-degree of freedom working arm and platform control system, etc. The overall structure of the system is shown in Fig. 1(a) and (b). The complex road condition mobile platform enables the work platform to enter the equipment area to ensure the practicability of the work platform. The lifting platform is mounted on the mobile platform, and the folding and telescopic structure can be used to transport the dedicated hanging grounding device to the designated working height. Multi-degree of freedom working arm is mounted at the end of the lifting work platform, its role is to clamp special insulation tools, cooperate with the lifting work platform to complete the work of hanging the ground wire. The platform control system uses remote control and video monitoring mode to realize the platform's remote control and the collection and remote display of platform data.

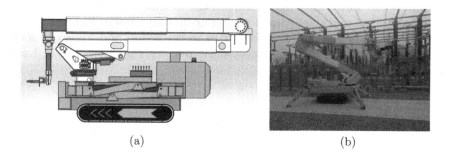

(a)   (b)

Fig. 1. Substation automatic hanging grounding wire working robot

3   Robot Control System

The substation automatic grounding wire working robot runs in the unstructured environment of substation, which has high working height and complex working environment. When using traditional remote control mode, the distance between operator and the working end is relatively long, and it is difficult to achieve the operation requirements of the system only by using video surveillance technology;

using fully autonomous control mode is difficult to achieve the control system because the control model is very complex and the robot needs to control many joints, and the cost is high, using a single autonomous control or master-slave remote control side is difficult to achieve the desired control effect.

In this paper, a robot control system for substation automatic grounding wire operation is proposed. The control system consists of two parts: robot motion control system and robot visual servo system [18]. The robot motion control system mainly realizes the kinematics modeling and analysis of the robot, and realizes the multi-axis open-loop motion control of the robot system. Robot visual servo system mainly uses binocular vision measurement technology to realize the dynamic acquisition of the three-dimensional position and attitude of the working object, and feeds the data back to the robot control system to realize the closed-loop control of the robot. The system adopts remote control + autonomous control mode during operation. When the robot is far from the target, remote control is used. When the robot is close to the target, it starts the autonomous mode and realizes the autonomous control of the robot's working process through the visual servo technology of the robot, which guarantees the control flexibility of the system and solves the problem of poor adaptability of the traditional control system.

3.1   Robot Motion Control System

The hardware connection of the system of the substation automatic grounding wire operation robot designed in this paper is shown in the figure Fig. 2. The motion control system is hydraulically driven. By controlling the opening of the hydraulic servo valve, the system realizes the motion control of the insulated lifting mechanism and the hydraulic manipulator. The system realizes the real-time acquisition of the motion information of the robot, and feeds the sensor information back to the hydraulic servo valve to realize the closed-loop hydraulic servo control of the system.

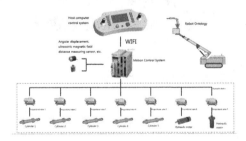

Fig. 2. Robot motion control system.

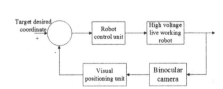

Fig. 3. Block diagram of the robot vision servo system.

The robot system has 11 degrees of freedom, including arm pitch, arm rotation, arm telescoping, platform rotation and platform pitch. The high power density hydraulic manipulator has 6 degrees of freedom, so the robot motion control model is very complex and it is difficult to realize. For this reason, this paper decouples the motion of live-working robot in substation into two independent motion control subsystems: one is the control subsystem of insulated lifting mechanism, which realizes the large-scale and fast movement of the robot system; the other is the motion control subsystem of hydraulic manipulator, which realizes the motion control of the manipulator. The combination of two independent subsystems can realize the effective fusion of large-scale fast motion and high-precision close-range motion of the robot system.

Since the control system of the two control subsystems has the same principle, this paper takes the manipulator control subsystem as an example to illustrate the operation process of the control system. As shown in the figure above, the system is a multi-axis closed-loop control system. The motion controller collects the angular and displacement sensor information of each axis, controls the action of the corresponding hydraulic servo valve of each axis by its output, drives the hydraulic motor or cylinder by the hydraulic servo valve to realize the servo control of each axis. The kinematics analysis module is the core module of the robot motion control system, which uses the D-H model. The forward/inverse kinematics analysis of the manipulator is carried out, and the tool coordinate system at the end of the manipulator is transformed into the coordinate system of each axis of the robot. The motion of the end position of the manipulator is transformed into the control quantity of each joint axis, and the motion control of the end position of the manipulator is realized through the coordinated control of multiple axes.

3.2   Robot Visual Servo System

As shown in Fig. 3, the visual servo system [19–21] of the robot consists of a robot control unit, a high voltage live working robot, a binocular camera and a visual positioning unit [22]. The system obtains the image of the work site by binocular stereo camera, and the position information of the equipment is obtained by the analysis and processing of the acquired image by visual positioning unit. The position information is received by the robot control unit, which realizes the closed-loop servo control of the substation automatic grounding wire operation robot.

In this servo system, the binocular camera needs to be calibrated in advance to eliminate the distortion of the camera itself and the level of the left and right camera poles.

Visual positioning unit is the most important part of the system, which converts the image signals collected by binocular camera into position signals that can be received by robot control unit.

This system uses binocular stereo vision correlation algorithm and target tracking algorithm to process image signal and acquire device position signal.

The specific implementation method will be described in detail in the next section.

Binocular Stereo Vision Algorithm. The binocular stereo vision algorithm is based on the parallax principle (as shown in Fig. 4). The baseline distance B is the distance between the left and right projection centers of the two cameras, and the camera focal length is f.

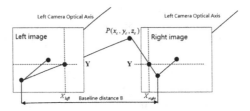

Fig. 4. Block diagram of the robot vision servo system.

Two cameras view the same feature points of space objects at the same time, The image of point $P(x_c, y_c, z_c)$ is captured on the "left eye" and the "right eye" respectively, and their image coordinates are:

$$P_{left} = (X_{left}, Y_{left})$$
$$P_{right} = (X_{right}, Y_{right})$$

Assuming that the two cameras are on the same plane, the image coordinates Y coordinates of the feature point P are the same, then the triangular geometry relation is used to obtain the following results:

$$\begin{cases} X_{left} = f\frac{x_c}{z_c} \\ X_{right} = f\frac{(x_c - B)}{z_c} \\ Y = f\frac{y_c}{z_c} \end{cases}$$

Then the parallax is: $Disparity = X_{left} - X_{right}$. The three-dimensional coordinates of the feature point P in the camera coordinate system can be calculated as follows:

$$\begin{cases} x_c = \frac{B * X_{left}}{Disparity} \\ y_c = \frac{B * Y_{left}}{Disparity} \\ z_c = \frac{B * f}{Disparity} \end{cases}$$

Therefore, as long as any point on the image plane of the left camera can find the corresponding matching point on the image plane of the right camera, the three-dimensional coordinates of the point can be determined. This method is a complete point-to-point operation. As long as there are corresponding matching points on the image surface, all points can participate in the above operation, and then obtain their corresponding three-dimensional coordinates.

The parallax of the left and right views is calculated by traversing all the similar point pairs in the left and right images. According to the above algorithm, the acquisition of three-dimensional point cloud information of the device is realized.

As mentioned above, it is difficult to satisfy the real-time requirement to process the whole image. In this paper, a target tracking algorithm is added before the stereo vision algorithm, which limits the stereo vision tracking algorithm to a limited area in the target tracking window to improve the real-time performance of the algorithm.

Target Tracking Algorithm. In this paper, CamShift algorithm is used for target tracking. CamShift algorithm can effectively solve the problem of target deformation and occlusion. It has low requirement for system resources and low time complexity. It can achieve good tracking effect under simple background. The implementation steps include color probability distribution calculation, MeanShift algorithm and CamShift target tracking.

Calculation of Color Probability Distribution. In order to reduce the influence of this change on tracking effect, image is first converted from RGB space to HSV space. Then, the statistical histogram of H component is made. The histogram data represent the probability of different H component values appearing in the image. The probability distribution of color is obtained by replacing the value of each pixel with the probability of H component appearing in the image.

MeanShift Algorithm. The MeanShift algorithm is a nonparametric method for density function gradient estimation. The algorithm locates the target by iteratively finding the extreme value of the probability distribution. The implementation process is as follows:

First select a tracking window in the image color probability distribution. Calculate the image zero-order M_{00}, first-order M_{10}, M_{01} and tracking window centroid in the tracking window, where $I(x, y)$ is the image equation, x is the image abscissa, y is the image ordinate, (x_c, y_c) is the tracking window centroid.

$$M_{00} = \sum_x \sum_y I(x,y) \, M_{10} = \sum_x \sum_y x I(x,y) \, M_{01} = \sum_x \sum_y y I(x,y)$$
$$x_c = M_{10}/M_{00} \quad y_c = M_{01}/M_{00}$$

Adjust the size of the tracking window, adjust the width to $s = \sqrt{M_{00}/256}$, length to $1.2\,s$. Move the center of the tracking window to the center of mass. If the moving distance is greater than the preset fixed threshold, repeat b, C and d until the moving distance between the center of the tracking window and the center of mass is less than the preset fixed threshold, or the number of cycles reaches a certain maximum and stops the calculation.

Camshift Target Tracking. Camshift algorithm is the extension of mean shift algorithm to continuous image sequence. It calculates the mean shift of all frames of video, and takes the result of the previous frame, i.e. the size and center of the search window, as the initial value of the search window of the next frame mean shift algorithm. In this way, the target tracking can be achieved by iteration.

In the case of complex background or interference of many pixels similar to the target color, because CamShift algorithm only considers the color histogram and ignores the spatial distribution of the target, it will lead to tracking failure, so Kalman filter is needed to filter the errors in the tracking process.

Kalman Filter. In the process of target tracking, the position, velocity and acceleration of the target are often noisy at any time. Kalman filter [23] utilizes the dynamic information of the target to eliminate the influence of noise and get a good estimation of the target position.

In this paper, the position and moving speed of the target window are taken as the observation values, and Kalman filter algorithm is used to estimate the observation values accurately, and then the results of target tracking are corrected to achieve accurate target tracking. Kalman filter can effectively reduce the error tracking in the process of target tracking, improve the accuracy of tracking, and then improve the effectiveness of the algorithm.

4   Experimental Data and Analysis

In order to verify the real-time performance of the proposed algorithm, 10 frames of images are extracted continuously from the video, and binocular operation of the whole image and the algorithm are used to calculate the running time. The results are shown in the following Figs. 5 and 6.

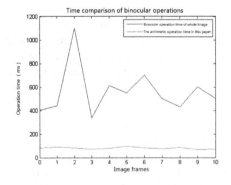 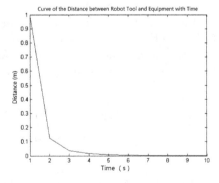

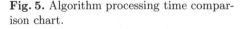

Fig. 5. Algorithm processing time comparison chart.

Fig. 6. Curve of the distance between robot tool and equipment with time

From the above figure, we can find that the whole binocular image algorithm runs a single frame image for a long time, mostly more than 400 ms, which can not meet the real-time requirements [24]. Moreover, because the time of the algorithm runs has a great relationship with the complexity of the image, the variance of the algorithm runs time is large, and the stability of the algorithm is poor.

Because of the target tracking algorithm adopted in this paper, the operation of the algorithm can be limited within the scope of the equipment, the amount of calculation is greatly reduced, the running time is relatively short, basically guaranteed within 100 ms, which can meet the real-time requirements of the algorithm, and because the image complexity is relatively stable within the scope of the equipment, the variance of the running time of the algorithm is small and the stability of the algorithm is greatly improved.

In order to verify the effectiveness of the proposed algorithm, the algorithm is applied to the high-voltage live working robot system, and the camera coordinate system coordinates are converted into the coordinates of the robot tool coordinate system. Analyse the relationship between the horizontal X (horizontal) direction, Y (vertical) direction and Z (distance) direction with time, and analyze the effectiveness of the algorithm.

This paper takes Z direction as an example to illustrate the effectiveness of the algorithm. It can be seen from the figure that the algorithm in this paper can ensure its rapid convergence. The distance has converged to 0 when the time is more than 6 s, thus ensuring the effectiveness of the algorithm.

5 Conclusion

The paper presents an intelligent control system for substation automatic grounding wire operation robot. The system obtains the position information of equipment by visual method, and feeds the position information back to the robot control system to form a servo control system. Experiments show that the system meets the real-time and effectiveness requirements of the application, can realize high-voltage live operation without any participation, improve the level of automation and operation efficiency, and realize the hanging work of high voltage grounding wire according to the different types of substations in different places. The system has high application and promotion value.

References

1. Lu, S., Ma, P., Qi, H., Li, B.: Research on high voltage electric power live line working robot. Autom. Electr. Power Syst. **17** (2003)
2. Li, C., Ren, Z., Bo, Y., An, Q., Li, J.: Image registration algorithm for high-voltage electric power live line working robot based on binocular vision. In: Society of Photo-Optical Instrumentation Engineers (2017)
3. Zhang, L., Liu, X., Lu, S., Ge, Z., Liu, C., Zhou, Y.: Kinematics of a climbing robot for intelligent wood inspection, pp. 3503–3506 (2017)
4. Zhang, Y., Tian, Y., Lu, S.: Design of a live maintenance mobile robot system for power substation equipment, pp. 882–887 (2017)

5. Gao, H., Liu, L., Tian, Y., Lu, S.: 3D reconstruction for road scene with obstacle detection feedback. Int. J. Pattern Recognit. Artif Intell. **32**(12), 1855021 (2018)
6. Dong, X., Lu, H., Wang, Y., Pan, H., Lu, S.: Kinematics analysis and simulation of the substation equipment water washing robot with hot-line, pp. 1–4 (2016)
7. Wang, Y.G., Yu, H.D., Xu, J.K.: Design and simulation on inspection robot for high-voltage transmission lines. Appl. Mech. Mater. **615**(25), 173–180 (2014)
8. Hu, Y., Chen, J.: The research of the automatic washing-brushing robot of 500 kV DC insulator string. In: Proceedings of ESMO 1993, IEEE 6th International Conference on Transmission and Distribution Construction and Live-Line Maintenance, pp. 411–424. IEEE (1993)
9. Lu, S., Gao, H., Lu, C., Wang, T.: Design of Chinese medical massage robot system, pp. 3882–3885 (2011)
10. Lu, S., Feng, L, Dong, J.: Design of control system for substation equipment inspection robot based on embedded Linux, pp. 1709–1712 (2008)
11. Huang, R., Zhang X., Huang, W., Wang, Z.: Substation live working robot system. In: International Conference on Applied Robotics for the Power Industry (2016)
12. Li J., Su, J., Fu, M., Lu, S., Dong, X.: Research and application of the water washing robot with hot-line working used in 220kV open type substation. In: International Conference on Applied Robotics for the Power Industry (2016)
13. Nakashima, M., Yano, K., Maruyama, Y., Yakabe, H.: The hot line work robot system "phase ii" and its human-robot interface "MOS". In: Proceedings 1995 IEEE/RSJ International Conference on Intelligent Robots and Systems. Human Robot Interaction and Cooperative Robots, vol. 2, pp. 116–123. IEEE (1995)
14. Sawada, J., Kusumoto, K., Maikawa, Y., Munakata, T., Ishikawa, Y.: A mobile robot for inspection of power transmission lines. IEEE Trans. Power Delivery **6**(1), 309–315 (1991)
15. Boyer, M.: Systems integration in telerobotics: case study: maintenance of electric power lines. In: Proceedings of IEEE International Conference on Robotics and Automation, vol. 2, pp. 1042–1047. IEEE (1996)
16. Tlale, N.S., De Villiers, M.: Kinematics and dynamics modelling of a mecanum wheeled mobile platform, pp. 657–662 (2008)
17. Taheri, H., Qiao, B., Ghaeminezhad, N.: Kinematic model of a four mecanum wheeled mobile robot. Int. J. Comput. Appl. **113**(3), 6–9 (2015)
18. Gimel'farb, G.L.: Intensity-based computer binocular stereo vision: signal models and algorithms. Int. J. Imaging Syst. Technol. **3**(3), 189–200 (1991)
19. Suetsugu, T., Matsuda, Y., Sugi, T., Goto, S., Egashira, N.: A visual supporting system for teleoperation of robot arm using visual servo control. In: Sice Conference (2014)
20. Fang H., Cui, X., Cui, L., Chen, Y.: An adapted visual servo algorithm for substation equipment inspection robot. In: International Conference on Applied Robotics for the Power Industry (2016)
21. Dong, G., Zhu, Z.H.: Position-based visual servo control of autonomous robotic manipulators. Acta Astronaut. **115**, 291–302 (2015)
22. Chen, L., Yang, R.Q., Xie, X.P.: An automatic vision based insulator positioning servo system of cleaning robot working under high voltage condition. Robot **26**(2), 161–165 (2004)
23. Gadzhiev, C.M.: Testing the covariance matrix of a renovating sequence under operating control of the kalman filter. Autom. Remote Control **57**(7), 1046–1052 (1996)
24. Lowe, D.G.: Distinctive image features from scale-invariant keypoints. Int. J. Comput. Vis. **60**(2), 91–110 (2004)

Application of Support Vector Machine Model Based on an Improved Elephant Herding Optimization Algorithm in Network Intrusion Detection

Hui Xu[✉], Qianqian Cao, Heng Fu, Chaochuan Fu, Hongwe Chen, and Jun Su

School of Computer Science, Hubei University of Technology, Wuhan, China
xuhui@mail.hbut.edu.cn

Abstract. In order to improve the accuracy of network intrusion detection, it is necessary to optimize Support Vector Machine (SVM) parameters. In view of the advantages of Elephant Herding Optimization (EHO) algorithm, such as simple control parameters and easy combination with other algorithms, this paper tries to optimize these parameters by using an Improved EHO (IEHO) algorithm. The IEHO-SVM algorithm is then proposed for parameters optimization, in order to improve the accuracy of network intrusion detection. The simulation experiment uses the KDD CUP99 data set for verification analysis. The experimental results show that, compared with the Particle Swarm Optimization (PSO)-SVM algorithm, Month-flame Optimization (MFO)-SVM algorithm and the basic EHO-SVM algorithm, the IEHO-SVM algorithm not only improves the global search ability of network intrusion, but also increases the accuracy rate of network intrusion detection by an average of 7.36%, 4.23% and 5.56% respectively, and reduces the false alarm rate by an average of 3.04%, 2.41% and 3.07% respectively, which aims at improving the efficiency of network intrusion detection.

Keywords: Network intrusion detection · Support vector machine · Parameter optimization · Elephant herding optimization algorithm · Particle swarm optimization algorithm

1 Introduction

With the increasing popularity of computer and network technology applications, intrusion detection can check the information in the computer network or computer system and analyze it to determine whether there are violations of security policies or computer system security in the system [1, 2]. In essence, network intrusion detection is a data classification problem, that is, the normal data and the intrusion data are separated by the detection model. The traditional network intrusion detection algorithms include Boyer-Moore (BM) algorithm, pattern matching algorithm, Quick Search (QS) algorithm, etc. These algorithms are single-mode network intrusion

© Springer Nature Singapore Pte Ltd. 2019
K. Knight et al. (Eds.): ICAI 2019, CCIS 1001, pp. 283–295, 2019.
https://doi.org/10.1007/978-981-32-9298-7_23

detection algorithms, which are difficult to meet the requirements of modern large-scale network security detection [3].

In recent years, with the development of artificial intelligence technology, the machine algorithm has been mature, and the artificial intelligence method can solve the high-dimensional, small sample and linear inseparable intrusion detection data. Intrusion detection models based on naive Bayes, support vector machines and neural networks have emerged [4, 5]. The Support Vector Machine (SVM) is a classifier for a limited sample data set and is insensitive to the dimensions of the data. It better overcomes the traditional machine learning algorithms such as neural networks, which are easy to fall into local optimization and over-fitting when training small sample network intrusion data. Therefore, SVM technology can detect intrusion data and classify data samples, so it is feasible to use SVM for intrusion detection [6, 7].

The performance of the network intrusion detection model based on SVM mainly depends on the value of its penalty factor C and kernel function parameters. In order to obtain better SVM parameters, scholars have proposed to use traditional grid method and gradient descent method, as well as genetic algorithm, Particle Swarm Optimization (PSO) and other parameter optimization methods, which all optimize SVM parameters to some extent and improve intrusion detection performance [8]. Elephant Herding Optimization (EHO) is a new meta-heuristic search algorithm based on elephant herding behavior, which is used to solve the global unconstrained optimization problem [9, 10]. It has been successfully applied to multi-level thresholds, support vector machine parameter optimization, scheduling problems and many other problems [11–13]. In order to improve the accuracy of SVM in network intrusion detection, this paper proposes a network intrusion detection method based on Improved Elephant Herding Optimization SVM to optimize SVM (IEHO-SVM).

2 Improved Elephant Herding Optimization (IEHO) Algorithm

2.1 Elephant Herding Optimization (EHO) Algorithm

In the basic EHO algorithm, the update operation is first performed to determine the search direction and the local search detail level of the algorithm, and then the separation operation is implemented. This process consists of two phases: the clan update operation and the separation operation.

Randomly initialize the elephant population, divide the elephant population into n clan, and each clan has j elephant individuals. In each iteration, the position of each elephant j is defined as Eq. (1).

$$x_{new,ci,j} = x_{ci,j} + \alpha \cdot (x_{best,ci} - x_{ci,j}) \cdot r \tag{1}$$

The position of the female matriarch $x_{best,ci}$ is defined as Eq. (2).

$$x_{new,ci,j} = \beta \times x_{center,ci} \tag{2}$$

The center of the elephant clan is defined as Eq. (3).

$$x_{center,ci,d} = \frac{1}{n_{ci}} \cdot \sum_{j=1}^{n_{ci}} x_{ci,j,d} \tag{3}$$

The elephant position with the worst fitness value is defined as Eq. (4).

$$x_{worst,ci} = x_{min} + (x_{max} - x_{min} + 1) \times rand \tag{4}$$

2.2   Improvement Strategy of Elephant Herding Optimization Algorithm

Since the elephant can only update the position according to the position of the patriarch, the local search ability is strong, but the global convergence is poor and easy to fall into the local optimum. Therefore, the levy flight strategy and the particle swarm search strategy are introduced to improve the global search ability and convergence speed of the algorithm [14, 15].

Introducing Levy Flight Strategy in the Updation Mechanism

Levy Flight [16] is a non-Gaussian randomization process that uses the frequent short-range search in Levy flight to search carefully around the current optimal solution to improve local search ability. The essence of Levy flight is a random step size whose position update formula is.

$$x_i^{t+1} = x_i^t + \alpha \oplus Levy(\lambda) \tag{5}$$

Where $i \in [1, \cdots, N]$ controls the size of the step size, $Levy(\lambda)$ is the random search path, and $\oplus$ is the point product.

$Levy(\lambda)$ represents the Levy distribution obeying the parameter λ, which satisfies the form of the power-law distribution, namely.

$$Levy(\lambda) \sim u = t^{-\lambda}, 1 < \lambda \leq 3 \tag{6}$$

Since the Levi distribution is very complicated, most of the current simulations using the Mantegna algorithm, the calculation formula of the Levi random number is.

$$Levy(\lambda) \sim \frac{\phi * u}{|v|^{1/\beta}} \tag{7}$$

In the EHO algorithm, the clan represents local search, and the location update of the individual in the algorithm will be affected by the location of the elephant with the best current fitness. If the current optimal individual is attracted by the local extremum, the algorithm has no effective mechanism to get rid of the constraint, which will lead to the algorithm being attracted to the local extreme value prematurely and falling into the local optimum. Inspired by Levy's flight strategy, this paper uses the levy flight behavior model to simulate the elephant's position update behavior, making full use of the levy flight short-range local search and the occasional long-distance jump search to expand the search range to avoid the elephant individual being attracted by local extremum. So as to make the elephant individuals move in a better direction and avoid the algorithm from premature convergence and falling into local optimal. The improved location update operation is,

$$x_{new,ci,j} = x_{ci,j} + Levy(\lambda) + \alpha \cdot (x_{best,ci} - x_{ci,j}) \cdot r \tag{8}$$

Introducing PSO Strategy in the Separation Mechanism

PSO is a population-based meta-heuristic intelligent optimization algorithm proposed by Kennedy et al. in 1995 according to the foraging behavior of birds [17]. The PSO algorithm is simple to understand, simple and easy to implement, fast convergence, and strong global search ability. It has been successfully applied to function optimization, neural networks, and intelligent optimization algorithm fusion. The speed and position update formula for the i-th particle is,

$$v_i^{t+1} = \omega \cdot v_i^t + c_1 \cdot r_1 \times (p_i^t - x_i^t) + c_2 \cdot r_2 \times (p_g^t - x_i^t) \tag{9}$$

$$x_i^{t+1} = x_i^t + v_i^{t+1} \tag{10}$$

In the EHO algorithm, male elephants leaving the clan perform global search. The larger their fitness value is, the larger the search range will be, which can improve the diversity of the population. However, in the basic EHO algorithm, the moving male elephant is the worst fitness function value in the clan. The basic EHO algorithm uses the Eq. (4) to replace the individual with the worst fitness value in the population. And the results are too random, which reduces the global search ability and the diversity of the population to some extent. Due to the simple parameter adjustment, fast convergence and wide range of global search capabilities of the PSO algorithm, the PSO algorithm can be embedded in the separation operation of the EHO algorithm. The specific updating strategy is: at each iteration, the male elephant with the worst fitness value in the population is regarded as each particle in the particle swarm. The PSO strategy is used to optimize the search space by using the inertia of the particles. And calculate the fitness function value of each particle, if a better fitness value is obtained, update $x_{worst,ci}$, otherwise, it will not be updated.

$$v_i^{t+1} = \omega \cdot v_i^t + c_1 \cdot r_1 (x_{best,ci}^t - x_{worst,ci}^t) + c_2 \cdot r_2 (x_{g,ci}^t - x_{worst,ci}^t) \tag{11}$$

$$x_i^{t+1} = x_{worst,ci}^t + v_i^{t+1} \tag{12}$$

The IEHO algorithm proposed in this paper mainly improves the updation mechanism and separation mechanism of the basic EHO algorithm. The flow chart of IEHO algorithm is shown in Fig. 1.

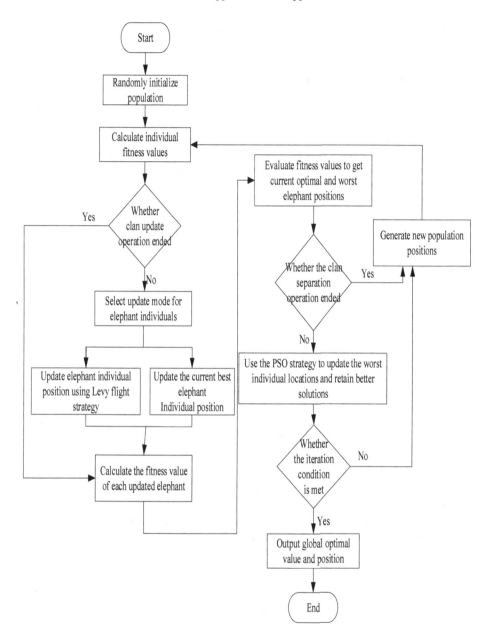

Fig. 1. Flow chart of IEHO algorithm

3   Optimized SVM Based on IEHO Algorithm for Network Intrusion Detection

3.1   Support Vector Machine (SVM)

SVM has better classification performance in the small sample classification problem [18]. The core idea of SVM is to replace a nonlinear mapping with a kernel function satisfying Mercer condition, so that the sample points in the input space can be mapped to a high-dimensional feature space, and be linearly separable in this space, and then construct an optimal hyperplane to approximate the ideal classification result [19]. The linearly separable training sample is set as $S = \{(x_i, y_i), i = 1, 2, \cdots, r\}, x_i \in R^d$, $y_i \in \{+1, -1\}$.

Then the classification hyperplane in the d-dimensional space is,

$$w.x + b = 0 \tag{13}$$

The problem of finding the optimal classification surface can be transformed into a constraint optimization problem, then,

$$\min \varphi(w) = \frac{1}{2}||x||^2$$
$$s.t. y_i[w.x_i + b] - 1 \geq 0, i = 1, 2, \cdots, r \tag{14}$$

The final classification function is,

$$f(x) = \text{sgn}(\sum_{i=1}^{r} a_i y_i(x_i, x) + b) \tag{15}$$

Where, the support vector is the sample corresponding to a_i which is not 0, and b is the offset vector, which can be solved by the constraint conditions after selecting a support vector.

In the case of linear inseparability, a relaxation variable ξ_i and a penalty factor C can be added to the constraint of Eq. (16).

$$\min \varphi(w, \xi) = \frac{1}{2}||x||^2 + c\sum_{i=1}^{r} \xi_i$$
$$s.t. y_i[w.x_i + b] - 1 + \xi_i \geq 0, i = 1, 2, \cdots, r, \xi_i \geq 0 \tag{16}$$

Where C is a constant and $C > 0$, which controls the penalty for the division of errors.

According to the functional theory, a suitable kernel function $K(x_i, x)$ can transform the nonlinear separable problem of the original space into a linear separable problem in the feature space. The kernel function of RBF is $K(x_i, x) = \exp(-\frac{||x_i - x||^2}{2\sigma^2})$, which has the advantages of fewer functions and better versatility. Therefore, the SVM can be established with the RBF core, and the classification decision function is,

$$f(x) = \text{sgn}(\sum_{i=1}^{r} a_i y_i K(x_i, x) + b) \tag{17}$$

According to the SVM regression theory, it can be seen that the penalty factor C and the σ value of the radial basis function have a great influence on the classification result. Therefore, in order to obtain the superior performance of the support vector machine, it is necessary to select the appropriate C and σ.

3.2 Optimization of SVM Intrusion Detection Parameters Based on IEHO Algorithm

The parameter selection of the SVM algorithm needs to be carried out in a large range, and the optimization process is easy to fall into the local optimum. The main factors affecting the support vector machine are: the choice of kernel function type, the kernel function parameters and the optimization of the penalty factor C to maximize the classification boundary and minimize the classification error. Since the parameters of the RBF kernel function are less easy to implement, the RBF function is selected as the kernel function of the SVM. In the classification problem, the penalty factor C also affects the classification accuracy of the intrusion detection. Therefore, in order to ensure the accuracy of intrusion detection, it is necessary to select an appropriate RBF parameter σ and a penalty factor C.

In view of the simple model and low parameters of the IEHO algorithm, it can effectively avoid the algorithm falling into local optimum and can perform large-scale global search, which effectively solves the problem of SVM parameter selection. In this paper, the IEHO algorithm is used to optimize the parameter σ of the RBF kernel function and the parameter penalty factor C of the SVM. The position of the elephant group is used to represent C and σ, and the correct rate of intrusion detection is used as the fitness function of the image group. Finding the optimal solution of the algorithm can search for the optimal parameters of the SVM and speed up the convergence of the SVM. The IEHO-SVM algorithm is used to establish an optimal network intrusion detection model to improve the detection rate of intrusion detection.

The IEHO-SVM algorithm flow chart is shown in Fig. 2.

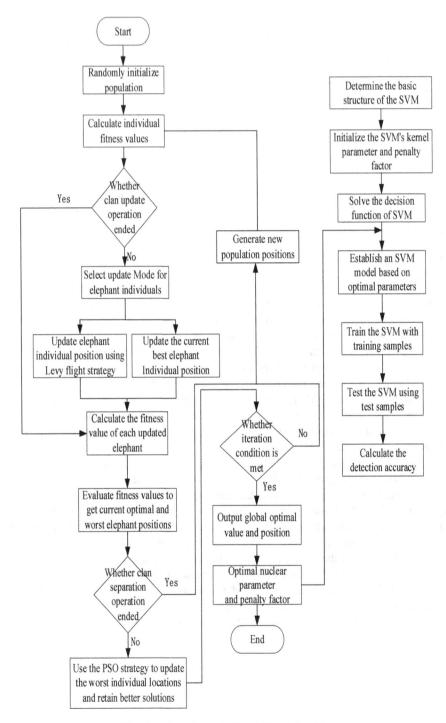

Fig. 2. Flow chart of IEHO-SVM algorithm

4 Simulation Experiments and Analysis

This paper uses the IEHO algorithm for SVM-based intrusion detection parameter optimization, and uses the commonly used evaluation indicators of intrusion detection to evaluate the performance of IEHO-SVM algorithm. The experimental platform uses CPU2.1 GHz, 2 GB memory, Windows7 operating system, simulation experiment on Matlab2014 platform, test analysis with JAVA software and weka software, weka is an open data mining work platform.

4.1 Data Source

The experimental data is selected from the KDD CUP99 intrusion detection data set, which includes four types of intrusion types: DOS, U2R, R2L, and Probe. Due to the large size of the complete KDD CUP99 dataset, which includes about 5 million records, each record includes 41 feature attributes, the last one is the flag attribute, except that Normal represents normal data, the rest are expressed as Abnormal [20]. In this paper, 12474 data were randomly selected from the 10% KDD CUP99 dataset, including 5012 training data sets and 7462 test data sets. The experimental data distribution is shown in Table 1.

Table 1. Distribution of experimental data sets

Category	Normal	Dos	Probe	R2L	U2R	Total
Training set	1632	1438	826	806	310	5012
Test set	2412	2046	1320	1246	438	7462

4.2 Data Preprocessing

Data preprocessing includes two stages: data digitization and normalization

(1) Data digitization: Since the data set contains data types (discrete type, numerical type) of each attribute, and SVM can only process continuous data between [0,1]. Therefore, it is necessary to standardize the attributes of textual data and discrete data, and use these standardized data to build SVM models for training.

(2) Normalization processing: Since the feature values of the intrusion data taken from the network are different units of measurement, the characteristic attributes of the samples are different. In order to solve the incomparability between data indicators, it is necessary to normalize the data.

$$X' = \frac{X - X_{min}}{X_{max} - X_{min}} \quad (18)$$

Where X is the original feature in the dataset, X_{max} is the maximum of the original features, X_{min} is the minimum of the original features, and X' is the normalized data.

4.3 Analysis of Experimental Results

In order to prove the effectiveness of the proposed IEHO-SVM algorithm in network intrusion detection, the EHO-SVM model, PSO-SVM model and MFO-SVM model are used as the comparison model under the same data set. In order to ensure the accuracy of the experiment, the test results were selected as the average of 20 experiments. In this paper, detection rate and false alarm rate are used as evaluation indicators.

The average detection results of different intrusion attack types based on the four algorithm are shown in Table 2, the corresponding optimal detection (accuracy) rate and false alarm rate are shown in Figs. 3 and 4 respectively. And Fig. 5 is a convergence graph of the four algorithm for detecting the optimal fitness value of different attack type data with the number of iterations under the same data set.

Table 2. Comparison of simulation results of different detection algorithms /%

Model	Evaluation index	Normal	Dos	Probe	R2L	U2R
PSO-SVM	Rate	81.01	82.92	83.78	80.18	82.55
	False	10.81	12.32	12.62	11.41	9.82
MFO-SVM	Rate	82.76	85.75	87.79	82.89	85.94
	False	10.54	11.22	11.87	9.44	10.77
EHO-SVM	Rate	89.81	81.69	82.01	78.40	87.58
	False	9.69	12.83	14.19	10.67	9.76
IEHO-SVM	Rate	94.05	87.36	89.53	85.16	91.14
	False	5.94	9.12	10.46	8.75	7.53

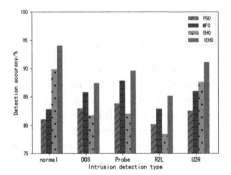

Fig. 3. Comparison of detection rates of different detection algorithms for different intrusion types

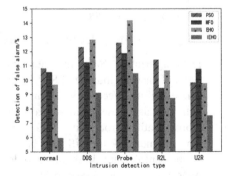

Fig. 4. Comparison of false alarm rate of different detection algorithms for different intrusion types

It can be seen from Table 2. that the IEHO-SVM algorithm obtains better detection results. The detection accuracy of the IEHO-SVM algorithm for Normal, Dos, probe, R2L, and U2R is 4.24%, 5.67%, 7.52%, 6.76% and 3.56% higher than the EHO-SVM algorithm, and the false alarm rate decreased by 3.75%, 3.71%, 3.73%, 1.92% and 2.23%, respectively. Indicating that IEHO algorithm effectively balances the local search and global search ability of the population, can effectively jump out of the local optimal solution, and has a good global search ability in SVM parameter optimization, which effectively improves the detection (accuracy) rate of SVM network intrusion detection and reduces the false alarm rate.

As can be seen from Figs. 3 and 4 that compared with the PSO-SVM algorithm, the MFO-SVM algorithm and the EHO algorithm, the network intrusion detection accuracy rate of the IEHO-SVM algorithm is significantly improved, among which for the Normal and U2R types, the detection rate of the IEHO-SVM algorithm is much higher than other comparison algorithms, and the false alarm rate also decreases to varying degrees. It can be seen that compared with EHO-SVM, PSO-SVM and MFO-SVM algorithms, IEHO-SVM algorithm has different degrees of improvement in the overall performance of network intrusion detection.

From the convergence curve of fitness values for four different attack types in Fig. 5, the convergence speed of the IEHO-SVM algorithm is generally faster than other comparison algorithms, and the fitness value is significantly better than other comparison algorithms. It shows that the IEHO algorithm has better performance in SVM parameter optimization, and verifies the effectiveness of the IEHO algorithm parameter optimization proposed in this paper.

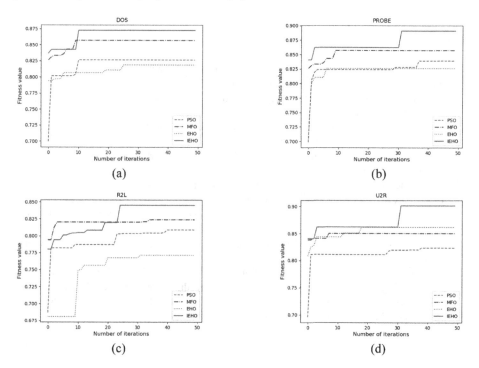

Fig. 5. Fitness value convergence curves of four different algorithms for different attack types

5 Conclusions

Since the parameters of SVM will affect the classification accuracy of intrusion detection, based on the shortcomings of traditional SVM which is easy to fall into local optimum when selecting parameters, the optimal SVM parameters can be obtained by using optimization algorithm. In this paper, the EHO algorithm is used to optimize the SVM parameters. In order to overcome the shortcomings of the basic EHO algorithm, that the convergence speed is too fast and easy to fall into the local optimum. An Improved Elephant Herding optimization algorithm based on the combination of particle swarm optimization strategy and Levy flight is proposed, and the optimization effect of IEHO algorithm is verified

In order to improve the accuracy of network intrusion detection, this paper proposes a network intrusion detection model based on IEHO-SVM algorithm. The IEHO algorithm is used to optimize the parameter penalty factor C and kernel parameters σ of the SVM to improve the classification accuracy of the SVM. The simulation experiment is carried out with KDD CUP99 dataset. The experimental results show that the IEHO-SVM model can obtain better SVM parameters, effectively improved the accuracy of SVM network intrusion detection and reduced the false alarm rate. Thereby establishing a better network intrusion detection model to meet the evolving network intrusion detection needs.

Acknowledgment. This work has been supported by the National Natural Science Foundation of China (No. 61602162, No. 61440024).

References

1. Hu, H.P., Chen, H.T., Huang, C.L., Tang, Y.: Research status and development trend of intrusion detection system. Comput. Eng. Sci. **23**(2), 20–25 (2001)
2. Denning, D.E.: An intrusion-detection model. IEEE Trans. Softw. Eng. **13**, 222–232 (1987)
3. Zhang, H.: Design of intrusion detection system based on a new pattern matching algorithm. In: Proceeding of 2009 IEEE International Conference on Computer Engineering and Technology, pp. 545–548. IEEE Press (2009)
4. Zhu, K., Zhang, Q.: Application of machine learning in network intrusion detection. J. Data Acquisition Process. **32**, 479–488 (2017)
5. Hodo, E.K., Bellekens, X., Hamilton, A.: Threat analysis of IoT networks using artificial neural network intrusion detection system. In: Proceedings of 2016 IEEE International Symposium on Networks, Computers and Communications, pp. 6865–6867. IEEE Press (2016)
6. Rao, X., Dong, C.X., Yang, S.Q.: Intrusion detection system based on support vector machine. J. Softw. **14**(4), 798–803 (2003)
7. Nskh, P., Varma, M.N., Naik, R.: Principle component analysis based intrusion detection system using support vector machine. In: Proceedings of 2016 IEEE International Conference on Recent Trends in Electronics, Information & Communication Technology, pp. 1344–1350. IEEE Press (2016)

8. Dong, S.K., Nguyen, H.N., Park, J.S.: Genetic algorithm to improve SVM based network intrusion detection system. In: Proceedings of 2005 IEEE International Conference on Advanced Information Networking and Applications, pp. 155–158. IEEE Press (2005)
9. Wang, G.G., Deb, S., Coelho, L.D.S.: Elephant herding optimization. In: Proceedings of 3rd IEEE International Symposium on Computational and Business Intelligence. IEEE Press (2015)
10. Deb, S., Gao, X.Z., Coelho, L.D.S.: A new metaheuristic optimisation algorithm motivated by elephant herding behaviour. Int. J. Bio-Inspired Comput. **8**(6), 394–409 (2017)
11. Tuba, E., Alihodzic, A., Tuba, M.: Multilevel image thresholding using elephant herding optimization algorithm. In: Proceedings of IEEE International Conference on Engineering of Modern Electric Systems. IEEE Press (2017)
12. Tuba, E., Stanimirovic, Z.: Elephant herding optimization algorithm for support vector machine parameters tuning. In: Proceedings of International Conference on Electronics, Computers and Artificial Intelligence, pp. 1–4 (2017)
13. Sarwar, M.A., Amin, B., Ayub, N., Faraz, S.H., Khan, S.U.R., Javaid, N.: Scheduling of appliances in home energy management system using elephant herding optimization and enhanced differential evolution. In: Barolli, L., Woungang, I., Hussain, O.K. (eds.) INCoS 2017. LNDECT, vol. 8, pp. 132–142. Springer, Cham (2018). https://doi.org/10.1007/978-3-319-65636-6_12
14. Qiantu, Z., Liqin, F., Yulong, Z.: Double subgroups fruit fly optimization algorithm with characteristics of levy flight. J. Comput. Appl. **35**(5), 1348–1352 (2015)
15. Yan-Hui, G., Ke-Jun, Z.: Hybrid PSO-solver algorithm for solving optimization problems. J. Comput. Appl. **147**(6), 261–265 (2011)
16. Senthilnath, J., Das, V., Omkar, S.N., Mani, V.: Clustering using levy flight cuckoo search. In: Bansal, J., Singh, P., Deep, K., Pant, M., Nagar, A. (eds.) Seventh International Conference on Bio-Inspired Computing: Theories and Applications (BIC-TA 2012). Advances in Intelligent Systems and Computing, vol. 202. Springer, India (2013). https://doi.org/10.1007/978-81-322-1041-2_6
17. Kennedy, J., Eberhart, R.: Particle swarm optimization. In: Proceedings of International Conference on Neural Networks, vol. 4, pp. 1942–1948 (2013)
18. Abhisek, U.: Support vector machine. Comput. Sci. **1**(4), 1–28 (2002)
19. Mill, J.: Support vector classifiers and network intrusion detection. In: Proceedings of 2008 IEEE International Conference on Fuzzy Systems. IEEE Press (2008)
20. Wu, J., Zhang, W., Ma, Y.: Data analysis and study on KDDCUP99 data set. In: Computer Applications and Software (2014)

Author Index

Printed in the United States
By Bookmasters